IEE TELECOMMUNICATIONS SERIES 43

Series Editors: Professor C. J. Hughes
　　　　　　　Professor J. D. Parsons
　　　　　　　Professor G. White

Telecommunications Signalling

Other volumes in this series:

Volume 1	**Telecommunications networks** J. E. Flood (Editor)
Volume 2	**Principles of telecommunication-traffic engineering** D. Bear
Volume 3	**Programming electronic switching systems** M. T. Hills and S. Kano
Volume 4	**Digital transmision systems** P. Bylanski and D. G. W. Ingram
Volume 5	**Angle modulation: the theory of system assessment** J. H. Roberts
Volume 6	**Signalling in telecommunications networks** S. Welch
Volume 7	**Elements of telecommunications economics** S. C. Littlechild
Volume 8	**Software design for electronic switching systems** S. Takamura, H. Kawashima, N. Nakajima
Volume 9	**Phase noise in signal sources** W. P. Robins
Volume 10	**Local telecommunications** J. M. Griffiths (Editor)
Volume 11	**Principles and practice of multi-frequency telegraphy** J. D. Ralphs
Volume 12	**Spread spectrum in communication** R. Skaug and J. F. Hjelmstad
Volume 13	**Advanced signal processing** D. J. Creasey (Editor)
Volume 14	**Land mobile radio systems** R. J. Holbeche (Editor)
Volume 15	**Radio receivers** W. Gosling (Editor)
Volume 16	**Data communications and networks** R. L. Brewster (Editor)
Volume 17	**Local telecommunications 2: into the digital era** J. M. Griffiths (Editor)
Volume 18	**Satellite communication systems** B. G. Evans (Editor)
Volume 19	**Telecommunications traffic, tariffs and costs: an introduction for managers** R. E. Farr
Volume 20	**An introduction to satellite communications** D. I. Dalgleish
Volume 21	**SPC digital telephone exchanges** F. J. Redmill and A. R. Valdar
Volume 22	**Data communications and networks 2nd edition** R. L. Brewster (Editor)
Volume 23	**Radio spectrum management** D. J. Withers
Volume 24	**Satellite communication systems 2nd edition** B. G. Evans (Editor)
Volume 25	**Personal & mobile radio systems** R. C. V. Macario (Editor)
Volume 26	**Common-channel signalling** R. J. Manterfield
Volume 27	**Transmission systems** J. E. Flood and P. Cochrane (Editors)
Volume 28	**VSATs: very small aperture terminals** J. L. Everett (Editor)
Volume 29	**ATM: the broadband telecommunications solution** L. G. Cuthbert and J.-C. Sapanel
Volume 30	**Telecommunication network management into the 21st century** S. Aidarous and T. Plevyak (Editors)
Volume 31	**Data communications and networks 3rd edition** R. L. Brewster (Editor)
Volume 32	**Analogue optical fibre communications** B. Wilson, Z. Ghassemlooy and I. Darwazeh (Editors)
Volume 33	**Modern personal radio systems** R. C. V. Macario (Editor)
Volume 34	**Digital broadcasting** P. Dambacher
Volume 35	**Principles of performance engineering for telecommunication and information systems** M. Ghanbari, C. J. Hughes, M. C. Sinclair and J. P. Eade
Volume 36	**Telecommunication networks, 2nd edition** J. E. Flood (Editor)
Volume 37	**Optical communication receiver design** S. B. Alexander
Volume 38	**Satellite communication systems, 3rd edition** B. G. Evans (Editor)
Volume 39	**Quality of service in telecommunications** A. P. Oodan, K. E. Ward and T. W. Mullee
Volume 40	**Spread spectrum in mobile communication** O. Berg, T. Berg, S. Haavik, J. F. Hjelmstad and R. Skaug
Volume 41	**World telecommunications economics** J. J. Wheatley
Volume 42	**Video coding: an introduction to standard codes** M. Ghanbari

Telecommunications Signalling

Richard Manterfield

The Institution of Electrical Engineers

Published by: The Institution of Electrical Engineers, London, United Kingdom

© 1999: The Institution of Electrical Engineers

This publication is copyright under the Berne Convention and the Universal Copyright Convention. All rights reserved. Apart from any fair dealing for the purposes of research or private study, or criticism or review, as permitted under the Copyright, Designs and Patents Act, 1988, this publication may be reproduced, stored or transmitted, in any forms or by any means, only with the prior permission in writing of the publishers, or in the case of reprographic reproduction in accordance with the terms of licences issued by the Copyright Licensing Agency. Inquiries concerning reproduction outside those terms should be sent to the publishers at the undermentioned address:

The Institution of Electrical Engineers,
Michael Faraday House,
Six Hills Way, Stevenage,
Herts. SG1 2AY, United Kingdom

While the author and the publishers believe that the information and guidance given in this work are correct, all parties must rely upon their own skill and judgment when making use of them. Neither the author nor the publishers assume any liability to anyone for any loss or damage caused by any error or omission in the work, whether such error or omission is the result of negligence or any other cause. Any and all such liability is disclaimed.

The moral right of the author to be identified as author of this work has been asserted by him/her in accordance with the Copyright, Designs and Patents Act 1988.

British Library Cataloguing in Publication Data

A CIP catalogue record for this book is available from the British Library

ISBN 0 85296 761 6

Printed in England by Short Run Press Ltd., Exeter

To Liz, Muriel and Dorien

Contents

1	**Introduction**	**1**
	1.1 The Changing Environment	1
	1.2 The Network Response	1
	1.3 Types of Network	2
	1.3.1 Telecommunications networks	3
	1.3.2 Internet technologies	3
	1.3.3 Broadcast/media	4
	1.3.4 Computing	4
	1.3.5 Management networks	4
	1.3.6 Evolution of networks	4
	1.4 Scope of this Book	5
	1.5 Signalling: The Heart of the Network	5
	1.6 Chapter Summary	6
2	**Principles of Signalling Systems**	**7**
	2.1 Introduction	7
	2.1.1 Scope	7
	2.1.2 Networks	7
	2.1.3 Signalling	7
	2.1.4 Terminology	8
	2.1.5 Types of signalling	8
	2.2 Channel Associated Signalling	9
	2.2.1 Principle	9
	2.2.2 Types of information	9
	2.2.3 Forms of channel associated signalling	10
	2.2.4 CAS disadvantages	10
	2.2.5 CAS application	11
	2.3 Common Channel Signalling	11
	2.3.1 Principle	11
	2.3.2 Circuit-related signalling	11
	2.3.3 Non-circuit-related signalling	12
	2.3.4 Messages	13
	2.3.5 Buffers	14
	2.3.6 Signalling activity	15
	2.3.7 Error detection/correction	16
	2.3.8 Specification	16

2.4	Advantages of Common Channel Signalling		16
	2.4.1 Harmony with control techniques		16
	2.4.2 Signalling terminations		17
	2.4.3 Overcoming limitations of CAS systems		17
	2.4.4 Speed of transfer		17
	2.4.5 User information		18
	2.4.6 Reliability		18
	2.4.7 Evolutionary potential		18
	2.4.8 Multipurpose platform		19
	2.4.9 CCS Network		19
2.5	Additional Requirements of CCS		19
	2.5.1 Reliability and security		19
	2.5.2 Speech continuity		20
	2.5.3 Processing overhead		20
2.6	Evolution of Signalling Systems		21
2.7	Types of CCS System		21
	2.7.1 International standards		21
	2.7.2 Specifications		21
	2.7.3 Access and inter-nodal differences		22
2.8	Modes of CCS Operation		24
	2.8.1 Associated mode		24
	2.8.2 Non-associated mode		24
	2.8.3 Quasi-associated mode		25
	2.8.4 Flexibility		26
2.9	Chapter Summary		26
2.10	References		28
3	**Architecture of CCS Systems**		**29**
3.1	Introduction		29
3.2	Architecture Requirements		29
	3.2.1 Evolution		30
	3.2.2 Flexibility		30
	3.2.3 Resource optimisation		30
	3.2.4 Openness		30
	3.2.5 Scope		30
	3.2.6 Operation		30
	3.2.7 Reliability		30
	3.2.8 Compatibility		31
	3.2.9 Signalling interworking		31
	3.2.10 Network interworking		31
	3.2.11 National use		31
	3.2.12 Degree of detail		31
	3.2.13 Clarity		31
	3.2.14 Legacy systems		31

		3.2.15	Speed of specification	32
	3.3	Architecture Principles		32
		3.3.1	Structure within a node	32
		3.3.2	Tiered structure applied between nodes	33
		3.3.3	Protocol	35
		3.3.4	Example of application	35
	3.4	Types of Architecture		36
		3.4.1	Four-level structure	36
		3.4.2	OSI layered model	36
		3.4.3	Layers for DSS1 and DSS2	36
		3.4.4	Layers for Internet Protocol	36
		3.4.5	Evolution	37
	3.5	Level Structure of CCSS7		37
		3.5.1	Level 1	37
		3.5.2	Level 2	37
		3.5.3	Level 3	38
		3.5.4	Message Transfer Part (MTP)	38
		3.5.5	Level 4	39
		3.5.6	Applying the level structure	39
	3.6	OSI Model		40
		3.6.1	General	40
		3.6.2	Layer 1 – Physical	42
		3.6.3	Layer 2 – Data Link	42
		3.6.4	Layer 3 – Network	42
		3.6.5	Layer 4 – Transport	42
		3.6.6	Layer 5 – Session	43
		3.6.7	Layer 6 – Presentation	43
		3.6.8	Layer 7 – Application Layer	43
	3.7	Applying the OSI Model to CCSS7		45
		3.7.1	Context	45
		3.7.2	Application to NCR signalling	46
		3.7.3	Signalling Connection Control Part (SCCP)	46
		3.7.4	Transaction Capabilities (TC)	46
		3.7.5	Application Parts	47
		3.7.6	Broadband ISDN User Part (B-ISUP)	47
	3.8	Overall Architecture of CCSS7		47
	3.9	DSS1 Layer Structure		49
		3.9.1	General	49
		3.9.2	DSS1 Layer 1 – Physical	51
		3.9.3	DSS1 Layer 2 – Data Link	51
		3.9.4	DSS1 Layer 3 – Network	52
		3.9.5	Primitives	52
	3.10	DSS2 Layer Structure		52
	3.11	Specification of CCS Systems		53

		3.11.1	Contents	53
		3.11.2	Avoiding ambiguity	54
		3.11.3	Options	54
		3.11.4	Speed	54
	3.12	Chapter Summary		55
		3.12.1	Architecture requirements	55
		3.12.2	Architecture principles	55
		3.12.3	Types of architecture	55
		3.12.4	Future architecture	56
		3.12.5	Way forward	57
	3.13	References		57
4	**CCSS7 Message Transfer Part**			**59**
	4.1	Introduction		59
	4.2	Architecture		59
		4.2.1	Structure	59
		4.2.2	Primitives	60
		4.2.3	Relationship with SCCP	60
	4.3	Functional Structure		60
	4.4	MTP Level 1: Signalling Data Link		62
	4.5	MTP Level 2: Signalling Link Functions		62
		4.5.1	General	62
		4.5.2	General Signal Unit format	62
		4.5.3	Error detection	64
		4.5.4	Error correction	64
		4.5.5	Basic Method of error correction	65
		4.5.6	Preventative Cyclic Retransmission Method of error correction	66
		4.5.7	Level 2 flow control	66
		4.5.8	Error monitoring	67
	4.6	MTP Level 3: Signalling Network Functions		67
		4.6.1	General	67
		4.6.2	Level 3: Signalling Message Handling Function	68
		4.6.3	Level 3: Signalling Traffic Management Function	70
		4.6.4	Level 3: Signalling Link Management Function	71
		4.6.5	Level 3: Signalling Route Management Function	72
	4.7	Chapter Summary		73
		4.7.1	General	73
		4.7.2	Primitives	73
		4.7.3	Level 1	73
		4.7.4	Level 2	73
		4.7.5	Level 3	74
	4.8	References		75

5 Signalling Connection Control Part — 77
- 5.1 Introduction — 77
- 5.2 Architecture — 77
  - 5.2.1 OSI Model — 77
  - 5.2.2 Structure — 78
- 5.3 Services — 78
  - 5.3.1 Service types — 78
  - 5.3.2 Service classes — 79
- 5.4 Primitives — 80
- 5.5 SCCP Formats — 81
  - 5.5.1 Principle of message structure — 81
  - 5.5.2 Message types — 82
  - 5.5.3 Parameters — 82
  - 5.5.4 Pointers — 83
  - 5.5.5 Overall message structure — 83
- 5.6 SCCP Procedures — 85
  - 5.6.1 Connection-Oriented procedures — 85
  - 5.6.2 Connection-Oriented with multiple sections — 86
  - 5.6.3 Connectionless procedures — 87
  - 5.6.4 Addressing — 87
- 5.7 Evolution — 88
- 5.8 Chapter Summary — 88
- 5.9 References — 89

6 CCSS7 Narrowband ISDN User Part — 90
- 6.1 Introduction — 90
- 6.2 ISUP Formats — 91
  - 6.2.1 Format principles — 91
  - 6.2.2 Examples of message formats — 92
  - 6.2.3 Format overhead — 95
  - 6.2.4 Message segmentation — 96
- 6.3 ISUP Procedures — 96
  - 6.3.1 Basic call set-up — 96
  - 6.3.2 Basic call release — 98
  - 6.3.3 Additional features — 99
- 6.4 ISUP Supplementary Services — 103
  - 6.4.1 Number identification — 103
  - 6.4.2 Call Offering — 104
  - 6.4.3 Call Completion — 106
  - 6.4.4 Multi-Party — 107
  - 6.4.5 Community of Interest — 108
  - 6.4.6 Charging — 109
  - 6.4.7 Additional Information Transfer — 110

	6.5	Chapter Summary	110
		6.5.1 Architecture	110
		6.5.2 Format	110
		6.5.3 Procedures	111
		6.5.4 Features	111
		6.5.5 Supplementary services	112
	6.6	References	112
7	**Transaction Capabilities**		**113**
	7.1	Introduction	113
	7.2	Users	113
	7.3	Network Service	114
	7.4	Portability	114
	7.5	Architecture	115
		7.5.1 OSI layers	115
		7.5.2 Intermediate Service Part (ISP)	115
		7.5.3 TCAP	116
		7.5.4 Component Sub-Layer	116
		7.5.5 Transaction Sub-Layer	117
	7.6	Primitives	117
		7.6.1 Component Sub-Layer	117
		7.6.2 Transaction Sub-Layer	118
	7.7	General TCAP Message Format	118
		7.7.1 Information Elements	118
		7.7.2 TCAP message structure	119
	7.8	Transaction Sub-Layer Format	121
		7.8.1 Unstructured Dialogue	121
		7.8.2 Structured Dialogue	121
	7.9	Component Sub-Layer Format	124
		7.9.1 Invoke	124
		7.9.2 Return Result	125
	7.10	Component Sub-Layer Procedures	125
		7.10.1 Single invocation	126
		7.10.2 Multiple invocation	126
		7.10.3 Segmentation	127
		7.10.4 Abnormal conditions	127
	7.11	Transaction Sub-Layer Procedures	128
		7.11.1 Unstructured Dialogue	128
		7.11.2 Structured Dialogue	128
	7.12	Chapter Summary	129
	7.13	References	130

8 Mobile Communications — 131
8.1 Introduction — 131
8.1.1 Evolution — 131
8.1.2 Technologies — 132
8.1.3 Signalling implications — 132
8.2 Network Structure — 133
8.2.1 Numbering — 133
8.2.2 Network elements — 133
8.2.3 Cellular structure — 135
8.3 MAP Services — 135
8.3.1 MAP service model — 135
8.3.2 MAP relationship with TC — 136
8.4 MAP Service Descriptions — 137
8.4.1 Common services — 137
8.4.2 Location services — 137
8.4.3 Handover services — 138
8.4.4 Call handling services — 141
8.4.5 Short Message Service (SMS) — 141
8.4.6 Supplementary Services — 141
8.4.7 Miscellaneous mobility services — 141
8.5 MAP Procedures — 144
8.5.1 Mobile station location updating — 144
8.5.2 Handover — 146
8.5.3 Outgoing Call Set-Up — 147
8.5.4 Incoming Call Set-Up — 148
8.5.5 Version compatibility — 149
8.5.6 Short Message Service — 150
8.5.7 Supplementary services — 151
8.6 Chapter Summary — 153
8.6.1 Introduction — 153
8.6.2 Numbering — 153
8.6.3 Network elements — 153
8.6.4 Services — 154
8.6.5 Procedures — 154
8.7 References — 155

9 Intelligent Network Application Part — 157
9.1 Introduction — 157
9.1.1 Distributed intelligence — 157
9.1.2 Centralised intelligence — 157
9.1.3 IN structure — 158
9.1.4 Modular approach — 159
9.1.5 Service invocation and control — 160
9.1.6 Signalling — 161

xiv *Contents*

9.2	IN Architecture	161
	9.2.1 Conceptual Model	161
	9.2.2 Implementation components	162
	9.2.3 INAP architecture	163
9.3	INAP Formats	165
9.4	INAP Procedures	165
	9.4.1 General	165
	9.4.2 Translation	166
	9.4.3 Translation with announcement	167
	9.4.4 Translation with user interaction	168
	9.4.5 Credit card calling	169
	9.4.6 Call gapping	170
9.5	Chapter Summary	171
	9.5.1 Introduction	171
	9.5.2 IN Structure	172
	9.5.3 Modularity	172
	9.5.4 Services	173
	9.5.5 INAP architecture	173
	9.5.6 Format	173
	9.5.7 Procedures	173
9.6	References	174

10 Management Aspects of CCSS7 — **175**

10.1	Introduction	175
10.2	CCSS7 Management Functions	175
	10.2.1 Scope	175
	10.2.2 OMAP	177
10.3	TMN Type of Management Functions	177
	10.3.1 Principles of TMN	177
	10.3.2 Architecture	178
	10.3.3 Managed objects	179
	10.3.4 Managed object specification	179
	10.3.5 Measurements	181
10.4	Application Type of Management Functions	181
	10.4.1 OMAP Model	181
	10.4.2 OMASE User	183
	10.4.3 OMASE primitives	185
	10.4.4 OMASE services	185
	10.4.5 OMASE parameters	185
	10.4.6 MTP Routing Verification Test (MRVT)	187
	10.4.7 MRVT messages	187
	10.4.8 MRVT procedures	187
	10.4.9 SCCP Routing Verification Test (SRVT)	188
	10.4.10 Circuit Validation Test (CVT)	188

Contents xv

	10.5	Chapter Summary	189
	10.6	References	190

11 DSS1 Physical and Data Link Layers — **191**
    11.1 Introduction — 191
    11.2 Physical Layer — 191
    11.3 Data Link Layer Function — 192
    11.4 Data Link Layer Formats — 193
        11.4.1 Flag — 193
        11.4.2 Address field — 193
        11.4.3 Identifiers — 194
        11.4.4 Control field — 194
        11.4.5 Information field — 195
        11.4.6 Check bits — 196
    11.5 Data Link Layer Procedures — 196
        11.5.1 Unacknowledged information transfer — 196
        11.5.2 Terminal Endpoint Identifier procedure — 197
        11.5.3 Terminal Endpoint Identifier Check procedure — 198
        11.5.4 Terminal Endpoint Identifier Removal procedure — 198
        11.5.5 Acknowledged information transfer — 199
    11.6 Chapter Summary — 201
    11.7 References — 202

12 DSS1 Network Layer — **203**
    12.1 Introduction — 203
    12.2 Format Principles — 203
        12.2.1 Protocol Discriminator — 204
        12.2.2 Call reference — 204
        12.2.3 Message type — 205
        12.2.4 Other information elements — 205
        12.2.5 Codesets — 207
    12.3 Examples of Message Formats — 209
        12.3.1 General — 209
        12.3.2 Set-up Message — 210
        12.3.3 Connect Message — 211
        12.3.4 Disconnect Message — 211
    12.4 Set-up Procedures for Circuit-Switched Calls — 211
        12.4.1 General — 211
        12.4.2 En bloc procedures with point-to-point working — 212
        12.4.3 En bloc procedures in conjunction with broadcast working — 214
        12.4.4 Overlap procedures — 215
    12.5 Call-Clearing Procedures for Circuit-Switched Calls — 216
    12.6 Other Features for Basic Circuit-Switched Calls — 217

	12.6.1	General	217
	12.6.2	Restart procedure	218
	12.6.3	Call re-arrangement procedure	218
	12.6.4	Error conditions	218
	12.6.5	User notification	218
12.7	Procedures for Packet-Data Calls		219
	12.7.1	General	219
	12.7.2	Circuit-switched access	219
	12.7.3	Packet-switched access – B channel	220
	12.7.4	Packet-switched access – D channel	220
12.8	User Signalling Bearer Service		221
	12.8.1	General	221
	12.8.2	Procedures	221
	12.8.3	Congestion control	222
12.9	Circuit Mode Multi-Rate Procedures		222
12.10	Supplementary Services		223
	12.10.1	General	223
	12.10.2	Keypad Protocol	223
	12.10.3	Feature Key Management Protocol	223
	12.10.4	Functional Protocol	224
	12.10.5	Message format	225
12.11	Chapter Summary		226
12.12	References		227

13 Private Networks — 229

13.1	Introduction		229
	13.1.1	General	229
	13.1.2	Signalling	230
13.2	PSS1 (QSIG) General		230
	13.2.1	Introduction	230
	13.2.2	Architecture	231
	13.2.3	Services	232
13.3	PSS1 Basic Call Procedures		233
13.4	PSS1 Supplementary Services		234
	13.4.1	Call-related supplementary services	234
	13.4.2	Active call supplementary services	234
	13.4.3	Call-independent supplementary services	235
13.5	PSS1 Message Format		235
13.6	GVNS – General		236
	13.6.1	Virtual network	236
	13.6.2	Access types	237
	13.6.3	GVNS features	238
	13.6.4	Operation and maintenance	238

13.7	GVNS Functional Model	238
	13.7.1 GVNS functions	238
	13.7.2 Model	239
	13.7.3 GVNS call types	240
13.8	GVNS Formats	240
	13.8.1 Forward GVNS parameter	240
	13.8.2 Backward GVNS parameter	241
13.9	GVNS Procedures	241
13.10	Chapter Summary	243
	13.10.1 Meeting the requirements	243
	13.10.2 PSS1 (QSIG)	243
	13.10.3 GVNS	244
13.11	References	245

14 Broadband Signalling Platform — 247

14.1	Introduction	247
14.2	Broadband Signalling Relations	247
14.3	Architecture	248
	14.3.1 UNI architecture	249
	14.3.2 NNI architecture	250
14.4	Platform Format Convention	250
14.5	Asynchronous Transfer Mode (ATM) Layer	251
	14.5.1 Cell structure	252
	14.5.2 Primitives	253
14.6	ATM Adaptation Layer (AAL)	253
	14.6.1 Segmentation and reassembly	254
	14.6.2 Blocking and deblocking	254
14.7	AAL Type 1 Services	255
	14.7.1 Services	255
	14.7.2 Architecture	255
	14.7.3 Primitives	255
	14.7.4 Functions	256
	14.7.5 Format	256
14.8	AAL Type 2	257
14.9	AAL Types 3/4 and 5	257
	14.9.1 Services	257
	14.9.2 Architecture	258
	14.9.3 Primitives	258
	14.9.4 Functions	258
	14.9.5 Format	260
14.10	Chapter Summary	261
14.11	References	262

15	**Broadband ISDN User Part for CCSS7**	**263**
15.1	Introduction	263
15.2	Architecture	263
15.3	Application of the Architecture	264
	15.3.1 Dynamic modelling	264
	15.3.2 Static modelling	265
15.4	B-ISUP Formats	266
	15.4.1 Principles	266
	15.4.2 Message format	266
	15.4.3 Message content	267
	15.4.4 Name codes	267
	15.4.5 Examples of message formats	268
	15.4.6 Message Segmentation	270
15.5	B-ISUP Procedures	270
	15.5.1 Call Control – Nodal Functions	270
	15.5.2 Compatibility – Nodal Functions	273
	15.5.3 Maintenance Control – Nodal Functions	273
	15.5.4 Application Service Elements (ASEs)	273
15.6	Chapter Summary	276
15.7	References	277
16	**Broadband Access Signalling, DSS2**	**279**
16.1	Introduction	279
16.2	Capabilities	279
	16.2.1 Connections	279
	16.2.2 ATM services	280
	16.2.3 Error recovery	280
	16.2.4 Narrowband interworking	280
16.3	Primitives	281
16.4	Message format	281
	16.4.1 General format	281
	16.4.2 Message type	282
	16.4.3 Message Length Field	283
	16.4.4 Variable Length Information Elements	283
16.5	Message Examples	285
	16.5.1 Set-up Message	285
	16.5.2 Release Message	286
16.6	Codesets	286
16.7	Call/Connection Procedures	286
	16.7.1 Message sequence – establishment	286
	16.7.2 Message sequence – release	287
	16.7.3 Connection identification	287
	16.7.4 Compatibility checking	288
	16.7.5 Support of N-ISDN Services	288

		16.7.6	Interworking of N-ISDN and broadband	289
	16.8	Chapter Summary		290
	16.9	References		291

17 Interworking of CCS Systems — 293

	17.1	Introduction		293
	17.2	Interworking Principles		293
		17.2.1	Constituents	293
		17.2.2	Primitive constituent	294
		17.2.3	Procedure constituent	295
		17.2.4	Format constituent	296
	17.3	Example of Interworking for a Basic Call		296
		17.3.1	General	296
		17.3.2	Basic call procedure constituent	297
		17.3.3	Basic call primitive constituent	299
		17.3.4	Basic call format constituent	301
	17.4	Example of Interworking for an Unsuccessful Basic Call		303
	17.5	Example of Database Access Call		304.
		17.5.1	Network structure	304
		17.5.2	Typical service	305
		17.5.3	Procedure constituent	305
		17.5.4	Format constituent	306
	17.6	Example of a Complex Database Access Call		307
	17.7	Chapter Summary		308
	17.8	References		309

18 Internet Protocols — 311

	18.1	Introduction		311
		18.1.1	General	311
		18.1.2	Terminology	311
	18.2	IP Signalling Architecture		312
		18.2.1	Structure	312
		18.2.2	Relationship with the OSI Model	313
		18.2.3	Protocols	313
	18.3	Address Resolution Protocol (ARP)		315
	18.4	Internet Protocol (IP)		315
		18.4.1	IP service	315
		18.4.2	IP formats	316
		18.4.3	Extension Headers	317
		18.4.4	Internet Control Message Protocol (ICMP)	318
	18.5	Transmission Control Protocol (TCP)		319
		18.5.1	Architecture	319
		18.5.2	Functions	319
		18.5.3	Formats	321

		18.5.4	Connection establishment	323
		18.5.5	Connection closure	324
		18.5.6	Interface to users	325
	18.6	User Datagram Protocol (UDP)		325
		18.6.1	Architecture	326
		18.6.2	Format	326
		18.6.3	Interfaces	327
	18.7	File Transfer Protocol (FTP)		327
		18.7.1	File transfer model	327
		18.7.2	Control Connection	328
		18.7.3	Data connection	330
	18.8	Simple Network Management Protocol (SNMP)		331
		18.8.1	Architecture	331
		18.8.2	General message format	332
		18.8.3	Polling Protocol Data Units	333
		18.8.4	Get-Response PDU	333
		18.8.5	Trap PDU	334
		18.8.6	Procedures	334
	18.9	Chapter Summary		335
		18.9.1	Common signalling principles	335
		18.9.2	IP signalling architecture	335
		18.9.3	The Address Resolution Protocol (ARP)	335
		18.9.4	The Internet Protocol (IP)	336
		18.9.5	The Internet Control Message Protocol (ICMP)	336
		18.9.6	The Transmission Control Protocol (TCP)	336
		18.9.7	The User Datagram Protocol (UDP)	337
		18.9.8	The File Transfer Protocol (FTP)	337
		18.9.9	The Simple Network Management Protocol (SNMP)	338
	18.10	References		338
19	**Conclusions**			**339**
	19.1	The Changing Environment		339
	19.2	Requirements		339
	19.3	Architecture		340
		19.3.1	Types of architecture	340
		19.3.2	Common factors	341
	19.4	Telecommunications and Internet Signalling		341
	19.5	Standards		342
	19.6	Vision		342
	19.7	Signalling: The Lifeblood		343
Appendix 1	**Networks**			**345**
	A1.1	Switches		345
		A1.1.1	Telephony	345
		A1.1.2	Broadband	346

				Contents	xxi

		A1.1.3	Mobile communications	346
		A1.1.4	Internet Protocol	346
	A1.2	Transmission Links		346
	A1.3	Calls		347

Appendix 2	**Channel Associated Signalling**			**351**
	A2.1	Introduction		351
	A2.2	Loop-Disconnect Signalling		351
		A2.2.1	Loop conditions	351
		A2.2.2	Dialled digits	352
		A2.2.3	Called customer answer	352
		A2.2.4	Calling customer cease	352
		A2.2.5	Called customer cease	353
		A2.2.6	Signals	353
		A2.2.7	Constraints	353
	A2.3	Long-Distance DC Signalling		354
	A2.4	Voice-Frequency Signalling		355
		A2.4.1	Frequencies	355.
		A2.4.2	Modes	355
		A2.4.3	Application	357
		A2.4.4	Signal reception	357
	A2.5	Outband Signalling		357
		A2.5.1	Frequency	357
		A2.5.2	Application	358
		A2.5.3	Modes	358
		A2.5.4	Signals	358
	A2.6	Multi-Frequency Inter-register Signalling		359
		A2.6.1	Registers	359
		A2.6.2	Selection signalling	359
		A2.6.3	Frequencies	359
		A2.6.4	Advantages	360
		A2.6.5	ITU-T signalling system R2	360
	A2.7	Signalling in Pulse Code Modulation Systems		362
		A2.7.1	General	362
		A2.7.2	30-channel PCM systems	362
		A2.7.3	24-channel PCM systems	363
	A2.8	References		365

Appendix 3	**Evolution of Signalling Systems**		**367**
	A3.1	Introduction	367
	A3.2	Step-by-step Switches	367
	A3.3	Separation of Control and Switch Block	367
	A3.4	Common Channel Signalling	368
	A3.5	Non-Circuit-Related Signalling	369

xxii *Contents*

Appendix 4	**ITU-T Signalling System No. 6**		**371**
	A4.1 Introduction		371
	A4.2 General Description		371
		A4.2.1 Transmission	371
		A4.2.2 Signal units	371
		A4.2.3 Exchange functions	372
	A4.3 Formatting Principles		373
		A4.3.1 General	373
		A4.3.2 Lone Signal Unit (LSU)	373
		A4.3.3 Multi-Unit Message (MUM)	374
		A4.3.4 Initial Addresss Message (IAM)	374
		A4.3.5 Signalling system control signals	375
		A4.3.6 Management signals	375
	A4.4 Procedures		376
		A4.4.1 Basic call set-up	376
		A4.4.2 Basic call clearing	377
		A4.4.3 Continuity check	377
		A4.4.4 Error control	378
	A4.5 Comparison of CCSS6 and CCSS7		378
	A4.6 Summary		379
	A4.7 References		381
Appendix 5	**CCSS7 Telephone User Part**		**383**
	A5.1 General		383
	A5.2 TUP Formats		383
		A5.2.1 Message format	383
		A5.2.2 Initial Address Message (IAM)	385
		A5.2.3 Address Complete Message (ACM)	385
		A5.2.4 Answer Signal (ANS)	386
		A5.2.5 Clear Forward (CLF) Signal	386
	A5.3 TUP Procedures		386
		A5.3.1 Basic call set-up	386
		A5.3.2 Call-release	388
		A5.3.3 Abnormal conditions	388
	A5.4 TUP Supplementary Services		389
		A5.4.1 General	389
		A5.4.2 Closed User Group (CUG)	390
		A5.4.3 Calling line identification	390
		A5.4.4 Called line identification	390
		A5.4.5 Redirection of calls	390
		A5.4.6 Digital connectivity	390
	A5.5 Summary		390
	A5.6 References		391

Appendix 6	**Inter-PABX Signalling**		**393**
	A6.1 Introduction		393
	A6.2 Private Network		393
	A6.3 Format		393
	A6.4 Procedures		394
		A6.4.1 Basic call	394
		A6.4.2 Supplementary services	395
	A6.5 Conclusion		396

Glossary — **397**

Abbreviations — **417**

Index — **425**

Acknowledgements

I should like to thank John Ziemniak and David Hilliard of Mentor Technology International for their support in writing this book. Thanks also apply to Professor Flood for his painstaking task of reviewing manuscripts.

My thanks are also extended to the many participants at the International Telecommunications Union (ITU) and other forums. It was both fun and hard work deriving recommendations and it was an honour to have worked with experts from around the world. Some figures and text have been reproduced from relevant recommendations with authorisation from the ITU. I chose the excerpts to explain the principles and the responsibility for accuracy, context, etc. is mine. The recommendations, forming the specifications of signalling systems, are available from the ITU, Place des Nations, Geneva, Switzerland.

Finally, thank you to all my friends and family for their support during this project. Leo and Star helped in their inimitable style.

Chapter 1
Introduction

1.1 The Changing Environment

Rapid and dramatic changes are taking place in today's society. Not least is the evolution of the Information Age, in which the role of the telecommunications, Internet, computing and broadcast/media industries is expanding to unprecedented levels. The amount of data being generated, and the desire to gain access to this information, are expanding beyond recognition. The Information Age will have an impact on society similar to the industrial revolution of the 1800s, transforming working practices and social behaviour.

Information Age customers are demanding increasing levels of value, choice and mobility and they are becoming much more global in their outlook. They expect greater influence over service provision and a higher degree of control over their services. Competition amongst suppliers of services is intense. Many joint ventures/mergers are taking place and the number of niche suppliers is expanding rapidly.

1.2 The Network Response

Responding to these ever more sophisticated needs is indeed a challenge. The key to providing services to customers is the establishment of a flexible, secure and high-functionality network.

In the past, network technologies and service provision have been inextricably mixed, resulting in slow and expensive evolution. Modifying a service, or introducing a new service, required changes to numerous network elements. A major drive over the last few years has been to separate the network infrastructure from the functions of service provision. The aim is to provide a flexible and secure network platform that can support a vast range of services, while remaining independent from the services. This separation has not yet been attained, but progress has been made and it remains a target for early achievement.

The intelligence to provide services in the past has invariably been lodged in networks (e.g. in a local exchange). With customer terminal equipment gaining processing power, the future will encompass a more balanced approach to service provision, with both networks and terminals taking an active role.

However, there will always be a need to cater for low-intelligence terminals (e.g. old telephones). Networks must therefore respond to both extremes of terminal processing power.

As competition in network operation intensifies, it is essential that costs are optimised and flexibility maximised. This is achieved by providing effective management capabilities. Although network management has been recognised in the past as essential for effective operation, even more focus is being placed on this aspect now. Service management will become a major contributor to effective communications in its own right. Advanced management capabilities will provide customers with much greater control over their own services.

Technologies are advancing swiftly. Such advances have provided the basis for much of today's communications ability. Digital techniques are giving greater processing power and higher levels of miniaturisation at lower prices. New switching and transmission technologies are providing the basis for an explosion in the amount of bandwidth available to customers, thus catering for the emerging multimedia and visual services.

1.3 Types of Network

The advanced capability to provide services depends on network provision. Today, several types of public network are operated to meet the increasingly demanding expectations of customers. Key types of network are illustrated in Figure 1.1. These networks originally operated on a national basis, with international communications conforming to strict regulation. As industries become more international in nature, and the level of business and personal travelling continues to expand, the networks are quickly becoming more global in nature.

Fig. 1.1 Types of network

1.3.1 Telecommunications networks

Telecommunications networks in the past provided telephony services, using analogue switches and transmission equipment. The services applied to fixed communications (i.e. with little capability for the mobility of users). The networks were based primarily on circuit switching, in which a physical circuit is provided when a call is requested and the circuit remains in place for the duration of a call.

As the need for data transfer grew, some Telecommunications Packet Data Networks were also provided, but these remained relatively small. Packet Data Networks operate on the basis of splitting a datastream into small parcels (packets) and routing each packet individually to its destination.

Telecommunications networks now operate with digital switches and transmission equipment. New nodes have been added to perform specific functions, e.g. remote databases providing specialised translation functions. Both circuit-switching and packet-switching technologies are encompassed. Digitally based networks have expanded their scope to facilitate:

- digital communication, using for example Integrated Services Digital Networks (ISDNs), adopting digital techniques from one user to another for data transfer, etc.;
- mobility, in the form of mobile telephones, debit card calling facilities, etc.;
- multimedia services, which combine computing, Internet and telecommunications techniques;
- high-bandwidth services.

Telecommunications networks continue to expand to provide communications between users and enable the transfer of information.

1.3.2 Internet technologies

The Internet has grown significantly over the past few years and is a major force in providing information transfer capabilities. Internet usage is booming and new services are being introduced continually. The Internet uses Internet Protocol (IP) technologies for information transfer. The term 'IP' refers to a method of transferring information (a protocol) within the Internet, but it is now also used as a general description of Internet technologies.

The Internet is based on a network of routers, search engines and servers/databases. Routers are similar to switches in a telecommunications network, routing calls according to user needs. However, routers operate on a completely different addressing scheme from telecommunications switches. Search engines analyse requests for information and direct a call to an appropriate database. Servers provide a capability for applications and databases hold vast quantities of information.

The Internet has evolved with unclear ownership and is designed to have simple management systems. This, in turn, results in a lower quality of service compared with telecommunications networks. To meet the increasingly sophisticated requirements of customers, especially the business community, more management capability will be included, but this will be with a cost penalty.

1.3.3 Broadcast/media

Broadcast/media industries, such as the television industry, have traditionally relied on the radio spectrum to distribute bandwidth, either on a terrestrial basis or by satellite. Cable-television operators originally entered the field in the guise of distributing television channels. They are expanding their scope to supply telecommunications and IP services, which are often more profitable.

The broadcast/media industry is now undergoing massive changes, with cable television operators vying for dominance with satellite and digital broadcasting in both the television and communications fields. Cable operators will continue to be a major form of competition in the provision of services and they will make use of telecommunications and IP networks.

1.3.4 Computing

The computing industry has evolved its focus from large mainframe devices to personal computers (PCs). PCs connected to networks and servers establish powerful information transfer capabilities. The processing capacity of PCs will continue to expand and they will provide services of a similar nature to those provided in telecommunications and IP networks.

1.3.5 Management networks

Networks are complex entities and, to bring out their true attributes, they need be managed effectively. Network management is a key factor in allowing network configurations to take place, to detect and rectify faults and to measure performance. Network management is part of an overall management approach that also includes service management and customer care systems. Management networks generally use packet data techniques as their primary transfer mechanism.

1.3.6 Evolution of networks

A common theme amongst the telecommunications, IP, broadcast/media and computing industries is the generation, storage, manipulation and transfer of information (data). In the past, the sectors followed their own paths. In the future, customers will demand convergence at a service level: services will be provided in a seamless manner, regardless of the network used to support the services.

All types of network will continue to evolve. Telecommunications networks will make more use of packet data techniques as, for example, asynchronous transfer mode (ATM) switching is introduced to cater for services requiring high bandwidth. Telecommunications networks will continue to improve mobility and the common aspects of fixed and mobile networks will encourage convergence as a cost reduction measure. The land-based broadcast networks (i.e. the ex-cable television networks) will join the telecommunications networks approach.

Internet technologies will continue to increase in influence and will provide a platform for services such as telephony over IP. The quality of service of IP will need to improve to meet the expectations of customers. At the same time, the cost of telecommunications provision will decrease. This will result in two network platforms capable of providing a foundation for services. Many people see telecommunications and IP as rivals in a bid to win supremacy. However, the maximum benefit to customers will be achieved by assuming that both types of network will co-exist, and the challenge will be to take maximum advantage from a balance of the two approaches.

Management networks will continue to increase in importance as cost reduction becomes paramount and users demand more control. They will expand significantly in their degree of functionality.

1.4 Scope of this Book

This book focuses on telecommunications signalling. However, signalling systems in IP technologies adopt very similar principles, e.g. the use of a tiered structure for specification. The book draws out these similarities and includes a chapter on IP signalling systems to provide an insight into IP technologies and to demonstrate the similarity with telecommunications in this subject area. Networks have a high level of management capability built into their procedures. These capabilities are described in this book. The book also describes the interface with the management network based on the Telecommunications Management Network (TMN) Architecture specified by the International Telecommunications Union (ITU).

1.5 Signalling: The Heart of the Network

It is an exciting time for the evolution of networks and services and there is a great deal yet to come. Signalling, which provides the ability to transfer control information, is at the heart of this revolution. This book provides a foundation for the principles of signalling and explains the great variety of systems in use and evolving. I hope that the importance of signalling becomes clear: effective signalling turns inert elements of equipment into a living, cohesive and powerful network that is the lifeblood of meeting the expectations of its customers.

1.6 Chapter Summary

Major changes are occurring in society and the evolution of the Information Age will have a significant impact upon working methods and social behaviour. A key response of networks to these changes is to separate the functions of network and service provision. The aim is to provide a flexible and secure network platform that can support a vast range of services.

The intelligence to provide services will encompass a balance between network and terminal equipment. Advanced management capabilities will provide customers and operators with much greater control over services. New switching and transmission technologies are providing the basis for an explosion in the amount of bandwidth available to customers, thus catering for the emerging multimedia and visual services.

The computing, broadcast/media, IP and telecommunications industries have a common goal to generate, store, manipulate and transfer information. The separate technological paths previously adopted by these industries will converge to meet the common requirements of the sectors.

All networks will evolve to provide digital communication, increase mobility (in the form of mobile telephones, etc.) and provide multimedia and other high bandwidth services.

Many people see telecommunications and IP technologies as rivals in a bid to win supremacy. However, the maximum benefit to customers will be achieved by assuming that both types of network will co-exist, and the challenge will be to take maximum advantage from a balance of the two approaches.

Signalling provides the ability to transfer control information and it is signalling that is at the heart of the Information Age revolution. This book provides a foundation for the principles of signalling in telecommunications and IP networks. Effective signalling turns inert elements of equipment into a living, cohesive and powerful network that is the lifeblood of meeting the expectations of its customers.

Chapter 2
Principles of Signalling Systems

2.1 Introduction

2.1.1 Scope

Chapter 1 described the dramatic changes that are taking place in communications. This Chapter takes a step into the detail and outlines the principles of signalling. The role of signalling is explained and the importance of bringing networks to life is demonstrated.

2.1.2 Networks

At a physical level, information transfer takes place over transmission links consisting of metal cables (e.g. copper cables), optical fibres or radio spectrum (e.g. mobile radio paths, satellite links). Each user needs to be able to communicate with all other users. This is achieved by providing switches that permit sharing of the transmission links. The combination of switches and transmission links forms the backbone of networks. Readers unfamiliar with the structure and terminology of networks are advised to study Appendix 1 which explains the role of switches and transmission links and introduces the concept of a call.

2.1.3 Signalling

So, what is signalling? Signalling is the lifeblood, the vitalising influence, of networks. Signalling transforms the inert network backbone from a passive group of elements into a live entity. It provides the bond that holds together the multitude of transmission links and nodes in a network to provide a cohesive entity. Through the flow of signalling information the network becomes a tremendously powerful medium that provides customers with an expansive communications ability.

Signalling provides the ability to transfer control information. Control information is any form of data required by the network or user terminal to meet customers' needs. The type of control information that is transferred varies greatly. One example is a customer request to establish a call. Another example is data needed to perform an intricate internal operation in the network. The control information can be passed (a) between customers, (b) between nodes and

(c) between customers and nodes. Signalling between nodes (or exchanges) is termed inter-nodal (or inter-exchange) signalling. Signalling between customers and nodes is termed access signalling.

It is signalling that provides the ability to transfer control information between customers and nodes and between nodes. It is signalling that allows a customer to indicate that a call is required. It is signalling that transfers the information to establish and release a call. It is signalling that transforms the foundation of the network from a passive group of elements into an active entity that can provide the required service to the customer.

So, we can say that signalling is critical to the successful operation of networks and provision of services. This chapter describes in further detail the types of signalling that exist and explains the principles that form the basis of modern signalling systems.

The critical nature of signalling and its role in the Information Age is driving its rapid evolution. Old signalling systems were simple mechanisms for transferring basic information. These are being replaced rapidly by efficient data-transfer highways. The ultimate objective is to provide an unimpeded transfer of information that forms the foundation for a wide range of services.

2.1.4 Terminology

Many terms have different meanings depending on the context in which they are being used. For example a 'user' can be a customer making a call within an overall communications context. In a network context, a user can be a computer program in the depths of a switching system. This form of inconsistency is apparent even in the international standards for signalling systems. This can be very confusing. The best advice is to remember that terms can have multiple meanings and that it is always necessary to determine the context in which a term is being used. With this advice, the principles of signalling systems can be established despite the confusing terminology.

This book adopts the terms used in international standards and every effort is made to explain the context of each term. The Glossary gives a brief explanation of key terms.

2.1.5 Types of signalling

Signalling information can be transferred in two ways, by Channel Associated Signalling (CAS) or Common Channel Signalling (CCS). In CAS systems, dedicated signalling capacity is provided for each traffic circuit. In CCS systems, signalling capacity is provided on a common basis and provided as and when necessary. An explanation of the principles of these two forms of signalling is the basis for the rest of this chapter.

2.2 Channel Associated Signalling

2.2.1 Principle

Consider two exchanges in a network that are connected by transmission links, as illustrated in Fig. 2.1.

Assume that the exchanges need to transfer information, e.g. to establish a speech path over a circuit within a transmission link. In Channel Associated Signalling (CAS), a dedicated means of transferring signalling information is provided for each circuit. Thus, in Fig. 2.1, each speech circuit (denoted by a solid line) has dedicated signalling capacity (denoted by a dotted line). The signalling information is either transferred over the circuit itself or dedicated signalling capacity is provided for each circuit within the transmission link.

The same principles apply to signalling systems used between customers and local nodes (access signalling). For each speech circuit from the customer to Exchange A, there is a dedicated signalling channel.

2.2.2 Types of information

CAS systems transfer three types of information, namely Line Signalling, Selection Signalling and Network Signalling.

Line Signalling conveys information that changes the status of the speech circuit to which the signal refers. Basic Line Signalling covers the set-up and release of a speech circuit. Examples of line signals are:

- Seize, indicating a request to use a particular speech circuit.
- Answer, indicating that the called customer has answered the call.
- Clear forward, indicating that the calling customer has released the call.
- Clear back, indicating that the called customer has ceased the call.

Selection Signalling is used to convey address information (e.g. the digits dialled by the calling customer to identify the called customer). The address information is used by the network to route the call to the called customer. Depending on the signalling system used, Selection Signalling may also include procedures to increase the speed of address transfer. One measure of the quality

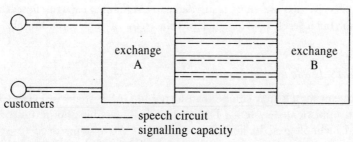

Fig. 2.1 Concept of channel associated signalling

of service perceived by the calling customer is the 'post-dialling delay', defined as the time between completion of dialling the address information and the receipt of a tone/announcement by the calling customer. The techniques used within the network for selection signalling have a great impact on the post-dialling delay.

Network Signalling covers those aspects of information transfer relating to network operation, e.g. maintenance and operational support information.

2.2.3 Forms of channel associated signalling

CAS is implemented in several forms, outlined in items (a) to (d) below. These are further expanded in Appendix 2. Further details of CAS systems are well documented.[1]

(a) In Loop Disconnect Signalling Systems, electrical conditions are used to send signals. For example, for an analogue circuit between a customer and a local exchange, an electrical loop is formed to allow speech to occur. Signals can be sent by establishing (making) and breaking this loop. Making the loop signals the intention to establish a call and breaking the loop indicates the release of the call. Trains of timed loop disconnects indicate digits being dialled by the calling customer.

(b) In Voice Frequency Signalling Systems, specific frequencies can be transmitted down analogue speech circuits in certain orders, or in timed pulses, to indicate a range of signals. These systems can use a single frequency or a combination of frequencies to define a signal.

(c) Outband Signalling Systems can be used in analogue transmission links carrying multiplexed circuits. Such transmission links often allocate 4000 Hz for each circuit but only use the range 300–3400 Hz for speech. A typical Outband Signalling System uses the frequency 3825 Hz to define signals.

(d) In digital transmission links, the circuits are typically 64 kbit/s streams. One method of transferring CAS information on digital transmission links is to encode the frequencies that have been described above into digital format and transmit that information down the speech path to which the information refers. Another method is to allocate one of the 64 kbit/s streams as a signalling channel and allocate a part of that signalling channel to each speech circuit within the transmission link. In this case, there is still dedicated signalling capacity for each speech circuit, but it is collected together and placed into a single signalling channel.

2.2.4 CAS disadvantages

The range of signals that can be transferred in CAS systems is severely limited, even for sophisticated systems. The systems are inflexible and optimised for use with old technologies. In addition, analogue CAS systems need to provide a signalling generator and a signalling terminator for each speech circuit at each exchange, resulting in an inherent cost penalty. These disadvantages are expanded further in Appendix 2.

2.2.5 CAS application

In general terms, CAS systems are designed for use in old-technology networks, in which exchanges use analogue techniques and transmission systems are primarily analogue (although some digital transmission systems are used between exchanges).

CAS systems performed a fine job in the days of analogue networks and limited requirements for services. Indeed, in access networks, CAS systems will continue to perform a basic role for many years. However, in today's environment of service explosion, digital networks and the need for great flexibility CAS systems cannot cope with more than basic requirements. For inter-nodal signalling (and high functionality in access networks), it is 'thank you' to CAS systems for a great job and over to CCS, explained in the following Section, for the future in communications.

2.3 Common Channel Signalling

2.3.1 Principle

In CCS systems, signalling capacity is provided in a common pool. The tie between the signalling path and corresponding circuits is removed. Information is transferred between users in packets termed 'messages'. These concepts allow CCS systems to be used in two ways. The first way is termed 'circuit related (CR)'. This reflects the original use of CCS to control the set-up, clear-down and operation of circuits. Thus a message in circuit-related signalling refers to a particular circuit.

The second way is termed 'non-circuit-related (NCR)'. NCR signalling takes advantage of the removal of the tie between the signalling path and circuits. It provides the capability to transfer information that is not related to a specific circuit. This opens up a significant opportunity to achieve the aim of unimpeded transfer of information between users. Users can be customers or network nodes. Here is the basis of a data highway for signalling.

2.3.2 Circuit-related signalling

For circuit-related signalling, capacity is provided in a common pool that handles the transfer of signalling information for numerous circuits. Signalling capacity is not reserved for each traffic circuit; it is allocated dynamically for use by each circuit as and when required. All signalling transfer relating to a transmission link takes place over a dedicated signalling channel, as shown in Fig. 2.2.

12 *Telecommunications Signalling*

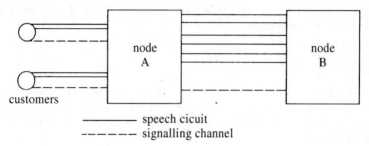

———— speech cicuit
— — — — signalling channel

Fig. 2.2   Circuit-related CCS

For inter-nodal signalling, Nodes A and B are connected by numerous circuits (e.g. speech circuits for telephony), denoted by solid lines. All the signalling that relates to the speech circuits is transferred between the nodes using the common signalling path (denoted by a dotted line). The common signalling path can be regarded as a pipe between the two nodes, typically operating at 64 kbit/s, into which all signalling information is funnelled.

Similarly, all signalling information pertaining to the speech circuits between each customer and Node A is transferred via the Access Signalling Channel. This is most appropriate to cases where a customer is connected to a local exchange by a digital line.

Because there is not a dedicated relationship between circuits and signalling, it is necessary to identify which traffic circuit a particular message refers to. This is achieved by including a circuit identifier in the message being transferred. This is explained further in Section 2.3.4.

2.3.3 *Non-circuit-related signalling*

In NCR signalling, the signalling channel is used as a general data transfer mechanism. This allows users to exchange information that is not related to circuit control. An example is given in Fig. 2.3.

The Local Node has circuits and a circuit-related signalling channel to the Transit Node. However, the Local Node also has an NCR connection to the Database. Note that the connection to the Database is a signalling connection only and is not related to any circuit. In this example, the Local Node needs spe-

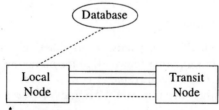

Fig. 2.3   Example of NCR signalling

cialised information from the database as part of the call set-up. The information can be requested and supplied using NCR signalling, allowing the call to proceed as normal.

As the link between the Local Node and the Database is a common pool of signalling capacity, it is necessary to identify to which transaction a particular message refers. This is achieved by allocating a Reference Number. This is explained further in the following section.

2.3.4 Messages

The transfer of signalling information is achieved by sending messages down the common signalling path. A message is a block of information that is divided into fields, each field defining a certain parameter. The structure of a message, including the fields and parameters, is defined by the specification of the signalling system.

The use of messages in CCS systems opens up a whole range of flexibility that is not present in CAS systems. Instead of being limited to a small number of meanings for signals, messages can be designed to cover a multitude of requirements. This approach is a major advantage of CCS systems. Numerous descriptions of messages are given in later chapters, but an example of a simple message is given in Fig. 2.4.

(a) Flag

In the example, Field 1 contains a 'flag', which is a unique code that identifies the start of the message. Hence, a node receiving numerous messages from the common signalling channel can detect when one message finishes and another message starts.

(b) Identifier Field

As all signalling information is funnelled into one signalling channel, it is necessary to identify to which traffic circuit or transaction a particular message refers. This is achieved by including information within the message itself and this is the function of Field 2 (Identifier).

For circuit-related signalling, each traffic circuit is allocated a unique number that identifies that traffic circuit. A message related to a particular circuit (e.g. an instruction to release a particular traffic circuit between two exchanges) is identified by the number of the circuit to which the information

field 4	field 3	field 2	field 1
check	information	identifier	flag

Fig. 2.4   Example of a simple message

refers. The Identifier Field in a circuit-related application is the number of the traffic circuit to which the message pertains.

For NCR signalling, a Reference Number is used to identify the relevance of a message. The Reference Number is allocated by nodes or customers' equipment from a pool of available numbers at the start of a call. The Reference Number is used for the life of the call. When the call is terminated, the Reference Number is returned to the pool.

(c) Information Field

The Information Field contains the heart of the signalling information that is being transferred. For example, the Information Field could be coded 'release', meaning that the traffic circuit indicated in Field 2 should be disconnected.

(d) Check bits

Field 4 contains check bits, which are generated by a known algorithm and which are used to ensure that the message has not been corrupted during its passage through the network.

2.3.5 Buffers

In CCS systems, the generation of messages is volatile. During some periods there is little signalling activity, whereas at other times there are too many messages to fit within the channel. To handle this situation, 'buffers' are provided at each end of a CCS link to store each message until the link is available. As messages are received for transfer, they are stored in the buffer and transmitted in a specified order. A typical order of transmission is first in/first out, resulting in messages being transmitted in the same order as they are received.

The principle of buffers is illustrated in Fig. 2.5.

At time T (part a of figure), Messages 2, 6 and 5 of Calls 2, 3 and 1, respectively, have been transmitted to line. The buffer is storing six messages in preparation for transmission. Message 5 of Call 2 is arriving at the buffer.

At time $T+1$ (part b of figure), the next time slot on the transmission link is available and Message 6 of Call 1 has been transmitted. The other messages in the buffer have moved up one place in the wait for transmission. Message 5 of Call 2, being the first new message to arrive at the buffer, has been allocated a slot for transmission after all the other messages in the buffer have been transmitted. Message 8 of Call 3 has arrived at the buffer and will be allocated a slot at time $T+2$.

When there are no messages to transmit, there is a need to maintain synchronisation of the signalling channel between two nodes. This is achieved by continuously transferring synchronisation information until a new message is ready for transmission.

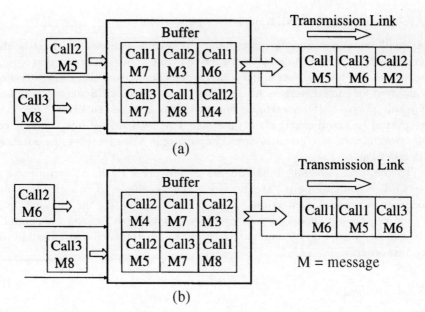

Fig. 2.5 Buffers

2.3.6 Signalling activity

To use a common signalling channel, CCS systems rely on the intermittent nature of signalling for traffic circuits.

For circuit-related signalling, to establish a call requires signalling between the calling and called users. During the conversation/data phase of the call, there is no requirement to transfer signalling information. When the call is completed, there is again a need to transfer signalling information to release the speech circuit. The signalling activity when setting-up and releasing a circuit is high. However, on average the signalling activity for a circuit is low because there is no signalling when calls are not being made and during the conversation phase of a call.

Hence, a single CCS channel can be used to handle numerous traffic circuits. The theoretical limit of the number of traffic circuits handled by a CCS channel is very high, but a typical practical value is 2000 traffic circuits.

The picture becomes more complex when NCR signalling activity is taken into account. NCR signalling can be intermittent (e.g. if it is used during call establishment to interrogate a database) or it can exhibit a high average signalling activity (e.g. if it is used to transfer large amounts of data). Hence, in a practical network, the calculation of signalling activity is an important element in designing the structure of the network and dimensioning signalling links.

2.3.7 Error detection/correction

A small error in a message can change its meaning dramatically. It is also important in an environment of software-controlled nodes to ensure that logical sequences are followed accurately. Error detection mechanisms are therefore employed for each message sent. The check bits in each CCS message are used for this purpose. If an error is detected, this triggers an error correction mechanism to retransmit the faulty message. Further details of detection/correction mechanisms are given in later chapters (e.g. Chapter 4 for the Message Transfer Part).

To improve signalling reliability on a network-wide basis, facilities (e.g. network reconfigurations) are provided. For example, if a signalling link between two nodes is interrupted, it is possible to re-route the messages over other transmission links and even via other nodes. These facilities are carefully defined to ensure that messages are received by the intended destination in the correct order.

2.3.8 Specification

CCS system specifications include the definition of 'formats' and 'procedures'. Formats define the structure of the messages used and the meaning of each field within the message. Procedures define the logical sequence in which messages can be sent.

The procedures of early versions of CCS circuit-related systems are very closely linked to the functions within nodes that control the set-up and release of calls. There is, therefore, a close relationship between CCS procedures and node call control. In later versions, attempts have been made to separate the two factors, so that signalling systems become more independent of the call control functions. Further details are given in Chapter 3.

2.4 Advantages of Common Channel Signalling

CCS has been adopted throughout the world in national and international networks. It has tremendous advantages over previous systems and these are summarised below:

2.4.1 Harmony with control techniques

In the early systems, exchanges could communicate, but in a limited and inflexible manner. Thus, the flexibility of call control was limited. In a CCS environment the objective is to allow uninhibited communication between node control functions (processors), thus tremendously broadening the scope and flexibility of information transfer.

For circuit-related applications, there is an aim to separate call control within a node and CCS. However, there is still a close relationship. Procedures for circuit-related CCS systems define the logical sequences of events that occur during call establishment and release, thus having a direct impact upon the call-control process. In this environment, CCS systems are designed to be in harmony with call control, achieving many more features than CAS.

In non-circuit-related (NCR) applications (e.g. Transaction Capabilities described in Chapter 7), there is much greater separation between CCS and call control. This results in even more flexibility in the signalling systems, thus allowing the call control applications to evolve freely without constraint.

2.4.2 Signalling terminations

Further advantages result from eliminating per-circuit signalling termination costs. These costs are inevitable in per-circuit CAS systems, but by funnelling all signalling information into a single common channel, only one signalling termination cost is incurred for each transmission link. There are cost penalties for CCS systems, e.g. the messages received by a node have to be analysed, resulting in a processing overhead. However, these cost penalties are more than outweighed by the advantages of increased scope of inter-processor communication and more efficient processor activity.

2.4.3 Overcoming limitations of CAS systems

CAS systems possess limited information-transfer capability due to:

(i) The restricted number of conditions that can be applied (e.g. the limited variations that can be applied to a DC loop or the limited number of frequency combinations that can be implemented in a voice-frequency system).

(ii) The limited number of opportunities to transfer signals (e.g. it is not possible to transmit voice-frequency signals during the conversation phase of a call without taking special measures).

Neither of these restrictions applies to CCS: the flexible message-based approach allows a vast range of information to be defined and the information can be sent during any stage of a call. Hence, the repertoire of CCS is far greater than CAS versions and messages can be transferred at any stage of a call without affecting the calling and called customers.

2.4.4 Speed of transfer

CCS systems transfer signals very quickly. A message used to establish a call in a CCS system can contain all the address digits in an information field. The message is delivered in a form suitable for modern processor-controlled nodes, thus resulting in fast route selection. This speedy signalling also permits the inclusion of far more information without an increase in post-dialling delay.

One of the problems that prompted the development of CCS systems was 'speech clipping' in the international network. In some CAS systems, it is necessary to split the speech path during call set-up to avoid tones being heard by the calling customer. This results in a slow return of the answer signal and, if the called customer starts speaking immediately after answer, then the first part of the statement by the called customer is lost. As the first statement is usually the identity of the called customer, this causes a great deal of confusion and inconvenience. CCS systems avoid the problem by transferring the answer signal quickly.

2.4.5 User Information

Techniques used in modern CCS systems can further improve the flexibility provided to customers. User-to-user signalling is a technique whereby messages can be transferred from one customer to another without undergoing a full analysis at each node in the network.

'End-to-end' signalling allows nodes (e.g. local exchanges) to transfer information to each other with minimum processing at intermediate nodes. Although forms of end-to-end signalling are possible using CAS systems, the technique can be more efficiently implemented with CCS systems.

2.4.6 Reliability

The processing ability of CCS systems allows a high degree of reliability to be designed into the signalling network. Error detection and correction techniques are applied, with a resulting high confidence in the transfer of uncorrupted information. In the case of an intermediate node failure, re-routing can take place within the signalling network, enabling signalling transfer to be continued. Although these features introduce extra requirements (see Section 2.5), the common channel approach to signalling allows a high degree of reliability to be implemented economically.

2.4.7 Evolutionary potential

A major restriction of CAS is the lack of flexibility, e.g. the ability to add new features is limited. One factor that led to the development of CCS was the increasing need to add new features and respond to new network requirements. Responses to new requirements in CCS can be far more rapid and comprehensive than for channel associated versions.

CCS systems are not just designed to meet current needs. They are designed to be as flexible as possible in meeting future requirements. One way of achieving this objective is to define modern CCS systems in a structured way, specifying the signalling system in a number of tiers. Chapter 3 describes the structure of modern CCS systems. This approach allows each tier to be optimised for various applications, while avoiding changes to other parts of the signalling system. The result is a flexible signalling system that can react quickly to evolving requirements. Some of these requirements were foreseen in the original designs of CCS systems, but many additional requirements will arise.

2.4.8 Multipurpose platform

A corollary of the structured approach is that modern CCS systems are not restricted to specific services (e.g. the establishment and clear-down of telephone calls). The ability of CCS systems to transfer general data, and the increased range of messages that can be transferred, mean that information related to any service can be handled. Hence, future services can be incorporated in a flexible and comprehensive manner. Changes to existing services can be implemented more quickly and at lower cost than with CAS systems.

2.4.9 CCS network

The separation of CCS from traffic circuits, and the direct interconnection of node processors, are early steps in establishing a cohesive CCS network to allow unimpeded signalling transfer between customers and nodes and between nodes in the network. The concept of a cohesive CCS network provides great flexibility and opens up the opportunity for the implementation of a wide range of network management, administrative, operations and maintenance functions. A major example of such a function is the quasi-associated mode of operation (described in Section 2.8.3). This mode of operation provides a great deal of flexibility in network security, reduces the cost of CCS on small traffic routes and extends the data-transfer capabilities for NCR signalling.

2.5 Additional Requirements of CCS

The introduction of CCS has many advantages, but additional requirements are introduced in three areas:

- reliability and security;
- speech continuity;
- processing overhead.

2.5.1 Reliability and security

A signalling channel carried on a 64 kbit/s link has the practical capacity to control approximately 2000 traffic circuits. Hence, the failure of an inter-nodal signalling link would cause the loss of a significant amount of traffic. For access signalling, the loss of the signalling link would mean isolation of the customer from the local node. It is therefore essential to take exceptional precautions to avoid such losses. On a message basis, error detection and correction mechanisms are adopted and these are described in Section 2.3.7.

Signalling security can also be improved by developing the signalling network itself. In the access network, it is possible to provide two signalling links to a customer (preferably on physically-diverse transmission links) and to transfer all signalling traffic to one link when the other link is interrupted. Similar arrangements can be made for inter-nodal signalling, with automatic reconfiguration of signalling paths, to maintain a signalling continuity in the event of the failure of a signalling link.

Using these techniques, security in the signalling network can be enhanced to meet the requirements for the unavailability of signalling. For example, the unavailability of signalling communication between two exchanges is specified as a maximum of 10 minutes per year.[2] The message and network techniques for improving security of signalling information are positive assets. However, there is a cost in implementing such techniques and these must be taken into account when assessing the advantages of CCS.

2.5.2 Speech continuity

CAS systems that use the speech path to transfer signalling information provide the inherent feature of checking the continuity of the speech path being established before conversation begins. If continuity is not achieved, the signalling transfer is not successful and the call is aborted or a further attempt is made to connect the call. This inherent continuity check is absent in CCS systems, due to the separation of the speech and signalling paths. Hence, if considered desirable, separate speech-continuity checks can be provided. However, many modern digital exchange and transmission systems do not require such a continuity check because of the general level of equipment reliability, and inherent self-checking features, in such systems.

2.5.3 Processing overhead

The flexible manner in which CCS systems are structured and the implementation of complex network management features mean that extra processing is necessary to operate CCS. Even the inherent concept of funnelling all signalling on a transmission link into a common signalling channel means that messages must be analysed to determine to which circuit (or transaction) they refer. However, this extra processing overhead is more than outweighed by the benefits of CCS systems.

2.6 Evolution of Signalling Systems

One of the major factors contributing to the advantages of CCS systems is the ability to establish a signalling network. This provides a flexible and secure foundation for service provision with clear interfaces to the call-control functions within nodes.

The evolution of signalling systems from basic CAS systems through to the CCS data highway networks of today is closely linked with the evolution of call control in evolving switch technologies. A brief description of the evolution of signalling systems from this perspective is given in Appendix 3.

2.7 Types of CCS System

2.7.1 International standards

The International Telecommunications Union (ITU-T) is responsible for defining international standards for signalling. ITU-T Signalling System Number 7 (CCSS7) is designed as an inter-nodal signalling system. Its designation derives from the previous organisation of ITU-T: it was the seventh system defined by the International Telegraph & Telephone Consultative Committee (CCITT) and it is still commonly referred to as CCITT No.7.

The Digital Subscriber Signalling System No. 1 (DSS1) is designed as an access signalling system for digital lines.

The bandwidth required by customers to fulfil their expectations is continually increasing. CCSS7 has a broadband ISDN User Part to cater for such high-bandwidth applications and access signalling uses the Digital Subscriber Signalling System No. 2.

The range and complexity of networks and services in telecommunications continues to increase. The growth in mobile communications is unprecedented and the application of signalling in this field is described in Chapter 8. Intelligent networks (IN) are becoming more important as a means of providing flexibility in networks and meeting the demands of service complexity. The signalling implications of INs are explained in Chapter 9.

Signalling System No. 6 (CCSS6) is an inter-nodal system that was defined during the early development of CCS. It is still in use, but is being replaced by CCSS7. CCSS6 is described in Appendix 4.

2.7.2 Specifications

ITU-T defines signalling systems in a series of 'Recommendations'. The objective is for all member countries to agree specifications that can be built by equipment manufacturers worldwide. International specifications of this nature reflect the ideas and views of experts throughout the world and allow customers in all countries to benefit from economies of scale in equipment development.

This book focuses upon the international standards and deliberately avoids details of national variants. In this way, the principles of CCS systems can be explained in an independent manner.

This book is not intended as an exhaustive account of CCS systems: that is left for the specifications. Although the specifications need to describe a vast range of failure modes and error conditions, these are largely omitted from this description to enable the reader to concentrate on the principles. Once the principles have been absorbed, the specifications will be much easier to read. In the same way, examples are given throughout the book, rather than an attempt to comprehensively design particular aspects of the signalling systems.

2.7.3 Access and inter-nodal differences

The first two CCS systems to be defined internationally, Signalling Systems Nos. 6 and 7, were inter-nodal systems for use within networks.

The objective of telecommunications networks is to provide fully flexible high-capability communication between customers. To achieve this objective not only requires evolution of the network but also requires corresponding abilities in links to customers. Hence, a parallel transition is taking place for access signalling systems, with the development of DSS1 and DSS2.

Access CCS systems must be able to interwork effectively with network CCS systems. Hence, there are many common aspects between the two systems and the principles described so far are applicable to both types of system. However, some differences arise between the design of the two systems as described below.

(a) Intelligence levels

Inter-nodal CCS systems continuously evolve to provide an increasing range of services. Thus, some nodes will operate later versions of CCS systems than others. However, the general level of intelligence is similar (e.g. each node can provide a basic service, even though some nodes are less complex than others). This means that assumptions can be made about the basic capability of each node in the network when designing an inter-nodal CCS system.

For access CCS systems, the capabilities of two communicating entities (e.g. a customer and a local node) can vary far more than for inter-nodal signalling systems. The node to which the customer is connected has at least the basic intelligence level discussed above. However, the capability of the terminal used by the customer can vary tremendously. The terminal can be a very simple telephone with a low level of intelligence or it can be a complex private automatic branch exchange (PABX) with as much intelligence as a network node.

Furthermore, a network operator is often unaware of the type of terminal being used by a customer. Hence, access CCS systems must take account of this wide variation in intelligence of customer terminals, e.g. by allowing simple terminals to ignore complex information without affecting the ability to establish a basic telephone call.

(b) Link efficiency

In inter-nodal CCS systems, a balance has to be achieved between the complexity of the call control at each node and the efficiency of the signalling links between the nodes. For example, in the ISDN User Part of CCSS7 described in Chapter 6, many of the commonly used messages include fields that do not explicitly state the name or length of the field. In these cases, the name and length of the field are derived from the type of message. This approach reduces the amount of information included in the message, thus allowing more messages to be carried on a signalling link. However, the approach can increase the level of processing required in a node because it is necessary to derive information from the message that is not provided explicitly.

In access CCS systems, there is less need to optimise the efficiency of the signalling link itself. This is because the signalling controls a limited number of circuits and there is less to gain by increasing the efficiency of the link at the expense of call-control complexity.

(c) Data levels

When establishing a call between two customers, signalling information can have significance to:

(i)   the link between the calling customer and the originating local node;
(ii)   the link between the called customer and the destination node;
(iii)   the network only;
(iv)   a combination of the elements of (i) to (iii).

Establishing a basic call, or invoking supplementary services, usually involves all forms of information, but the largest amount of information usually applies to items (i) and (ii). Hence, access CCS systems have to be capable of handling such large amounts of data on a per-call basis.

(d) Management

A CCS system for use in networks needs to exhibit extensive network-control mechanisms: these need to control both traffic circuits and signalling channels. Hence, CCSS7 contains complex network-control techniques. Although access signalling systems require management functions to allow operation of the access link, the management techniques are generally simpler than those adopted in the network.

(e) Terminals

Local nodes are not usually aware of the type of terminal connected by customers to access links. Indeed, numerous terminals can be connected to one access link. The protocols for the access link therefore have to take into account the prospect of multiple terminals being connected to one link.

2.8 Modes of CCS Operation

CCS systems can operate in a number of modes. To understand the modes of operation, it is first necessary to explain some appropriate terminology. A node in a telecommunications network that operates CCS is termed a 'signalling point' (SP).

Any two SPs with the possibility of signalling communication are said to have a 'signalling relation'. The realisation of the signalling relation is by sending signalling messages between the two nodes. The path taken by the signalling messages is determined by the mode of operation. Hence, the mode of operation determines how signalling messages are routed between SPs. The mode of operation can be 'associated', 'non-associated' or 'quasi-associated'.

2.8.1 Associated mode

In the associated mode of operation, the signalling messages pertinent to a particular signalling relation are transferred over transmission links directly connecting the relevant SPs. Fig. 2.6 represents the associated mode of operation, with Nodes A and B having a signalling relation. The signalling link directly connects the two nodes.

2.8.2 Non-associated mode

In the non-associated mode of signalling, the messages pertinent to a particular signalling relation are not transferred over transmission links directly connecting the relevant SPs. Instead, the messages are transferred using intermediate (or tandem) SPs. In an extreme case of non-associated signalling,

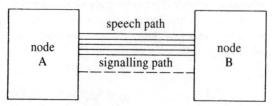

Fig. 2.6   Associated mode of operation

each message between two nodes could take a random route, with no preferred route being predetermined by the network.

2.8.3 Quasi-associated mode

In practical networks, a specific form of non-associated signalling is used, termed 'quasi-associated'. In the quasi-associated mode of operation, the path taken by a message through the signalling network is predetermined by information assigned by the network.

In Fig. 2.7, Nodes A and B have a signalling relation and are interconnected by speech paths. However, the signalling path used to implement the signalling relation is via Node C (and not directly between Nodes A and B). Hence, Fig 2.7 is an example of the non-associated mode of operation. However, the figure is also an example of the quasi-associated mode of operation because the message routing through Node C is predetermined by the network. In this case, Node C is termed a 'signal transfer point (STP)'.

The quasi-associated mode of operation can be used as a back-up in the case of signalling link failure. For example, in Fig. 2.8, Nodes A and B usually operate in the associated mode, with a signalling link directly connecting the two signalling points. However, in the case of failure of the A-B signalling link, the A-C-B signalling link can be used to control the speech paths between Nodes A and B.

The quasi-associated mode of operation can also be used to reduce the overhead cost of CCS on a small number of speech paths. In Fig. 2.9, Nodes A and C and Nodes C and B are connected by a large number of speech paths. Nodes A and B are connected by a small number of speech paths.

In this case, the small number of speech paths between Nodes A and B can be supported by quasi-associated signalling (routed A-C-B), thus avoiding the allocation of the signalling overhead to the A-B speech paths.

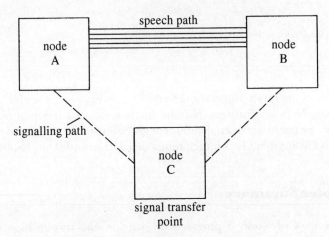

Fig. 2.7 Quasi-associated mode of operation

26 *Telecommunications Signalling*

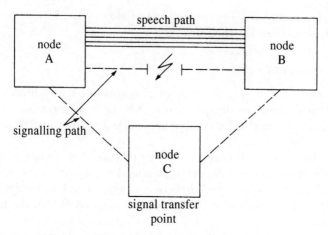

Fig. 2.8 *Use of quasi-associated mode of operation for back-up*

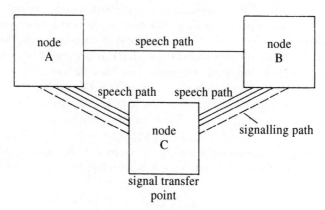

Fig. 2.9 *Use of quasi-associated mode for small routes*

2.8.4 Flexibility

These modes of operation illustrate the great flexibility and powerful nature of CCS systems. Network features, like the quasi-associated mode of operation, greatly enhance the range of applications of CCS. Such features provide a major step towards CCS systems becoming unimpeded data-transfer mechanisms.

2.9 Chapter Summary

The backbone of networks is provided by switches and transmission links. In circuit switching, a circuit is dedicated to a call for the whole period of a call. In

ATM switches, information is split into cells and the cells are transmitted over the transmission links. Each transmission link is shared on a real time basis with many users and a virtual circuit is provided when there is information to transfer. IP networks use routers, search engines and databases.

Signalling is the vitalising influence, the lifeblood, of networks. Signalling transforms the inert network backbone into a live entity and provides customers with an expansive communications ability.

There are two types of signalling system for telecommunications networks. In Channel Associated Signalling (CAS) systems, signalling capacity is provided on a dedicated basis for each traffic circuit. In Common Channel Signalling (CCS) systems, signalling capacity is provided in a common pool and is allocated on a dynamic basis, as and when required.

Appendix 2 describes six categories of CAS system. However, modern networks adopt CCS systems and this book concentrates on describing these systems. International standards have been derived for the use of CCS systems in both national and international networks. These standards are described in this book to avoid reference to national variants.

Appendix 4 gives a brief description of ITU-T Signalling System No. 6 (CCSS6), but the focus of the book is on the more modern systems, namely ITU-T Signalling System No. 7 (CCSS7) and the Digital Subscriber Signalling Systems No. 1 (DSS1) and No.2 (DSS2). CCSS7 is the modern CCS system for inter-nodal signalling and DSS1/2 are its counterparts for access signalling.

CCS systems use a common signalling channel to carry signalling information for numerous circuits. The signalling information is carried in messages. Each message is delineated by a flag and the transaction to which a message refers is defined by an identifier. In circuit-related applications, the identifier used is the circuit number of the appropriate traffic circuit. In non-circuit-related applications, the identifier used is a reference number that is independent of traffic circuits. Each message contains an information field defining the meaning of the message. Check bits are used to detect message corruption during transmission. Messages are stored in buffers to await the opportunity to be transmitted. Error detection and correction procedures are applied to provide a secure transfer of information.

The major advantages of CCS systems are that they are compatible with modern software-controlled networks, they overcome the limitations of CAS systems (particularly increasing the repertoire of messages that can be sent) and they exhibit evolutionary potential. Extra requirements are the need for a high level of reliability, the need to ensure traffic-circuit continuity and an increase in processing overhead.

Inter-nodal and access CCS systems have many aspects in common and such commonality is encouraged to reduce the complexity of interworking. However, the environments in which DSS1 and CCSS7 perform are different. Thus, differences in characteristics arise, e.g. DSS1 must be able to handle a wide variety of customer terminals.

CCS systems can operate in a number of modes. The usual modes are the associated mode and the quasi-associated mode. In the associated mode, signalling messages are transferred over transmission links directly connecting relevant signalling points. In the quasi-associated mode, signalling messages are routed via signal transfer points. The quasi-associated mode introduces a very flexible element into network design, e.g. quasi-associated signalling can be used as a back-up method of routing signalling messages if disruption occurs to the normal signalling link.

The critical nature of signalling and its role in the Information Age is driving its rapid evolution. The ultimate objective is to provide efficient data-transfer highways with an unimpeded transfer of information that forms the foundation for a wide range of services.

2.10 References

1 WELCH, S: Signalling in telecommunications networks' (Peter Peregrinus, 1981)
2 ITU-T Recommendation Q.706: 'Message Transfer Part Signalling Performance' (ITU, Geneva)

Chapter 3
Architecture of CCS Systems

3.1 Introduction

When defining modern signalling systems it is essential to adopt a structured approach for three main reasons:

- The complexity of modern signalling systems would make a specification exceedingly cumbersome if produced in a monolithic form.
- Signalling systems need to continue to evolve to match increasingly demanding requirements of customers and networks. A structured approach provides flexibility to handle new services and changes to existing services.
- A structured specification allows a disciplined approach to design and development, thus easing the process of implementation.

The term 'Architecture' is used to describe the structured approach to the specification of CCS systems. The architecture of CCS systems is the key to their flexibility and evolutionary capability. The requirements for an effective architecture are outlined in Section 3.2.

Ideally, all modern signalling systems should have been specified using a common architecture, with variances only occurring to handle specific requirements. For example, CCSS7 and DSS1 could have used a common architecture, with the functions for CCSS7 optimised for inter-nodal signalling and the functions for DSS1 optimised for access signalling.

Unfortunately, various architectures have been adopted as technologies have evolved and as long-term objectives have changed. The types of architecture are outlined in Section 3.4. However, the various architectures use a common set of principles and these are explained in Section 3.3. The architectures for CCSS7, DSS1 and DSS2 are described in the following sections.

3.2 Architecture Requirements

A large number of factors influence the architecture adopted for CCS systems. The important requirements are given below.

3.2.1 Evolution

A high degree of potential for evolution is required, thus catering for as many future requirements as possible. The objective is to achieve unimpeded communication between users of the signalling network. Implementing new concepts should not involve major changes to the signalling system.

3.2.2 Flexibility

Modern CCS systems are very complex and involve a large number of interacting parts. This results in difficulty in making changes to the system, because changing one part of a system can result in numerous changes to other parts. A key requirement is to minimise the impact of this interaction and permit the optimisation of parts without affecting other parts.

3.2.3 Resource optimisation

Networks are expensive to provide and operate. It is therefore essential that network elements are used efficiently. From a signalling perspective, this means efficient use of the signalling network itself, as well as facilitating efficient use of the other network elements.

3.2.4 Openness

Signalling system specifications should be as 'open' as possible, i.e. provide users with access to the widest range of applications possible and make interfaces standard so that they can be implemented by a range of suppliers.

3.2.5 Scope

In the past, internationally specified signalling systems catered only for telephony. Modern signalling systems need to cater for a variety of applications including telephony, data, ISDN, multimedia and mobile services.

3.2.6 Operation

In the increasingly competitive marketplace an important aspect of cost reduction is to provide a high level of automation and management control.

3.2.7 Reliability

As signalling is critical to the satisfactory operation of networks, it is essential that reliability standards are of the highest quality. Loss of a signalling link can have a major impact upon a large number of customers. In particular cases, faults can spread through the network and result in network-wide failure.

3.2.8 Compatibility

CCS systems must be able to match the changing requirements placed upon the network. This continual updating process results in new versions of CCS systems being implemented. These new versions must be compatible with previous versions that have already been implemented.

3.2.9 Signalling interworking

Significant costs can be incurred if different signalling systems are developed without consideration of interworking. For example, DSS1 in the access part of the network must interwork with CCSS7 in the inter-nodal part of the network.

3.2.10 Network interworking

The number of network operators is increasing in response to more widespread liberalisation of national telecommunications networks. This increases the level of interworking between networks. These network interworking costs need to be kept to a minimum.

3.2.11 National use

In the past, internationally specified signalling systems were designed for use in the international network, resulting in expensive signalling interworking requirements at the boundary of the international and national networks. Signalling systems must be developed for use in both national and international networks to reduce interworking costs.

3.2.12 Degree of detail

If a signalling system is defined in too much detail, the innovation of the manufacturer is constrained. If not enough detail is given, different implementations will not work with each other. A means of specification is required that assists the achievement of the correct balance.

3.2.13 Clarity

The specification must facilitate clear and unambiguous definition. Comprehensive definition of all conditions must be achieved, including the procedures for failure conditions.

3.2.14 Legacy systems

A signalling system must be able to interwork with equipment already existing in the network. Evolving CCS systems must be able to interwork satisfactorily with diverse national systems (e.g. old CAS systems).

32 *Telecommunications Signalling*

3.2.15 Speed of specification

In the past, standards for signalling systems have taken a great deal of time to develop. This has resulted in equipment developments to interim standards, thus negating the advantage of economies of scale that applies to an international standard. It also results in reluctance to derive new standards because there is a wide variety of conditions from which networks need to evolve. The architecture must therefore allow the speedy specification of evolving standards.

3.3 Architecture Principles

Various architectures have been adopted to specify modern signalling systems. Despite the variances, a common set of principles has been applied.

3.3.1 Structure within a node

Each architecture is based on defining the signalling system in a series of tiers, as illustrated in Fig. 3.1. The term 'tier' is not used in international standards; these use the terms 'layer' and 'level'. However, the terms are applied in different ways according to the context, as explained below. The term 'tier' is used here as a common terminology for the various architectures to assist in establishing the principles of architecture.

In this example architecture, the signalling functions of each node are categorised into Tiers X, Y and Z. Each tier contains a prescribed set of functions.

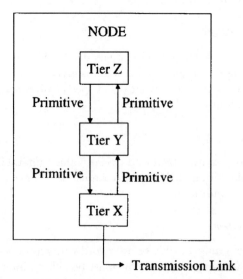

Fig. 3.1   Tiered structure within a node

Each tier is self-contained and can be optimised and/or modified without affecting the other tiers.

The interfaces between tiers are closely defined. Within a node, information can pass between tiers only by the use of units of information termed 'primitives'. Primitives are defined in terms of the information content that they carry. Information that varies according to the type of communication (e.g. a telephone number) is carried in 'parameters' (information fields) within each primitive.

Each tier, in conjunction with its lower tiers, provides a 'service' to the tier above. In Fig. 3.1, Tier X provides a service to Tier Y. Tiers Y and X provide a service to Tier Z.

To illustrate the principles, assume that the functions in Tier X determine the physical and electrical conditions that apply to the transmission link. Tier X provides a Transmission Service to Tier Y. This permits Tier Y to focus on its own functions. In a similar way, Tier Y (in conjunction with Tier X) provides a Routing Service to Tier Z. This permits information provided by Tier Z to be routed to another node.

3.3.2 Tiered structure applied between nodes

Now consider two nodes that need to transfer information, as shown in Fig. 3.2.

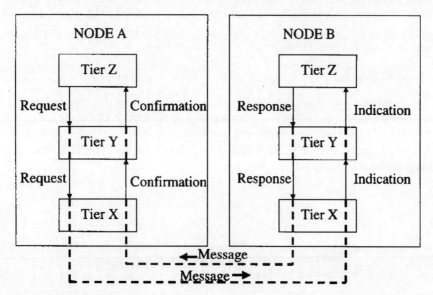

Fig. 3.2 Information transfer

The functions within a tier at Node A can communicate with the functions of the same tier at Node B by making use of the services provided by the lower tiers. For example, Tier Z in Node A can communicate with Tier Z in Node B by using the services offered by the respective Tiers Y and X.

Information transfer between tiers is by means of primitives. The general format of primitives is illustrated in Fig. 3.3. The service provider field identifies the tier(s) providing the service. The Generic Name describes the action that should be performed by the addressed tier, e.g. set-up a call. The Specific Name indicates the direction of the primitive flow and is described below. Parameters are the elements of information that need to be transmitted between two tiers (e.g. the telephone number to which a call should be established).

Two forms of primitive are available for use between nodes, namely extensive and unitdata. The extensive form applies when two nodes communicate and the initiating node expects an answer to a request. The unitdata form applies when a node sends information and does not expect a response. Communication between nodes can include both forms of primitive, depending on the type of service required by each tier.

(a) Extensive form

The extensive form of primitives is shown in Fig. 3.2, in which Node A initiates an action. Within Node A, a Request Primitive is used by a higher layer to request activities to be performed by a lower layer. Within Node B, a request from Node A is carried by an Indication Primitive. The response within Node B is carried by a Response Primitive. The response from Node B is carried within Node A as a Confirmation Primitive. The terms Request, Indication, Response and Confirmation are the Specific Names shown in Fig. 3.3. Similar primitives apply for Node B initiating an action: in this case, Node B would initiate the action with a Request Primitive.

(b) Unitdata form

The unitdata form of primitive is illustrated in Fig. 3.4. In this case, a Unitdata Request Primitive is used to initiate an action in Node A. A Unitdata Indication Primitive is used to convey the request within Node B. As Node A does not expect a reply, the Unitdata Primitive does not include Response and Confirmation Specific Names.

Service Provider	Generic Name	Specific Name	Parameter

Fig. 3.3 General format of primitives (Source: ITU Recommendation Q.70)

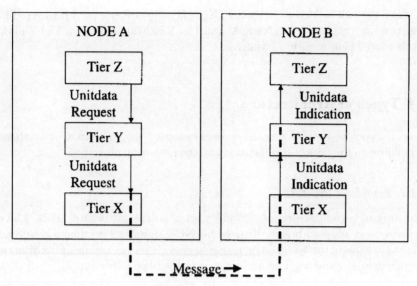

Fig. 3.4 Information transfer: Unitdata

3.3.3 Protocol

An aim of the structure is to separate the functions performed in an entity from the flow of signalling information between entities. The term 'protocol' is used to describe the procedures and the format of the signalling between a tier in one entity and the corresponding tier in another entity. The term peer-to-peer protocol is often used to describe such communication between corresponding tiers.

3.3.4 Example of application

To illustrate the use of the tier structure, assume that Node A in Fig. 3.2 needs to release a call that has been established between Nodes A and B. Tier Z in Node A is responsible for recognising the need to release the call. Thus, Tier Z formulates appropriate information and passes it, using a Request Primitive, to Tier Y. Tier Y is responsible for routing a CCS message from Node A to Node B. Thus, Tier Y amalgamates the primitive from Tier Z with its own routing information and passes the result to Tier X, again using a Request Primitive. Tier X is responsible for using an appropriate transmission link and carrying the information from Tiers Z and Y to Node B.

The result of the information supplied by Tiers X, Y and Z within Node A is a message suitable for transfer over the transmission link.

On receipt of the message, the Tiers in Node B perform corresponding functions to those at Node A, resulting in Tier Y passing an Indication Primitive to Tier Z requesting the release of the call.

When the call is released, Tier Z in Node B issues a Response Primitive. This information is transferred to Node A, and Tier Y in Node A informs Tier Z of the result using a Confirmation Primitive.

3.4 Types of Architecture

Several types of architecture have been adopted for modern CCS systems, depending on the time of specification and the long-term objectives.

3.4.1 Four-level structure

The original architecture of CCSS7 uses a 4-level structure.[1] This 4-level structure was adopted before the need for NCR signalling became widespread. The main focus of the 4-level structure is to provide services while optimising the use of network resources. This architecture is described in Section 3.5.

3.4.2 OSI layered model

The International Standards Organisation (ISO) uses a 7-layer Open Systems Interconnection (OSI) Protocol Reference Model.[2] This was originally applied to inter-processor communications, but it can also apply to any other form of communication. The main focus of the OSI Model is to provide a high degree of openness (i.e. providing interfaces that can be implemented by numerous manufacturers).

CCSS7 uses the OSI Model architecture for its NCR capabilities, e.g. the Intelligent Network Application Part (INAP) and the Mobile Application Part (MAP). The OSI Model architecture is described in Section 3.6.

3.4.3 Layers for DSS1 and DSS2

DSS1[3] uses a 3-layer structure to separate the physical, link and network aspects of access signalling. DSS2[4] is specified at the equivalent of the DSS1 Layer 3, but DSS2 uses an ATM Adaptation Layer (Chapter 14) as a Layer 2.

The relationship between the DSS1 layers and the OSI layers is still under study, but these two models do not align at this stage. There is therefore great scope for confusion in the common use of the term 'layer'. Similar comments apply for DSS2. The DSS layers are described in Sections 3.9 and 3.10.

3.4.4 Layers for Internet Protocol

The Internet Protocol (IP) uses a 4-layer structure. The IP layers have a close relationship with the OSI layers and are described in Chapter 18.

3.4.5 Evolution

Many changes in technology and focus are taking place, including:

- New requirements are arising to perform signalling for broadband (i.e. using high bandwidth) services.
- Advances in technology have provided improved processing power, memory capacity and quality of performance.
- Software advances have provided structured programming techniques, high-level languages, etc.
- A change of focus is occurring from network optimisation towards a higher degree of openness.

With the changes above occurring, the seven layers in the OSI Model represent a comprehensive and disciplined approach to architecture. The model is the most suitable reference for future signalling systems.

CCSS7 will continue to use the 4-level structure for those parts already specified in that manner, e.g. the MTP. Newer parts of CCSS7, e.g. Transaction Capabilities (Chapter 7) are already specified using the OSI approach. Future CCSS7 parts will also adopt the OSI approach.

The relationship between the DSS layers and the OSI Model layers will be clarified in the future to ensure as much clarity as possible.

3.5 Level Structure of CCSS7

The early circuit-related functions of CCSS7 were defined using four tiers, termed 'levels'. The 4-level structure is illustrated in Fig. 3.5.

3.5.1 Level 1

Any node with the capability of handling CCSS7 is termed a 'signalling point'. The direct interconnection of two signalling points uses one or more 'signalling link(s)'. Level 1 defines the physical, electrical and functional characteristics of the signalling link. Defining such characteristics within Level 1 means that the rest of the signalling system (Levels 2 to 4) can be independent of the transmission medium adopted.

3.5.2 Level 2

Level 2 defines the functions that are relevant to an individual signalling link, including error control and link monitoring. Thus, Level 2 is responsible for the reliable transfer of signalling information between two directly connected signalling points. If errors occur during transmission of the signalling information, it is the responsibility of Level 2 to correct the errors.

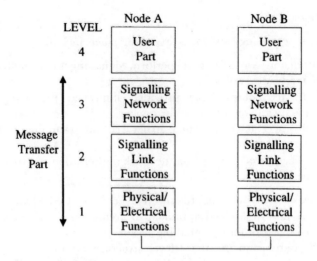

Fig. 3.5 Level structure of CCSS7

3.5.3 Level 3

The functions that are common to more than one signalling link, i.e. signalling network functions, are defined in Level 3. These functions include message handling and signalling network management.

When a message is transferred between two nodes, there are usually several routes that the message can take, including via a signal transfer point (STP). The message handling functions are responsible for the routing of the messages through the signalling network to the correct node. Signalling network management functions control the configuration of the signalling network, e.g. if a node within the signalling network fails, Level 3 of CCSS7 can reroute messages and avoid the node that has failed.

3.5.4 Message Transfer Part (MTP)

Levels 1 to 3 constitute the Message Transfer Part (MTP). The MTP is responsible for transferring information in messages from one signalling point to another. The MTP does not understand the meaning of the messages being transferred, but it controls a number of signalling message, link and network management functions to ensure correct delivery of messages. This means that the messages are delivered to the appropriate node in an uncorrupted form and in the sequence that they were sent, even under failure conditions in the network. Details of the MTP are given in Chapter 4.

3.5.5 Level 4

Level 4 contains the 'User Parts'. These parts define the meaning of the messages transferred by the MTP and the sequences of actions for a particular application (e.g. an ISDN service).

If new requirements arise that had not been foreseen previously, the relevant user part can be enhanced (or a new user part derived) without modifying the transfer mechanism or affecting other user parts.

Although CCSS7 is specified as a signalling system, Level 4 specifies a number of call-control functions. Later versions of the User Parts, e.g. the ISDN User Part (ISUP), attempt to separate the control of a call (Call Control Part) from the control of a connection (Bearer Connection Control Part) to provide a more structured approach. The Call Control Part handles those aspects of the User Part that relate purely to the call, thus being independent of the traffic circuit used. The Bearer Connection Control Part handles the functions for controlling connections. This approach reaps the benefits of a more rigorous structured specification and the resulting User Part is termed the 'ISDN Signalling Control Part (ISCP)'.

User Parts specified at an earlier stage, e.g. the Telephone User Part, are implemented without this separation of call and connection control.

The ISUP was specified to operate in a narrowband environment (generally ≤ 2 Mbit/s). The abbreviation N-ISUP is sometimes used to highlight this narrowband application. Details of the ISUP are given in Chapter 6 and the Telephone User Part is explained in Appendix 5.

3.5.6 Applying the level structure

(a) General

The application of the level structure is illustrated in Fig. 3.6.

Nodes A and B are directly connected by speech circuits (denoted by the solid lines connecting the respective switch blocks). A signalling link is also available between Nodes A and B (denoted by the dotted line). It is shown that Level 4 (the User Part) is closely associated with the control function of the node.

When the control function of Node A needs to communicate, it requests the Level 4 functions to formulate an appropriate message. Level 4 requests the MTP to transport the message to Node B. Level 3 of the MTP analyses the request and determines the means of routing the message to Node B. The message is then transported via Levels 1 and 2. At Node B, Level 3 of the MTP recognises that the message has arrived at the correct node and passes the message to Level 4. Level 4 in Node B then interacts with the control function to determine the appropriate action and response.

40 *Telecommunications Signalling*

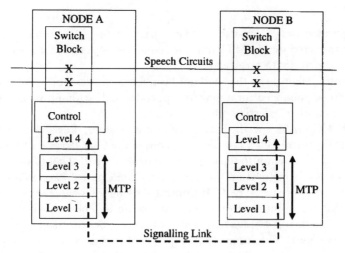

Fig. 3.6   Applying the 4-level structure

If problems arise in the transmission process between Nodes A and B, causing message corruption, the Level 2 functions are responsible for detecting the corruption and retransmitting the information. If the signalling link between Nodes A and B is not available (e.g. failure of the link), the Level 3 functions are responsible for rerouting the information through the signalling network to Node B.

Using these techniques, Nodes A and B can send each other appropriate messages until the need to communicate on a particular transaction ceases.

(b) Signal transfer point

Fig. 3.7 shows the more general case of Nodes A and B communicating using quasi-associated signalling. In this case, Node C acts as a Signal Transfer Point (STP). Hence, the MTP in Node A routes the message to Node C. As part of the STP function, Level 3 of the MTP in Node C recognises that the message needs to be onward routed to Node B. Level 4 in Node C is not involved in the message transfer, thus avoiding unnecessary Level 4 processing.

3.6 OSI Model

3.6.1 General

The OSI Model uses seven tiers called 'layers'. The 7-layer structure is illustrated in Fig. 3.8. In an OSI environment, a 'user' is a person or computer needing to communicate with another person or computer.

Architecture of CCS Systems 41

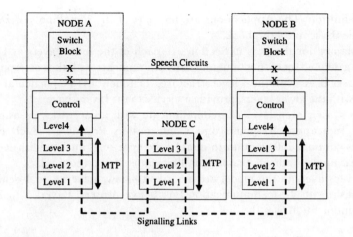

Fig. 3.7 STP working in the 4-level structure

The seven layers of the OSI Model provide the ability for Users A and B to communicate. The 'Applications' at Users A and B are the activities that each user wishes to perform, e.g. the procedures, programmes and functions for performing an action. The term 'Application Processes' is also used.

Applications sit above the communications capabilities in the model. Applications make use of the communications capabilities to exchange information, but they remain separate from the communications capabilities. The communications capabilities of the model are a key part of modern signalling

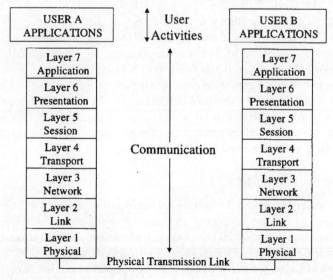

Fig. 3.8 OSI Model

system definition. The Applications are not part of the signalling systems and are outside the scope of this book.

As explained for the tiers in Section 3.3, each of the seven layers at User A performs a defined set of functions and can communicate with the corresponding layer in User B. As with the 4-level structure, the functions in a particular layer, together with the lower layers, provide a service to the layer above.

Within a user, each layer can communicate with adjacent layers using the Request, Indication, Response and Confirmation Primitives. Each of the primitives contains information to qualify the type, e.g. an Establish Request Primitive is used to indicate that a signalling connection is required.

The seven layers in the model are described below. Layers 1 to 4 deal with the establishment of a communication link between users. Layers 5 to 7 deal with communication over the link.

3.6.2 Layer 1 – Physical

Layer 1 relates to the physical transfer of a bit stream over a transmission medium. It provides the interface to the transmission medium.

3.6.3 Layer 2 – Data Link

Layer 2 is responsible for overcoming deficiencies in the transmission medium by, for example, the provision of error detection and correction techniques. Hence, errors in transmission (e.g. from bursts of noise) causing corruption of messages are overcome. Layer 2 applies between two directly connected nodes.

3.6.4 Layer 3 – Network

Layer 3 transfers data through a network from one user to another user and is responsible for analysing address digits and routing accordingly. Layer 3 communication can either be 'connection oriented' or 'connectionless'. In the connection-oriented case, a relationship needs to be established between two users to ensure co-ordination of the data that is exchanged, e.g. to guarantee that messages sent by one user are delivered in the same order to the other user.

The term 'connectionless' is used when no guarantee is given that messages will have a particular relationship, e.g. delivery of messages in the order of sending them is not guaranteed by the Network Layer.

3.6.5 Layer 4 – Transport

Layer 4 provides the ability to establish a transparent transfer of information from one user to another, relieving the users of involvement in the means of achieving the data transfer. Hence, Layer 4 allows the establishment of direct communication between two users. Layer 4 can also be used to enhance some of the functions of Layer 3, e.g. to provide a higher quality of service than normally provided by Layer 3.

3.6.6 Layer 5 – Session

The Session Layer is the lowest layer that deals with direct communication between the two users, as opposed to the establishment of the communication link. The Session Layer defines the type of interaction to be used between the two users, including the nature and timing of interactions. For example, the communication could be two-way simultaneous (i.e. both users able to communicate simultaneously), two-way alternate (i.e. both users able to communicate, but one at a time) or one-way (only one user able to send information).

3.6.7 Layer 6 – Presentation

One of the aims of the OSI Model is to allow users to adopt different data syntaxes and still be able to communicate with each other. Layer 6 translates the syntax of the data being transferred between that used by one user and that used by the other user. If both users adopt the same syntax, this function is not required.

3.6.8 Layer 7 – Application Layer

(a) Principle

The Application Layer (Layer 7) specifies the type of communication needed by the users. The Application Layer is part of the communications capability of the OSI Model. It should not be confused with the Applications (or Application Processes) themselves. Layer 7 is the bridge between User Applications and the other OSI layers.

It would be desirable for Layer 7 to provide a common set of functions that could be used by all Applications. However, in practice, the range of Applications is immense and it is necessary to provide Application-Specific Functions in addition to the common set. It is also necessary to be able to add Application-Specific Functions to handle future needs.

(b) Structure

To cater for these demanding requirements, Layer 7 is structured in a modular manner. The structure is illustrated in Fig. 3.9.

Consider an Application in one node wishing to communicate with a corresponding Application in another node. An Application Entity (AE) defines the functions that the Applications use to perform this communication. An AE can be regarded as a software program defined to perform a set of functions.

The basic component of an AE is an Application Service Element (ASE). An ASE defines a function, or a group of functions. An ASE can be regarded as a sub-program within the AE. An example of an ASE is the Association Control Service Element (ACSE). The ACSE is responsible for establishing a logical connection between the Application Layers in two nodes. The logical connection

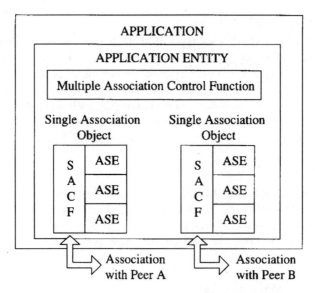

Fig. 3.9 OSI Layer 7 structure (Source: ITU Recommendation Q.1400) SACF = Single Association Control Function ASE = Application Service Element

allows the Application Layers to exchange protocols and is termed an 'Association'.

The Single Association Control Function (SACF) defines the rules governing the use of the ASEs. For example, a rule could be adopted that the ACSE is always the first ASE to be applied, thus ensuring that a logical connection is established between Application Layers before other functions are commenced.

The Single Association Object (SAO) is a collection of ASEs and SACF rules used by an Application to communicate with a peer over a single association.

The Multiple Association Control Function (MACF) performs a similar function to the SACF, but in this case it applies rules to multiple communications between applications. So, if it is necessary to establish more than one association, the MACF provides the rules for correct performance.

(c) Example ASEs

OSI has standardised several ASEs, allowing signalling system designers to pick from a range of ready-specified ASEs for a particular form of communication. There are two ASEs of special interest.

The ACSE, given as an example above, sets up and releases an Association between Applications. An ACSE is always provided to meet the needs of application communication.

The Remote Operations Service Element (ROSE) provides a framework for invoking remote procedures and receiving the results of the remote procedures.

For example, an Application Entity in a node in the network might need to communicate with a corresponding entity in a database during the set-up of a call to obtain additional routing information. The ROSE can be used for this purpose, using 'Operations' to carry the requests and responses. Operations are classified according to the type of response expected, e.g. report:

- success or failure of a request;
- failure of a request only;
- success of a request only.

Operations can be synchronous (in which a reply is needed before other operations are initiated) or asynchronous (in which further operations can be initiated before a reply is received).

A form of ROSE is implemented in, for example, the Transaction Capabilities Application Part of CCSS7 described in Chapter 7.

Information is transferred between Layer 7 peers in information elements termed Application Protocol Data Units (APDUs).

3.7 Applying the OSI Model to CCSS7

3.7.1 Context

The use of the OSI Model in CCSS7 depends upon the context in which the model is applied, as illustrated in Fig. 3.10. From a customer context, the

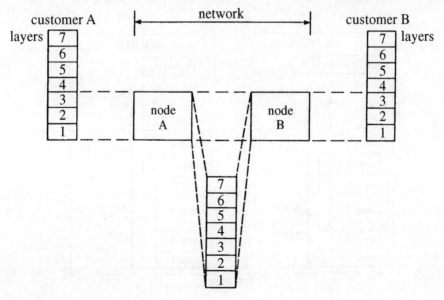

Fig. 3.10   Context of OSI Model

network is perceived as providing the functions of Layers 1 to 3. The functions of Layers 4 to 7 are provided by the customers and their terminals.

When the users are Applications within nodes, the model is applied between the nodes and the network provides the full range of functions of Layers 1 to 7.

3.7.2 Application to NCR signalling

The application of the OSI Model to the non-circuit-related (NCR) aspects of CCSS7 is illustrated in Fig. 3.11. The constituent parts are described in the following sections.

3.7.3 Signalling Connection Control Part (SCCP)

Although the MTP provides a comprehensive transfer technique, it does not provide all the functions necessary to fulfil an OSI Network Service (i.e. the full functions of Layers 1 to 3 of the OSI Model). The Signalling Connection Control Part (SCCP)[5] is defined to enhance the functions of the MTP to meet the OSI Layer 3/4 interface requirements and provide a Network Service. Details of the SCCP are given in Chapter 5.

3.7.4 Transaction Capabilities (TC)

Transaction Capabilities[6] (TC) is a generic non-circuit-related protocol. It encompasses Layers 4 to 6 and part of Layer 7 of the OSI Model. TC is divided into the Transaction Capability Application Part (TCAP) and the Intermediate Service Part (ISP). TCAP, together with functions provided by its users,

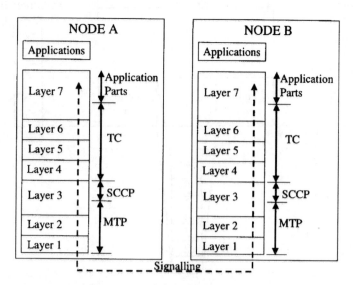

Fig. 3.11 Applying the OSI Model to CCSS7 NCR

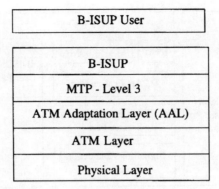

Fig. 3.12   Model for B-ISUP

provides the functions of Layer 7. The ISP provides the functions of Layers 4 to 6.

TC is designed to be portable, i.e. TC can use any transport mechanism offering a network service. This can be the combination of the MTP and SCCP, but it can also use any other transport mechanism compliant with the OSI Model. In a similar way, many different types of Application can use TC to establish communication with other Applications.

3.7.5 Application Parts

Application Parts are the users of TC and examples are the Mobile Application Part[7] (MAP) explained in Chapter 8, the Intelligent Network Application Part[8] (INAP) explained in Chapter 9 and the Operations, Maintenance and Administration Application Part (OMAP) explained in Chapter 10.

3.7.6 Broadband ISDN User Part (B-ISUP)

The B-ISUP[9] is the equivalent to the ISUP applied in a broadband environment (generally > 2 Mbit/s). The B-ISUP adopts the structure outlined for Layer 7 of the OSI Model and Layers 4-6 are regarded as transparent. Several configurations are possible. If an ATM Layer (Chapter 14) is used, the B-ISUP makes use of Level 3 of the MTP, the ATM Adaptation Layer and the ATM Layer as illustrated in Fig. 3.12.

3.8 Overall Architecture of CCSS7

The overall architecture of CCSS7 is shown in Fig. 3.13. It is built up from the original 4-level structure and the OSI 7-layer structure. The Telephone User Part (TUP) and Data User Part (DUP) are not shown for clarity.

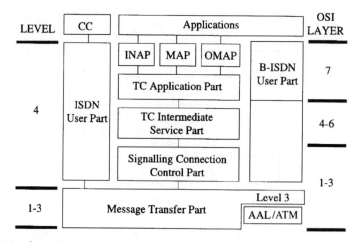

Fig. 3.13   Overall architecture of CCSS7

CC = Call Control   AAL = ATM Adaptation Layer

The MTP and ISUP conform to the level structure, with the MTP forming Levels 1–3 and the ISUP forming Level 4. The MTP is a transport mechanism and the ISUP allows the node to set up and release traffic circuits for speech and circuit-switched data.

The MTP and the SCCP are equivalent to Layers 1 to 3 of the 7-layer model and offer a network service to Transaction Capabilities (TC). TC and SCCP allow a node to invoke non-circuit-related functions when required.

TC forms Layers 4 to 6 and some functions of Layer 7. The users of TC (e.g. INAP, MAP and OMAP) complete the functions of Layer 7 and form a bridge to the respective Applications.

The B-ISUP provides the Layer 7 functions of the OSI Model. Level 3 of the MTP and the ATM Adaptation Layer are used as the transport mechanism.

Fig. 3.14 shows an example of using the CCSS7 Architecture. Nodes A and B are interconnected by traffic circuits (denoted by solid lines). The ISUP of Node A receives a request to set up a speech circuit. Node A recognises that the call requires specialised routing information and the control function requests TC to determine the required routing translation from the remote network database. The messages are transferred using the MTP and SCCP, without setting up a speech circuit to the database.

The translator in the database provides the required routing translation and returns the information to the control function of Node A. With the new information, the control function of Node A can now request the ISUP to set up a speech circuit to Node B. This is achieved by the ISUP in Node A communicating with the ISUP of Node B. Further detail of a similar example is given in Chapter 17.

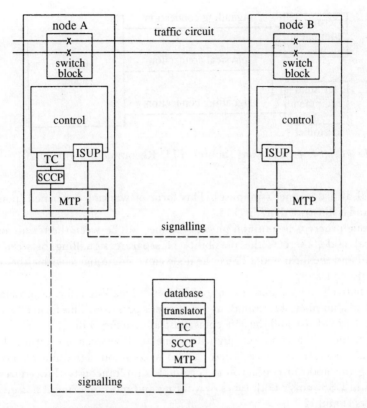

Fig. 3.14 Example of CCSS7 architecture usage

3.9 DSS1 Layer Structure

3.9.1 General

The ITU-T Digital Subscriber Signalling System No. 1 (DSS1) is designed as an access signalling system in a narrowband ISDN. DSS1 signalling information between users and the network is carried on a signalling channel termed a D-Channel. A D-channel is either a 16 kbit/s channel (used to control two 64 kbit/s channels), or a 64 kbit/s channel (used to control multiple 64 kbit/s channels).

The architecture of DSS1 uses three tiers, termed 'layers'. Unfortunately, the layers of DSS1 and those of the OSI Model do not align at this stage. Work to determine the relationship between these layers has not yet been conducted. The layers for DSS1 are explained in following Sections.

DSS1 must handle a variety of circumstances at customers' premises. For example, a local node might need to send signalling information to a specific

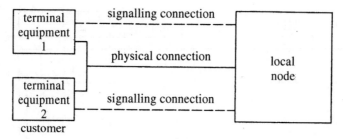

Fig. 3.15   Point-to-point working (Source: ITU Recommendation Q.920)

terminal at a customer's premises. This form of working is termed 'point-to-point' and is illustrated in Fig. 3.15.

Although there is a common physical connection between the terminals and the local node, there is the possibility of separate signalling between each terminal and the local node. Thus, the node can communicate with either of the terminals separately.

Alternatively, a local node might need to send signalling information simultaneously to a number of terminals at a customer's premises. This form of working is termed 'broadcast' and the concept is illustrated in Fig. 3.16.

Again, there is a common physical connection between the terminals and the local node. In this case, there is also a common signalling connection, allowing the node to contact both terminals simultaneously. Procedures and formats in DSS1 cover both types of working and more information is given in Chapters 11 and 12.

DSS1 is also required to handle the transfer of packet data[10] in dedicated 'packet data networks'. This is a form of data transfer in which data is divided into blocks called 'packets'. Each packet is routed independently through the network, the blocks being recombined when they have been received at the intended destination. In this way, transmission capacity within the packet network is allocated to a call only when required to transmit a packet. Between successive packets relating to a call, the transmission capacity is used to convey packets for other calls. Details of packet data networks are beyond the scope of

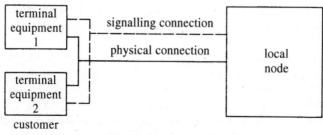

Fig. 3.16   Broadcast working (Source: ITU Recommendation Q.920)

this book, but the means by which DSS1 is specified to handle packet data is covered in Chapter 12.

3.9.2 DSS1 Layer 1 – Physical

Layer 1 of DSS1 defines the physical, electrical and functional characteristics of the transmission link. The physical layer provides the physical connection for the transmission of bits and allows the transfer of messages generated by DSS1 Layers 2 and 3. It therefore performs similar functions to the OSI Model Layer 1 and the CCSS7 Level 1. Further information is given in Chapter 11.

3.9.3 DSS1 Layer 2 – Data Link

When functions within Layer 3 at a customer's premises need to communicate with the network, an Association (or connection) is established to allow information transfer to occur. Such Associations are termed 'Data Link Connections'. DSS1 Layer 2 provides the ability to establish and control one or more Data Link Connection(s) on a D channel.

The functions of Layer 2 include sequence control (to maintain the sequential order of messages across a Data Link Connection) and the detection and correction of errors in messages transmitted on the data link. Information is transmitted in 'frames'.

The points at which DSS1 Layer-2 services are offered to Layer 3 are termed 'Service Access Points', each point being identified by a 'Service Access Point Identifier (SAPI)'. Many customer terminals can be associated with a SAPI and, to identify a particular terminal, each terminal is allocated a number called a 'Terminal End-Point Identifier (TEI)'. The point-to-point form of working can be implemented by selecting the TEI of the appropriate terminal. The broadcast form of working is implemented by selecting a TEI reserved for the purpose.

Two types of operation are defined for the Data Link Layer. These are 'Acknowledged Operation' and 'Unacknowledged Operation'. In Acknowledged Operation, each frame transmitted by DSS1 Layer 2 is numbered. This allows the Data Link Layer to acknowledge each frame that is received. If errors are detected, or a frame is missing, retransmission of the frame occurs. Flow control procedures exist in acknowledged operation to limit the number of frames being transmitted. Flow control avoids overloading equipment at the node or customer's premises, thus improving the success rate of frame transmission. Acknowledged operation is applicable only to point-to-point working.

In Unacknowledged Operation, DSS1 Layer 3 information is transmitted in frames that are not numbered and DSS1 Layer 2 does not provide an acknowledgement of each frame received.

3.9.4 DSS1 Layer 3 – Network

DSS1 Layer-3 functions are responsible for controlling the establishment and release of calls and connections (both circuit-switched and packet). The functions support the control of basic calls (e.g. basic telephony) and those involving supplementary services.

DSS1 Layer 3 is also responsible for providing transport capabilities in addition to those defined in Layer 2. It is DSS1 Layer 3 that generates and interprets messages that are transported by Layer 2. This involves the administration of call references used for call control and ensuring that the services provided by Layer 3 are consistent with the requirements of the customer.

Functions provided by DSS1 Layer 3 include:

- routing messages to an appropriate destination (typically a local node);
- conveying user-to-user information, with or without the establishment of a circuit-switched connection;
- multiplexing multiple calls onto a single Data Link Connection;
- segmenting (i.e. splitting) and reassembling (i.e. re-combining) messages to allow their transport by the Data Link Layer;
- detecting errors in the procedures defined for Layer 3, interpreting and reacting to errors detected by the Data Link Layer and detecting errors in messages provided by the Data Link Layer operating in the unacknowledged method;
- ensuring that messages are delivered to the destination customer in the same order as they are generated by the originating customer.

3.9.5 Primitives

The Primitive Types used by DSS1 depend on the type of operation used. In Acknowledged Operation, the Request, Indication, Response and Confirmation Primitives are used, conforming to the general description outlined in Section 3.3. The Primitive Types are qualified by 'Establish' and 'Release' to indicate the need to set up and clear down a DSS1 Layer 2 connection. A Connection Reference is used to identify a particular connection.

In Unacknowledged Operation, the Unitdata Primitive is used. In this case, the Unitdata Primitive is sent without a Connection Reference: hence, a connection is not established between peer layers. DSS1 Layer 3 sends a Request Unitdata Primitive to Layer 2 to request the transfer of information to another entity. A receiving entity passes the Primitive Unitdata Indication from Layer 2 to Layer 3 to present information received by the Data Link Layer.

3.10 DSS2 Layer Structure

The Digital Subscriber Signalling System No.2 (DSS2) is designed for use in the access part of broadband (i.e. high bandwidth) ISDNs. Many aspects of

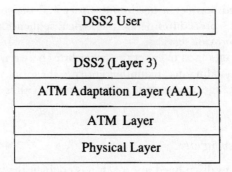

Fig. 3.17   DSS2 layers

DSS2 are similar to those for DSS1 and this Section highlights some of the differences that apply in a broadband context.

DSS2 signalling information between users and the network is carried on a virtual channel that is allocated for signalling purposes and is separate from the traffic-carrying virtual channels.

The architecture of DSS2 is based on Layers, but these do not align with the layers in the OSI Model and the work to determine their relationship has not yet been conducted. Several configurations are possible, but if ATM is used, the layers for DSS2 are summarised in Fig. 3.17.

The Physical Layer (Layer 1) is similar to that for DSS1. However, DSS2 Layer 2 needs to cater for the application of broadband technologies, particularly the use of a packet-oriented approach to traffic. Thus, DSS2 can make use of the ATM and ATM Adaptation Layers, in a similar way to the B-ISUP for CCSS7.

DSS2 Layer 3 has similar functions to Layer 3 of DSS1. DSS2 Layer 3 functions are responsible for controlling the establishment and release of calls and virtual connections. The functions support the control of basic calls and those involving supplementary services. Further details of DSS2, including procedures and formats, are given in Chapter 16.

3.11 Specification of CCS Systems

3.11.1 Contents

The specification of modern CCS systems is based upon the tier approach described earlier. Although details vary according to the type of tier adopted, the requirements in Section 3.2 are met by the architectures adopted. In the context of CCS specifications, the customers and nodes in the network can be termed 'entities'. CCS systems are defined by specifying the:

- functions performed in each tier within an entity;
- primitives used between adjacent tiers within an entity – by defining the primitives, the interface between adjacent tiers can be held stable even if the functions performed by one tier change;

- procedures used between a tier in one entity and the corresponding tier in another entity – procedures define the logical sequence of events and flows of messages to provide services;
- format of messages used to enact the procedures between a tier in one entity and the corresponding tier in another entity – the format defines the general structure of messages and specific codings of fields within messages.

3.11.2 Avoiding ambiguity

The adoption of this disciplined approach has enabled the production of high-quality specifications for modern CCS systems. Many experts throughout the world have contributed, resulting in CCS systems that are flexible in meeting the needs of customers. However, despite all these measures, it is still difficult to produce a specification that can be implemented by manufacturers and network operators and yet be completely unambiguous. The stringent test takes place when the signalling between two entities built by different manufacturers is put into operation.

The important point is to learn from practical experience and feed the results of this experience back into the specifications, thus continuously improving the definitions of the CCS systems. It is of interest to note that the majority of the problems encountered during field trials of systems have been due to design anomalies, rather than specification ambiguities. This reflects the high quality of specifications that have been attained.

3.11.3 Options

When it is difficult to reach agreement on how to specify a particular feature, 'options' can be specified that give alternative methods of implementing the feature. Too many options complicate the specification, increase the complexity of interworking and represent an inability to achieve a true standard, thus diluting the benefits of a common specification. If options are essential, they must be clearly identified and the implications of interworking between different implementations must be fully evaluated.

3.11.4 Speed

New customer requirements arise continually and CCS systems must respond to the challenge of supporting them. Changes to specifications can be minimised by adopting flexible architectures, but there will be occasions when new features, or new ways of implementing features, are required. Network operators must respond quickly to customer needs and, to avoid adoption of varying interim systems, the derivation of standards must be as fast as possible.

3.12 Chapter Summary

3.12.1 Architecture requirements

Modern CCS systems are specified in a structured manner to ensure that they can flexibly meet customer requirements and exhibit evolutionary potential. A structured approach eases the specification of complex systems and assists in the design and development processes. The structured approach to specification is known as the architecture of the signalling system.

To allow flexibility and evolution, the architecture must allow one part of the signalling system to be modified without affecting other parts. The signalling system must cater for a range of services and it must be able to handle a multitude of interworking cases. The architecture must also support the speedy production of unambiguous specifications in a manner that optimises network resources and provides open interfaces.

3.12.2 Architecture principles

Ideally, all modern signalling systems should have been specified using a common architecture, with variances occurring only to handle specific requirements. Unfortunately, the architectures adopted have differed as technologies have evolved and as long-term objectives have changed.

However, the various architectures use a common set of principles and these are based on dividing signalling functions into a number of tiers. The tiers consist of self-contained groups of functions. Each tier (in conjunction with lower tiers) offers a service to higher tiers. Communication between tiers is by means of primitives.

Specifications are derived by defining the:

- functions performed in each tier within an entity;
- primitives used between adjacent tiers within an entity;
- procedures used between a tier in one entity and the corresponding tier in another entity;
- format of messages used to enact the procedures between a tier in one entity and the corresponding tier in another entity.

3.12.3 Types of architecture

There are three basic architectures used to specify CCS systems:

- early circuit-related versions of CCSS7 use a 4-Level structure;
- the OSI Model uses a 7-Layer approach;
- DSS1 and DSS2 use a 3-Layer structure, but the DSS Layers are not the same as the OSI Model Layers.

In addition, IP technologies use a 4-layer structure with a close relationship to the OSI layers. The IP structure is described in Chapter 18.

Levels 1 to 3 of CCSS7 form the Message Transfer Part (MTP), which is responsible for transferring messages from one signalling point to another, even in the event of link and network failures. Level 4 contains the User Parts, which define the meaning of the messages transferred by the MTP and the logical sequence of events for a particular application.

The Signalling Connection Control Part (SCCP) of CCSS7 is defined to provide the MTP with additional functions to meet the OSI Model Network Service.

Transaction Capabilities (TC) of CCSS7 is a protocol designed to carry non-circuit-related data between nodes in the network. TC forms Layers 4–6 and part of Layer 7 of the OSI Model. Application Parts complete the Layer 7 functions, including the Intelligent Network Application Part (INAP), the Mobile Application Part (MAP) and the Operations, Maintenance and Administration Application Part (OMAP).

The Broadband ISDN User Part (B-ISUP) of CCSS7 is specified at Layer 7 of the OSI Model.

DSS1 is specified in terms of three Layers, but they are not the same as the layers in the OSI Model. This leads to confusion and it needs to be made clear which type of layer is meant.

DSS1 Layer 1 covers the physical and electrical conditions of the transmission link. DSS1 Layer 2 establishes Data Link Connections to convey signalling information. Layer-2 operation can be in an Acknowledged or Unacknowledged Form. To cater for a variety of terminal arrangements at customers' premises, Layer 2 allows for a point-to-point form of working (in which communication with a particular terminal is required) and a broadcast form of working (in which signalling information is transmitted to all terminals).

DSS1 Layer-3 Functions are responsible for controlling the establishment and release of calls and connections (both circuit-switched and packet). The functions support the control of basic calls and those involving supplementary services.

DSS2 also uses a 3-layer structure. Layers 1 and 3 are similar to those for DSS1. Layer 2 is based on the ATM Adaptation Layer to cater for the packet-oriented nature of broadband networks.

3.12.4 Future architectures

Future developments in CCS systems will use the OSI Model as the framework for architecture. Work will need to be carried out on existing systems that do not align with the OSI Model, e.g. DSS1 and DSS2, to ensure that an understanding of the relationship is derived.

3.12.5 Way forward

The international specification process makes use of experts throughout the world to produce high quality standards. It is very difficult to achieve unambiguous specifications and it is essential that feedback from practical implementations is used to achieve continuous improvement. Options allowed in specifications should be kept to a minimum. It is essential that international specifications are produced as fast as possible to obviate the need for interim implementations to meet evolving customer requirements.

3.13 References

1. ITU-T Recommendation Q.700: 'Introduction to ITU-T Signalling System No 7' (ITU, Geneva)
2. International Standards Organisation IS7498: 'Information Processing Systems—Open Systems Interconnection—Basic Reference Model'. ITU-T Recommendation X.200, 'Reference Model of Open Systems Interconnection for ITU-T applications' (ITU, Geneva)
3. ITU-T Recommendations Q.920, Q.921, Q.930 and Q.931: 'Digital Subscriber Signalling System No. 1' (ITU, Geneva)
4. ITU-T Recommendation Q.2931: ' Digital Subscriber Signalling System No. 2' (ITU, Geneva)
5. ITU-T Recommendation Q.711: 'Functional description of the Signalling Connection Control Part' (ITU, Geneva)
6. ITU-T Recommendation Q.771: 'Functional description of the Transaction Capabilities' (ITU, Geneva)
7. ITU-T Recommendation Q.1001: 'General aspects of Public Land Mobile Networks' (ITU, Geneva)
8. ITU-T Recommendation Q.1201: 'Principles of Intelligent Network Architecture' (ITU, Geneva)
9. ITU-T Recommendation Q.2761: 'Functional Description of the B-ISUP' (ITU, Geneva)
10. ITU-T Recommendations X.25 and X.75: 'Packet Mode Interface' and 'Packet Switched Signalling System' (ITU, Geneva)

Chapter 4
CCSS7 Message Transfer Part

4.1 Introduction

The Message Transfer Part[1] (MTP) is responsible for the transfer of signalling messages between signalling points. The MTP does not understand the meaning of the messages being transferred: its job is to deliver the messages in a flexible and secure manner. The aim is to deliver the information within messages without loss, without duplication, free of errors and in a pre-arranged sequence. Procedures are included that react to network and system failures, thus providing a high level of security.

The specification of the MTP is structured to allow flexible implementations that can be optimised in practical networks. It is an intelligent transfer mechanism that can reconfigure and control signalling traffic to overcome failures in the network. The users of the MTP define the meaning of the messages and high-level procedures.

The MTP is responsible for a very high level of reliability for successful conveyance of messages. To achieve the high targets set by the ITU-T[2], error detection and error correction techniques are employed, as well as actions to control the signalling network. Some significant performance targets are:

- undetected errors; less than one in 10^{10};
- loss of messages; less than one in 10^7;
- out-of-sequence delivery to higher levels; less than one in 10^{10}.

4.2 Architecture

4.2.1 Structure

The 4-Level Structure of CCSS7 is described in Chapter 3. The MTP consists of Levels 1 to 3 of that structure, as illustrated in Fig. 4.1 for the MTP serving Node A. The users of the MTP can be any entity meeting the MTP interface requirements and are typically CCSS7 User Parts or the SCCP.

Level 1 defines the physical, electrical and functional characteristics of the transmission path for signalling. Level 2 defines the functions and procedures for the transfer of signalling messages between two nodes over a signalling data link

60   *Telecommunications Signalling*

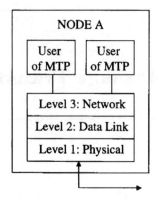

Fig. 4.1   MTP levels

(the physical link between the two nodes). Level 3 provides network functions, including comprehensive routing and management capabilities. The levels are described in greater detail in subsequent sections.

4.2.2 *Primitives*

The MTP primitives are summarised in Table 4.1.

The Originating Point Code (OPC) is the unique identity of the node sending a message. The Destination Point Code (DPC) is the unique identity of the node to which a message is sent. The Signalling Link Selection Field and Service Information Octet are described in Section 4.5.

4.2.3 *Relationship with SCCP*

When the transfer of information has to meet the OSI Model Network Service outlined in Chapter 3, then a combination of the MTP and the Signalling Connection Control Part (SCCP) is required.

4.3 Functional Structure

The functions of the MTP applied between two nodes are shown in Fig. 4.2. In this case, the users in Nodes A and B are using the MTP to exchange messages. The functions performed at each level are described in Sections 4.4 to 4.6.

Table 4.1 MTP primitives

Name	Description	Type	Typical parameters
MTP-Transfer	Used between Levels 3 and 4 to provide the Message Transfer Service	Request, Indication	Originating Point Code Destination Point Code Signalling Link Selection Service Information Octet User Data
MTP-Pause	Indicates to the user the inability to provide the MTP Service to the specified destination	Indication	Affected DPC
MTP-Resume	Indicates to the user the ability to provide the MTP Service to the specified destination	Indication	Affected DPC
MTP-Status	Indicates to the user the partial inability to provide the MTP Service to the specified destination	Indication	Affected DPC Cause

(Source: ITU-T Recommendation Q.701)

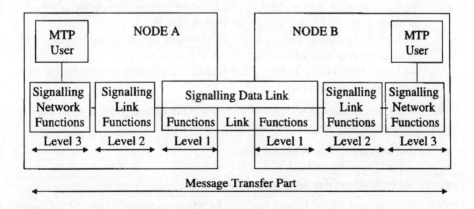

Fig. 4.2 MTP functions

4.4 MTP Level 1: Signalling Data Link

Level 1 of the MTP[3] defines the physical, electrical and functional characteristics of the transmission path for signalling. It is normally a 64 kbit/s path within a PCM system, but other forms of transmission (including analogue) can be used. The provision of a Level 1, defining the interface to the transmission medium, means that the higher levels (Levels 2–4) are independent of the type of transmission medium used.

4.5 MTP Level 2: Signalling Link Functions

4.5.1 General

Level 2 of the MTP[4] relates to the link between two nodes, defining the functions and procedures for the transfer of signalling messages. The combination of Levels 1 and 2 provides a Signalling Link for reliable transfer of signalling messages. The Level-2 functions provide a framework for the information transferred over each link and consist of:

- Signal Unit delimitation and alignment (using flags, etc.);
- error detection procedures;
- error correction procedures;
- error monitoring;
- flow control.

4.5.2 General Signal Unit format

Signalling information transferred between signalling points is divided into messages termed 'Signal Units'. These Signal Units vary in length according to the type of information being transferred. There are three types of Signal Unit:

(i) the Message Signal Unit (MSU), which is used for transferring signalling information supplied by a user (a User Part or SCCP);
(ii) the Link Status Signal Unit (LSSU), which is used to indicate and monitor the status of the Signalling Link;
(iii) the Fill In Signal Unit (FISU), which is used when there is no signalling traffic to maintain link alignment.

The three types of Signal Unit have a very similar format. The type of Signal Unit is identified by a length indicator (LI) as follows:

- $LI = 0$ indicates a Fill In Signal Unit;
- $LI = 1$ or 2 indicates a Link Status Signal Unit;
- $LI > 2$ indicates a Message Signal Unit.

The most comprehensive Signal Unit is the Message Signal Unit (MSU) and its format is shown in Fig. 4.3.

CCSS7 Message Transfer Part   63

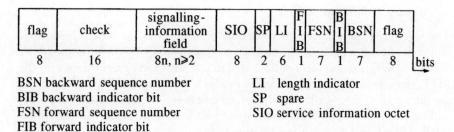

BSN backward sequence number
BIB backward indicator bit
FSN forward sequence number
FIB forward indicator bit

LI length indicator
SP spare
SIO service information octet

Fig. 4.3   Format of message signal unit
(Source: ITU-T Recommendation Q.703)

The MSU is divided into a number of fields, with a specified number of bits allocated to each field. Hence, the format of the MSU defines each of the fields within the message and allocates a meaning to each bit within the message. The exception to this is the Signalling Information Field, which is defined by higher-level functions.

(a) Flag

The Flag acts as a Signal Unit Delimiter, showing the start and end of each Signal Unit. The Flag is a unique 8 bit pattern that is not used for any other purpose. In typical implementations, the end Flag of one MSU acts as the beginning Flag of the next MSU. The pattern is coded 01111110. To avoid the Flag being imitated by another part of the Signal Unit, the node transmitting the MSU inserts a zero after every sequence of five consecutive ones occurring in any position in the MSU (apart from the Flag). This zero is deleted at the receiving end of the Signalling Link after the Flags have been detected.

(b) Sequence Numbers

The Backward Sequence Number, Backward Indicator Bit, Forward Sequence Number and Forward Indicator Bit are used in the error-correction mechanism described in Sections 4.5.5 and 4.5.6.

(c) Length Indicator

The Length Indicator (LI) gives the length of the Signal Unit. An LI value of greater than two indicates that the Signal Unit is an MSU.

(d) Service Information Octet

The Service Information Octet (SIO) defines the user that is appropriate to the message. For example, the SIO can indicate that the message is relevant to the ISDN User Part, the SCCP, etc.

(e) Signalling Information Field

The Signalling Information Field (SIF) consists of up to 272 octets, the formats and codes being defined by the user. Early designs of CCSS7 have a maximum Signalling Information Field of 62 octets, in line with early specifications for the MTP, but the larger size is essential in today's environment. The SIF contains the information that needs to be transferred between the MTP users at two signalling points. Thus, the MTP is not aware of the contents of the SIF, except for the Routing Label, which is information that is used for routing messages in the signalling network (see Section 4.6.2). Apart from such routing information, the MTP merely transfers the information in the SIF from the MTP user of one node to the MTP user of another node.

4.5.3 Error detection

Error detection is performed by means of 16 check bits provided at the end of each Signal Unit. The check bits are derived by the node sending the Signal Unit. The check bits are generated by applying a complex polynomial $(x^{16}+x^{12}+x^5+1)$ to the information in the Signal Unit. The polynomial is chosen to optimise the detection of bursts of errors during transmission.

The check bits transmitted are the ones complement of the resulting 16 bit field, i.e. the ones are changed to zeros and *vice versa*. This change is performed to minimise the risk of faulty operation of the receiving-node equipment.

The check bits are analysed at the receiving node according to a corresponding algorithm. If consistency is not found, then an error has been detected and the message is discarded. Discarding an MSU in this way invokes the error correction mechanism.

4.5.4 Error correction

There are two error correction methods defined for CCSS7. The Basic Method is appropriate for links with one-way propagation delays of less than 15 ms and the Preventive Cyclic Retransmission Method is appropriate for links with one-way propagation delays of >15 ms.

Using these techniques, messages that have been corrupted (e.g. due to error bursts on the transmission medium) are retransmitted in sequence. Level 3 is unaware of any problems if the error correction mechanism is successful. In this case, the messages are delivered to the user without loss or duplication. If persistent faults occur, Level 3 is informed so that management action can be taken. An example of such an action is re-routing messages via different Signalling Links. The Basic Method of error correction is described in Section 4.5.5 and the Preventive Cyclic Retransmission Method is described in Section 4.5.6.

4.5.5 Basic Method of Error Correction

The Basic Method of error correction is a Non-Compelled, Positive/Negative Acknowledgement, Retransmission Error-Correction System. Non-Compelled means that messages are sent once only, unless they are corrupted during transfer. Positive/Negative Acknowledgement means that each message is acknowledged as being received, with an indicator to explain whether or not the message is corrupted. Error correction is by means of Retransmission. The functions involved in the error-correction mechanism are shown in Fig. 4.4.

When the MTP User of Node A needs to send signalling information to the MTP User of Node B, the relevant information is passed (via Level 3) to Level 2 of Node A. Level 2 of Node A is provided with a transmission buffer and a retransmission buffer. The transmission buffer is used to store the MSU before sending it on the Signalling Link. Hence, the transmission buffer acts as a store until Signalling Link capacity is available to send the MSU. The retransmission buffer keeps a copy of the MSU in case of corruption during the transfer of the MSU to Node B.

Each MSU contains a Forward Sequence Number (FSN), a Forward Indicator Bit (FIB), a Backward Sequence Number (BSN) and a Backward Indicator Bit (BIB). When the Signalling Link is acting normally, the FIB is set to a particular value (e.g. zero) and the BIB is set to the same value (zero).

When an MSU is received by Level 2 from Level 3 at Node A, it is entered into the transmission buffer. The transmission buffer acts as a queue, working on the principle that the first MSU received is the first to be transmitted. When the Signalling Link is free, and it is the turn of the MSU in the example to be sent, the MSU is allocated an FSN based on the last FSN plus 1 (modulo 128). The MSU is then transmitted to Node B. A copy is also entered into the retransmission buffer.

At the receiving buffer at Node B, the FSN is compared with the expected value (last FSN + 1). If the FSN is the expected value, the MSU is passed to Level 3 for processing. The FSN value is copied into the BSN field and the BIB is left unaltered. The values of BSN and BIB indicate a positive acknowledgement

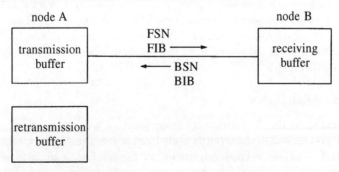

Fig. 4.4   Error-correction functions

to Node A. Upon receipt of the correct BSN and BIB values at Node A, the MSU is deleted from the retransmission buffer.

If the comparison of FSNs at Node B shows a discrepancy (e.g. due to the error detection mechanism discarding an MSU), the value of the BIB is inverted, (i.e. changed to value 1), thus denoting a negative acknowledgement to Node A. In this case, the BSN is made the value of the last correctly received FSN. Upon receipt of a negative acknowledgement at Node A, the transmission of Signal Units is interrupted and the MSUs in the retransmission buffer are retransmitted in the original order. The value of the FIB is inverted (i.e. changed to value 1) so that the FIB and BIB are once again the same value.

4.5.6 *Preventive Cyclic Retransmission Method of error correction*

This method is a Positive Acknowledgement, Cyclic Retransmission, Forward Error Correction System. This means that negative acknowledgements are not used, the mechanism relying on the absence of a positive acknowledgement to indicate corruption of the message.

Forward error correction is achieved by a programmed cycle of retransmission of unacknowledged MSUs. Each Signal Unit contains an FSN and BSN (as for the basic method), but the FIB and BIB are not examined and are permanently set to value 1. When no new MSUs are available for transmission, a cyclic retransmission of all MSUs remaining in the retransmission buffer is commenced. The original FSNs are maintained during the retransmission. If a new MSU arrives, the cyclic retransmission is stopped and the new MSU is transferred with an FSN of the last new MSU plus 1 (modulo 128). If no further new MSUs are received, the cyclic retransmission is recommenced.

An uncorrupted MSU is positively acknowledged by means of the receipt at Node A of a BSN equal to the allocated FSN. After positive acknowledgement, the relevant MSU is discarded from the retransmission buffer and is no longer available for retransmission.

One disadvantage of this system is that the sending transmission and retransmission buffers could become overloaded. To prevent message loss, a procedure called Forced Retransmission is adopted. The number of MSUs, and the number of octets in the retransmission buffer, are monitored continuously. If either parameter reaches a pre-set threshold, new MSUs are not accepted and priority is given to cyclic retransmission of MSUs in the retransmission buffer. The retransmission cycle is continued until the two activating parameters fall below the specified thresholds.

4.5.7 *Level 2 flow control*

If the Level 2 of the MTP detects congestion at a receiving node, a Busy Indication is returned to the sending node in a Link Status Signal Unit (LSSU). Positive and negative acknowledgements of Signal Units are also withheld. LSSUs are returned periodically to indicate continued congestion. When congestion has abated, the receiving node resumes acknowledging MSUs. This indicates to the sending node that procedures are to revert to normal.

4.5.8 Error monitoring

Level 2 includes functions to monitor the error rate on the Signalling Link. The measure for an operating Signalling Link is the Signal Unit Error Rate Monitor. The aim is to estimate the number of errors occurring on the link to determine its acceptability for operation. Three parameters are measured:

- the number of consecutive Signal Units received in error;
- the number of Signal Units received in error per block of 256 Signal Units;
- the number of octets received before a correctly checking Signal Unit is received.

Unacceptable error rates are reported to Level 3.

4.6 MTP Level 3: Signalling Network Functions

4.6.1 General

Chapter 2 describes the drive for CCS systems to be able to provide unimpeded communication between nodes. For maximum flexibility, separate signalling networks are required. It is Level 3 that establishes the foundation of a signalling network with comprehensive routing and management capabilities[5]. The use of Signal Transfer Points (STPs) enhances the flexibility of the MTP.

Level 3 is responsible for those functions that are appropriate to a number of Signalling Links. Level 3 functions are responsible for the reliable transfer of signalling information from one node to another, even in the case of signalling and traffic network failures. The MTP can reconfigure message routing to ensure the delivery of messages to the correct destination without corruption or duplication.

Level 3 functions are categorised into two main areas, namely Signalling Message Handling and Signalling Network Management. These are illustrated in Fig. 4.5.

Signalling Message Handling covers the transfer and routing of messages. This is described in Section 4.6.2. Signalling Network Management refers to the management of the signalling network and covers the control of the signalling network itself. The Signalling Network Management Functions can be divided into:

(i) Signalling Traffic Management, which covers the reconfiguration of signalling traffic in response to changes in network status (see Section 4.6.3);
(ii) Signalling Link Management, which controls the Signalling Links (see Section 4.6.4);

68 *Telecommunications Signalling*

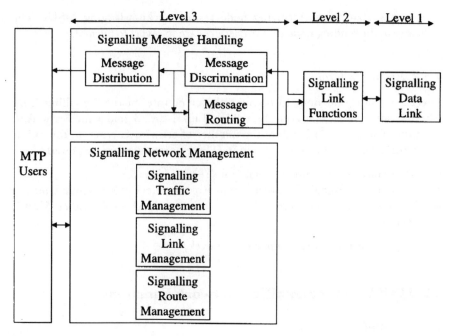

Fig. 4.5   Categories of network function

(iii) Signalling Route Management, which covers the distribution of information on the signalling network status (see Section 4.6.5).

4.6.2 Level 3: Signalling Message Handling Function

This function defines how Signal Units are routed. Each Signalling Point within a signalling network is identified by a Point Code, which is a 14 bit code or address that is unique within the signalling network. The Destination Point Code (DPC) identifies the Destination Signalling Point of a message and the Originating Point Code (OPC) identifies the Originating Signalling Point of a message.

The Message Handling Function has three categories of function: Message Routing is the process of selecting a Signalling Link for an outgoing message. Message Discrimination determines whether or not an incoming message is intended for this particular Signalling Point. Message Distribution is the process for choosing which user or Level 3 Function should receive an incoming message.

(a) Message Routing

For an outgoing message, Level 4 sends signalling information to the Routing Function. The information includes a Routing Label, which in turn includes a

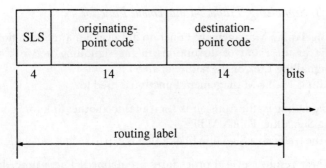

Fig. 4.6 Structure of Routing Label

DPC. The structure of the Routing Label is shown in Fig. 4.6. By analysing the DPC, the Routing Function can determine to which node the Signal Unit should be sent.

If two or more Signalling Links to the required destination exist, a load-sharing activity can be performed over the links. In this case, the Signalling Link Selection (SLS) Field is used to identify the chosen Signalling Link and hence permit load sharing. The SLS consists of the 4 bits following the OPC.

(b) Message Discrimination

For an incoming message, a Signal Unit is received by the Message Discrimination Function from Level 2. The DPC is analysed to determine if the Signal Unit is destined for the receiving Signalling Point or for another Signalling Point.

If the message is destined for the receiving Signalling Point, the message is passed to the Message Distribution Function. If the message is destined for another Signalling Point, the message is passed from the Message Discrimination Function to the Message Routing Function for onward transfer to the appropriate DPC. In this case, the Signalling Point performing the analysis of the DPC is acting as a Signal Transfer Point. One important aspect in this case is that the message is re-routed without Level 4 needing to know. Thus, the Signal Transfer Point working avoids Level 4 processing overhead on each message.

(c) Message Distribution

For an incoming message from the Message Discrimination Function, the Message Distribution Function chooses the appropriate user to receive the delivery. This is determined by analysing the Service Information Octet.

4.6.3 Level 3: Signalling Traffic Management Function

The Signalling Traffic Management Function of Level 3 maintains a flow of signalling traffic in the event of disruptions in the signalling network, e.g. the failure of a Signalling Link or the failure of an STP.

The Signalling Traffic Management Function is used to:

(i)   divert signalling traffic from a link (or route) to another link (or route);
(ii) restart a Signalling Point's MTP;
(iii) reduce the traffic to a Signalling Point that is experiencing congestion.

To achieve these results, several procedures are defined. These procedures are Changeover, Changeback, Forced Rerouting, Controlled Rerouting, MTP Restart, Management Inhibiting and Signalling Traffic Flow Control.

(a) Changeover

The Changeover Procedure ensures that signalling traffic is diverted from one Signalling Link to an alternative Signalling Link, as quickly as possible. A typical example of when the changeover procedure is initiated is when a Signalling Link fails. The changeover is implemented without message loss, duplication or mis-sequencing. This is achieved by implementing procedures to ensure that messages in the retransmission buffer of the unavailable Signalling Link are transferred to the alternative link. The changeover is implemented to avoid interrupting signalling traffic that already exists on the alternative Signalling Link.

(b) Changeback

The Changeback Procedure reverts the routing of signalling messages back to the conditions prevailing before changeover. Procedures are included to control the message sequence to ensure that messages are not lost or mis-sequenced. Changeback is initiated when the conditions causing changeover have been rectified, e.g. when a Signalling Link is restored.

(c) Forced Rerouting

Forced Rerouting is initiated at a signalling point when a Signalling Route becomes unavailable. This is indicated by the reception of a Transfer Prohibited Message. The Forced Rerouting Procedure stops the transmission of signalling traffic on the link set(s) pertaining to the unavailable route and stores the messages in a Forced Rerouting Buffer. An alternative route is determined and the contents of the Forced Rerouting Buffer are transmitted on the new route.

(d) Controlled Rerouting

The Controlled Rerouting Procedure restores the optimal signalling routing while minimising mis-sequencing of messages. It is used, for example, to reverse the routing imposed by the Forced Rerouting Procedure.

(e) MTP Restart

MTP Restart is used when a signalling point has been isolated from the network for some time, e.g. due to failure of the signalling point. In this case, the signalling point needs to check that its routing information is still valid. A central part of the Restart Procedure is the exchange of network status information between the restarting MTP and the adjacent nodes. When an adjacent node has completed sending updated information to the restarting node, the adjacent node sends a Traffic Restart Allowed (TRA) Message. When the restarting node has the required information from all adjacent nodes, it sends each adjacent node a TRA Message indicating that the procedure has been completed.

(f) Management Inhibiting

The Management Inhibiting Procedure is used to facilitate maintenance or testing. The procedure does not cause a change of status at Level 2, thus leaving the link available to send maintenance and test messages. The inhibiting condition is usually removed by means of an Uninhibiting Procedure.

(g) Signalling Traffic Flow Control

Signalling Traffic Flow Control is used to limit signalling traffic at its source when the signalling network is unable to transfer all signalling traffic offered by users. This can occur when:

- A failure in the signalling network results in Route Set unavailability.
- Congestion of a Signalling Link or signalling point results in a situation where reconfiguration is not appropriate.
- A user is unable to handle messages delivered by the MTP.

4.6.4 Level 3: Signalling Link Management Function

This function is used to set up and maintain Signalling Links (and certain aspects of Signalling Link Sets). The Signalling Link Management Function specifies several procedures, including: Signalling Link Activation and Deactivation, Signalling Link Restoration and Link Set Activation.

(a) Signalling Link Activation and Deactivation

Signalling Link Activation is the process of making a Signalling Link ready to carry traffic. It involves establishing alignment on the Signalling Link and performing a test to ensure correct functioning. Signalling Link Deactivation is the process of taking a link out of service.

(b) Signalling Link Restoration

Signalling Link Restoration is similar to Signalling Link Activation, but it applies to the re-introduction of service on a Signalling Link after failure.

(c) Link Set Activation

Two forms of Link Set Activation are specified. The Normal Activation Procedure applies when a Link Set is being brought into service for the first time or when a Link Set is being restarted. The Normal Activation Procedure is commenced on as many Signalling Links as possible. An Emergency Procedure is also specified when the Normal Procedure is not considered fast enough or when it is not possible to communicate with the Signalling Point at the remote end of the Link Set. The Emergency Procedure uses a less comprehensive test set and lower timer values.

4.6.5 Level 3 : Signalling Route Management Function

A Signalling Route is a collection of Signalling Links connecting two Signalling Points. It may be a direct route, in which Signalling Links are directly connected, or an indirect route, in which signalling is via one or more STPs.

The Signalling Route Management Function is used to distribute information about the signalling network status in order to Block (i.e. prevent access to) Signalling Routes. There is also an Unblock Function. The procedures include Transfer Prohibited, Transfer Allowed and Signalling Route Set Test.

(a) Transfer Prohibited and Transfer Allowed

The Transfer Prohibited Procedure is initiated by an STP. The aim is to notify one or more adjacent signalling points that messages should not be routed via the STP to a particular destination. An example is when an STP recognises that a particular destination is inaccessible. In this case, the Transfer Prohibited Procedure prevents messages being sent to the STP unnecessarily. A Transfer Allowed procedure is used to remove the prohibited status.

(b) Signalling Route Set Test

The Signalling Route Set Test Procedure is initiated by a signalling point to test whether or not signalling traffic towards a particular destination can be routed via an STP. The Signalling Route Set Test Message contains the current route status of the destination, as understood by the sending signalling point. On receipt of the message, the STP compares the status of the destination specified in the received message with the actual status. If they are the same, no action is taken. If they are different, the result is returned to the signalling point.

4.7 Chapter Summary

4.7.1 General

The Message Transfer Part (MTP) consists of Levels 1 to 3 of the 4-level structure adopted for CCSS7. The MTP is responsible for transferring messages between Signalling Points. The objective of the MTP is to transfer messages without loss or duplication and deliver them to the intended destination in an error-free condition and in the sequence in which they were transmitted. The MTP enables a signalling network to meet demanding message transfer requirements, even during fault conditions.

4.7.2 Primitives

The interface between the MTP and its users employs the MTP-Transfer, MTP-Pause, MTP-Resume and MTP-Status Primitives.

4.7.3 Level 1

Level 1 of the MTP defines the physical, electrical and functional characteristics of the transmission link. Higher levels can thus be independent of the transmission system adopted to carry the signalling system.

4.7.4 Level 2

Level 2 of the MTP defines the functions pertinent to a single Signalling Link. Information is transferred in Signal Units that can be Message Signal Units (carrying User Information), Link Status Signal Units (reflecting the status of the Signalling Link) and Fill In Signal Units (used to maintain synchronisation). The most comprehensive Signal Unit is the Message Signal Unit (MSU), consisting of:

- flags delimiting the MSU;
- sequence numbers and indicators used for error correction;
- a Length Indicator;
- the Service Information Octet, indicating the appropriate User;
- the Signalling Information Field containing the User Information;
- check bits used for error detection.

Two forms of error correction mechanism are defined. The Basic Method is a Non-Compelled, Positive/Negative Acknowledgement, Retransmission Error Correction System. The Preventive Cyclic Retransmission Method is a Positive Acknowledgement, Cyclic Retransmission, Forward Error Correction System.

Level 2 also performs flow control to overcome congestion and error monitoring procedures to ensure that the quality of links remains acceptable.

4.7.5 Level 3

Level 3 of the MTP defines the functions to provide a cohesive and comprehensive signalling network. Level 3 Functions are categorised into Signalling Message Handling Functions and Signalling Network Management Functions.

(a) Signalling Message Handling Functions

The Signalling Message Handling Functions are used to route messages to the correct destination. The Message Routing Function uses a Routing Label to route outgoing messages to the Destination Signalling Point (SP).

The Message Discrimination Function accepts incoming messages and determines if they are intended for the receiving SP or another SP. For those messages intended for another SP, the Message Discrimination Function sends them to the Routing Function for onward transmission. Those messages intended for the receiving SP are passed to the Message Distribution Function for delivery to the appropriate User.

(b) Signalling Network Management Functions

This defines a range of features and facilities to control the flow of signalling messages through the network.

Signalling Traffic Management Functions are used to divert signalling traffic from a link (or route) to another link (or route), restart a signalling point's MTP or reduce the traffic to an SP that is experiencing congestion. Several procedures are defined to achieve these results, including Changeover, Changeback, Forced Rerouting, Controlled Rerouting, MTP Restart, Management Inhibiting and Signalling Traffic Flow Control.

Signalling Link Management Functions provide a means to establish and maintain the capability of Signalling Links and certain aspects of Signalling Link Sets. In the event of Signalling Link failures, the Signalling Link Management Functions control the actions aimed at restoring full capability. The Signalling Link Management Functions specify several procedures, including:

- Signalling Link Activation and Deactivation;
- Signalling Link Restoration;
- Link Set Activation.

Signalling Route Management Functions are used to distribute information about the signalling network status in order to Block (i.e. prevent access to) and Unblock Signalling Routes. The procedures include:

- Transfer Prohibited and Transfer Allowed;
- Signalling Route Set Test.

In summary, the MTP provides a flexible and secure method for transferring messages from one user to another. The MTP forms the basis for establishing a signalling network.

4.8 References

1 ITU-T Recommendation Q.701: 'Functional Description of the MTP' (ITU, Geneva)
2 ITU-T Recommendation Q.706: 'MTP Signalling Performance' (ITU, Geneva)
3 ITU-T Recommendation Q.702: 'Signalling Data Link' (ITU, Geneva)
4 ITU-T Recommendation Q.703: 'Signalling Link' (ITU, Geneva)
5 ITU-T Recommendation Q.704: 'Signalling Network Functions and Messages' (ITU, Geneva)

Chapter 5
Signalling Connection Control Part

5.1 Introduction

The evolution towards more use of non-circuit-related signalling requires a comprehensive platform upon which information can be transferred. The Signalling Connection Control Part (SCCP)[1] of CCSS7 is specified to provide the means to establish logical (i.e. non-circuit-related) signalling connections. It also provides a capability to transfer elements of information termed 'Network Service Data Units' (NSDUs). The SCCP is defined in terms of the Open Systems Interconnection (OSI) 7-Layer Model.

5.2 Architecture

5.2.1 OSI Model

Although the MTP is a comprehensive transfer mechanism, it was specified before the OSI Model was applied to telecommunications signalling systems. The MTP adequately covers Layers 1 and 2 of the OSI Model, but a number of functions need to be added to the MTP to provide adequate Layer 3 functions. For flexibility in future network environments, it is important that CCSS7 fits within the OSI 7-Layer Model approach for non-circuit-related applications.

It is the SCCP that provides functions in addition to those of the MTP in order to offer an OSI Network Service. The combination of the MTP and the SCCP is called the Network Service Part (NSP). By offering a Network Service, the NSP can be utilised by a variety of users.

Chapter 3 describes how the SCCP fits into the overall architecture of CCSS7 and provides the interface between Layers 3 and 4 of the OSI Model.

5.2.2 Structure

The structure of the SCCP is illustrated in Fig. 5.1.

The SCCP has four main functional groups. The Connection-Oriented Control Function establishes and releases SCCP Connections between nodes and facilitates data transfer.

The Connectionless Control Function facilitates data transfer between nodes without establishing a Signalling Connection.

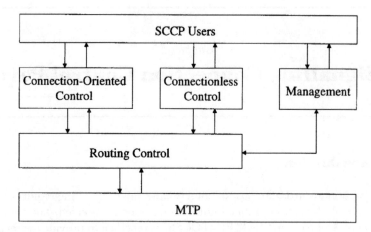

Fig. 5.1 SCCP structure

The Routing Control Function routes messages according to the Called Party Address. For messages received from the MTP, the Routing Control Function distributes the message to the appropriate Control Function or, if the message is not for this node, routes the message to another node. For messages generated by the SCCP User, the Routing Control Function delivers the message to the MTP for onward transfer to another node.

The Management Function provides capabilities to handle congestion or failure of the SCCP, the SCCP user or the signalling route.

5.3 Services

5.3.1 Service types

The SCCP offers both Connection-Oriented and Connectionless Services.

In Connection-Oriented Services, a Relationship (Connection) is established between two communicating nodes before data transfer begins. The signalling connection is released upon completion of the information transfer. A connection is established by exchanging Local Reference Numbers, i.e. numbers allocated by each node to identify to which communication a message refers. In this case, any data transmitted between the nodes includes the Local Reference Numbers, thus identifying the appropriate connection.

The information in Connection-Oriented Services is transferred in blocks called Network Service Data Units (NSDUs), which can be up to 255 octets in length. For longer data streams, the information is split (Segmented) into blocks of 255 octets at the sending node so that each block can be transmitted separately. The blocks are then put back together (Reassembled) at the receiving node.

In Connectionless Services, the information is transferred without establishing a specific connection. In this case, information is routed by the SCCP, but confirmation of receipt is not part of the service. Again, a Segmentation and Reassembly capability is provided for long information streams.

5.3.2 Service classes

There are four classes of service:

- Class 0, Basic Connectionless;
- Class 1, In-sequence Connectionless;
- Class 2, Basic Connection-Oriented;
- Class 3, Connection-Oriented with Flow Control.

Class 0 is a Connectionless Service. In Class 0, each NSDU is transported from the sending SCCP to the receiving SCCP in an independent manner using the MTP. Thus, NSDUs at the receiving node might not be in the same sequence as they were sent. This restriction applies under normal operating conditions, as well as under fault conditions.

Class 1 is also a Connectionless Service. Class 1 is similar to Class 0, but a limited sequencing mechanism is included. This allows the sending node to request that the NSDUs be delivered to the receiving node in the same sequence as they are sent. The sequencing is performed by the MTP in response to the SCCP selecting a consistent Signalling Link Selection (SLS) field. This technique works under normal operating circumstances but, under fault conditions in the network, the lack of a connection can still result in mis-sequenced messages.

Classes 2 and 3 are Connection-Oriented Services that exchange NSDUs between nodes on the basis of a specific connection. Classes 2 and 3 can establish temporary connections (i.e. on demand by a user) or permanent signalling connections that are established by management action. Temporary Connections have a Connection Establishment Phase, a Data Transfer Phase and a Connection Release Phase.

In Class 2, NSDUs may be transferred in both directions during the Data Transfer Phase of the connection. Class 3 complements Class 2 by the inclusion of:

- flow control (i.e. the ability to control the rate of message transfer);
- expedited data transfer, in which certain data is given priority for transfer and can override the congestion procedures;
- additional capabilities for the detection of message loss or mis-sequencing.

Protocols used to deliver the Service Classes are termed Protocol Classes. Protocol Classes 0–3 correspond to the Service Classes 0–3.

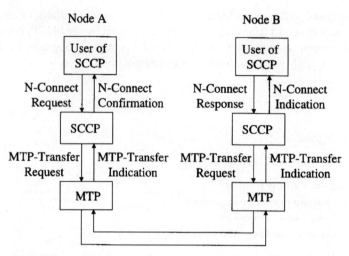

Fig. 5.2 SCCP N-Primitives

5.4 Primitives

The SCCP defines primitives for (*a*) interfaces to higher layers and (*b*) the interface to the MTP. Fig. 5.2 illustrates the use of the primitives by considering a request to establish a connection by the SCCP user in Node A. This is initiated by the User issuing an N-Connect Request Primitive. The SCCP in turn passes an MTP-Transfer Request Primitive to the MTP.

Table 5.1 Examples of SCCP N-primitives

Generic name	Specific name	Applies to	Description
N-Connect	Request, Indication, Response, Confirmation	Connection Oriented, Temporary	Requests the set-up of a connection
N-Data	Request, Indication	Connection-Oriented, Temporary & Permanent	Contains User data to be transferred between nodes
N-Disconnect	Request, Indication	Connection Oriented, Temporary	Initiates release of the connection
N-Unitdata	Request, Indication	Connectionless	Contains User data to be transferred between nodes

Table 5.2   Typical parameters

Primitive	Typical parameters	Description
N-Connect	Called Address	e.g. Called Point Code
	Calling Address	e.g. Calling Point Code
	Quality of Service	Negotiates Protocol Class and flow control window size
	Importance	Gives priority to a message during congestion
N-Data	User Data	Information being transferred
N-Disconnect	Originator	User of Network Service
	Reason	Gives reason for disconnection or connection refusal
N-Unitdata	Called Address	As above
	Calling Address	As above
	Sequence Control	Requests Class 1 Service
	Return Option	indicates to discard or return NSDU when encountering transport problems
	User Data	As above

In this case, the MTP Primitives do not include the Response and Confirmation Types. Hence, the MTP-Transfer Request and Indication Primitives are used to return the result to Node A. Examples of primitives between the SCCP User and the SCCP are given in Table 5.1. Examples of the parameters that are included within SCCP User Primitives are given in Table 5.2.

5.5 SCCP Formats

5.5.1 Principle of message structure

SCCP Messages are carried in the Signalling Information Field (SIF) of Message Signal Units (MSUs), as described in Chapter 4. The structure of an SCCP Message[2] is illustrated in Fig. 5.3.

The Routing Label is the standard version described in Chapter 4. The Message Type Code consists of one octet and it is mandatory for all SCCP messages. The Message Type Code uniquely defines the function and format of the SCCP message. The remainder of the message consists of parameters which, for a given Message Type, can be mandatory or optional. The mandatory parameters can be of fixed or variable length.

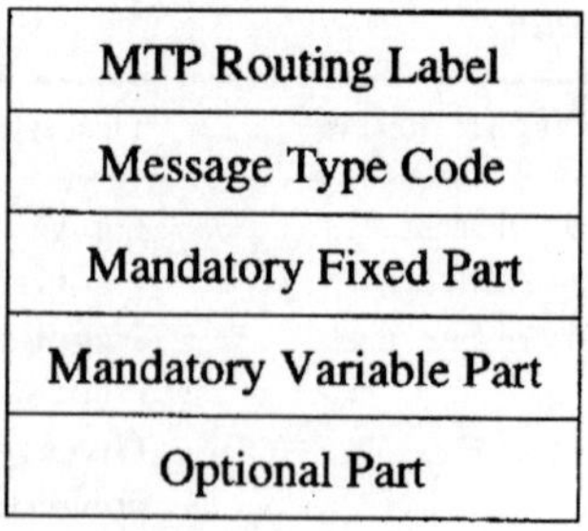

```
┌─────────────────────────┐
│   MTP Routing Label     │
├─────────────────────────┤
│   Message Type Code     │
├─────────────────────────┤
│   Mandatory Fixed Part  │
├─────────────────────────┤
│  Mandatory Variable Part│
├─────────────────────────┤
│     Optional Part       │
└─────────────────────────┘
```

Fig. 5.3   SCCP message structure (Source: ITU-T Recommendation Q.713)

5.5.2 Message types

Examples of message types are given in Table 5.3.

5.5.3 Parameters

Each message contains a number of parameters that complement the information contained in the message type. In general, each parameter consists of a name, a length indicator and a data field, as shown in Fig. 5.4.

Table 5.3   Examples of message types

Message type code	Service class	Description
Connection Request (CR)	2 & 3	indicates a request by one node to set up a connection, with given characteristics, to another node
Connection Confirm (CC)	2 & 3	confirms that a connection has been established between two nodes in response to a CR message
Data Form 1	2	used to pass data transparently between nodes
Data Form 2	3	used to pass data transparently between nodes
Released	2 & 3	indicates that the signalling connection is being released by either node
Release Complete	2 & 3	indicates completion of connection release
Unitdata	0 & 1	transfers data in connectionless mode

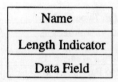

Fig. 5.4   Parameter structure

The Name uniquely identifies the parameter and is coded as a single octet. The Length Indicator specifies the length of the parameter and the Data Field contains supporting information. However, not all of these fields are included in all parameters. Parameters can be Fixed Mandatory, Variable Mandatory or Optional.

Fixed Mandatory Parameters must be present in a given Message Type and they are of fixed length. The position, length and order of Fixed Mandatory Parameters are uniquely defined by the Message Type, so the Parameter Names and Length Indicators are not included in the message.

Variable Mandatory Parameters must be present in a given Message Type, but they are of variable length. The name of the parameter is implicit in the Message Type and hence the Parameter Name is not included in the message.

Optional Parameters may or may not occur in any particular Message Type. Each Optional Parameter includes the Name and a Length Indicator before the Data Field carrying the parameter contents.

5.5.4   Pointers

SCCP messages can be long and complex. To assist in processing, Pointers are used within a message to indicate the location of certain parts of the message. A Pointer is one or two octets in length and indicates the number of octets between the Pointer itself and the start of the parameter being identified. Pointers are used to indicate the start of parameters in the Mandatory Variable Part of the message and also to indicate the start of the Optional Part of the message.

5.5.5   Overall message structure

The overall message structure is illustrated in Fig. 5.5.

To illustrate the principles of SCCP formatting, consider a Connection Request Message that is used to establish a connection when using an SCCP Connection-Oriented Service. An example of such a message is shown in Fig. 5.6.

The Message Type Code indicates that the message is a Connection Request. The Message Type is followed by four parameters. The first parameter is a Fixed Mandatory Parameter called Source Local Reference, which indicates the Reference Number that the Originating SCCP has allocated to identify messages relevant to a particular connection.

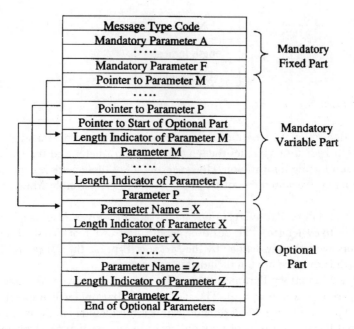

Fig. 5.5 Overall message structure

parameter 4	parameter 3	parameter 2	parameter 1	message type
O	V	F	F	
calling-party address	called-party address	protocol class	source-local reference	connection request

F - fixed mandatory
V - variable mandatory
O - optional

Fig. 5.6 Example of message

The second parameter is also Fixed Mandatory and it is called Protocol Class, reflecting the Class of Service requested.

The third parameter is Variable Mandatory and it is termed Called Party Address, indicating the identity of the SCCP to which the message is being sent. This parameter includes a Length Indicator to show the number of address digits included in the data field of the parameter.

The fourth parameter is Optional and is called Calling Party Address, indicating the identity of the SCCP sending the message. This parameter includes a Length Indicator and a Name.

5.6 SCCP Procedures

SCCP procedures[3] are described for both the Connection-Oriented and Connectionless classes of service.

5.6.1 Connection-Oriented procedures

An example of the message sequence for Connection-Oriented Services is shown in Fig. 5.7. In this example, higher-layer functions in Node 1 need to communicate with corresponding functions in Node 2. SCCP1 receives a request from a higher-layer function within Node 1 to establish a connection with SCCP2. SCCP1 analyses the Called Party Address (i.e. the address of SCCP2)

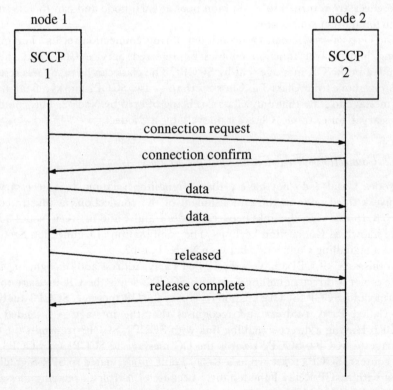

Fig. 5.7   Connection-Oriented procedures

and the result of the analysis is that the connection should be established over an appropriate signalling link, via the MTP, to Node 2.

A Connection Request (CR) Message is sent to SCCP2 via the MTP. Upon receipt of the CR Message at Node 2, the MTP delivers the CR Message to SCCP2. Analysing the Called Party Address, SCCP2 recognises that the CR Message has reached its intended destination and that a connection needs to be established with SCCP1. A Connection Confirm (CC) message is returned to SCCP1.

When the CR and CC Messages have been exchanged, the signalling connection has been established and data transfer can occur. When data transfer is complete, either SCCP1 or SCCP2 can initiate a release procedure by sending a Released (RLSD) Message. Receipt of the RLSD Message by a node is confirmed by returning a Release Complete (RLC) Message.

During connection establishment, Source and Destination Local References are assigned to the connection. The Source Local Reference is chosen by SCCP1 (from a pool of numbers) and the Destination Local Reference is chosen by SCCP2. The combination of these Local References then acts as the Reference Number to identify uniquely the SCCP connection. The Local References are mandatory fields in SCCP Messages. Upon release of the connection, the Local References are returned to a common pool at each node and can then be used again for another connection.

The class of service can be negotiated during connection set-up. The originating higher-layer function chooses a preferred Service Class and this is included in the CR message sent by SCCP1. The class can be made less restrictive (e.g. move from Class 3 to Class 2). In this case, SCCP2 marks a field in the CC message to state that only Class 2 is being offered by Node 2. This could be necessary if, for example, Class 3 is unavailable at Node 2.

5.6.2 *Connection-Oriented with multiple sections*

If Nodes 1 and 2 do not have a direct-signalling relation, it is necessary to involve a third node in the establishment of the connection, as illustrated in Fig. 5.8. In this case, the links between Nodes 1 and 3 and between Nodes 3 and 2 are known as Connection Sections. The combination of Connection Sections forms a Signalling Connection between Nodes 1 and 2.

In this case, SCCP1 analyses the Called Party Address and, recognising that there is not a direct signalling link with SCCP2, sends the CR message to an intermediate SCCP (SCCP3). Upon receipt of the CR message, SCCP3 analyses the Called Party Address and recognises that the message is intended for SCCP2. Having a direct signalling link with SCCP2, SCCP3 transmits the CR message to SCCP2. SCCP2 returns the CC message to SCCP1 via SCCP3. In this context, SCCP3 is known as a Relay Point (abbreviated to SPR-Signalling Point with SCCP Relay Functionality) because of its role in relaying messages between SCCPs 1 and 2.

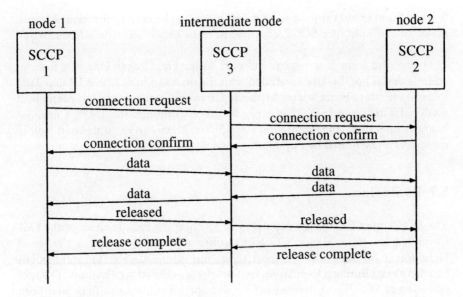

Fig. 5.8 Multiple sections

In some cases, a coupling of the two Connection Sections is necessary to correlate set-up information and permit data transfer. In this case, SCCP 3 performs the Coupling Function.

5.6.3 *Connectionless procedures*

In Connectionless Services, data is transferred in Unitdata Messages without establishing a connection. In this case, the higher-layer functions in the Originating Node request the SCCP to send an NSDU by Protocol Class 0 or 1. The Originating SCCP analyses the Called Party Address and determines that a message needs to be sent to a subsequent node. Upon receipt of the Unitdata Message at the subsequent node, an analysis of the Called Party Address determines that the message has reached its intended destination and the information within the NSDU is delivered to the appropriate higher-layer function.

If an intermediate SCCP is required to route the Unitdata Message, this is recognised by analysing the Called Party Address. The result of the analysis indicates the node to which the message should be routed. If the Unitdata Message is delivered to a node that cannot route the message further, procedures exist either to discard the message or to return the message to the Originating Node.

5.6.4 *Addressing*

The SCCP recognises two forms of address for routing purposes. The first form is the Destination Point Code (DPC) of the required node, together with a Sub-

System Number that is used to route messages to the correct function within a node. In this case, the SCCP does not need to implement special translation facilities.

The second form is termed the Global Title. The Global Title is a form of address that is not directly associated with the destination SCCP. A Global Title therefore needs to be translated by the SCCP to facilitate routing. An example of a Global Title is an ISDN Customer Number. Although the ISDN Customer Number identifies the Destination Node, it needs to be translated before messages can be routed by the SCCP.

5.7 Evolution

The SCCP gives CCSS7 the capability of meeting the requirements of the OSI Layer 3/4 Interface. Hence, CCSS7 can offer a Network Service to a range of higher-layer functions. In the short term, this allows CCSS7 the tremendous benefit of establishing a foundation for non-circuit-related applications. The specification of SCCP was derived before widespread implementations had been achieved. Thus, the implementation of standard versions of SCCP throughout the world will help to reduce costs, avoiding the development costs of many different solutions.

In the longer term, the benefits will be even greater. By providing a Network Service, the combination of the MTP and SCCP can be used to provide a transfer capability for a range of higher-layer protocols conforming to the OSI 7-Layer Model, irrespective of whether or not the higher-layer protocols are specified as part of CCSS7. The possibility will exist to mix and match the various protocols available to network operators, providing great flexibility in implementation and allowing technical solutions to be optimised to meet specific environments. The prospects are fascinating.

5.8 Chapter Summary

The Signalling Connection Control Part (SCCP) defines the functions, additional to the MTP, to meet the Layer 3/4 boundary requirements of the OSI 7-Layer Model. Thus, the combination of the MTP and SCCP provides a Network Service for higher layers. The SCCP is used in non-circuit-related applications.

The primitives for the Layer 3/4 boundary of the OSI Model are defined as N-Primitives in line with the principles outlined in Chapter 3.

The SCCP provides four classes of service to higher layers. Classes 0 and 1 are Connectionless Services in which data is transferred without the establishment of a signalling connection. In Classes 2 and 3, a signalling connection is established before data is transferred.

SCCP Messages are carried in the Signalling Information Field (SIF) of the MTP. SCCP Messages consist of:

(i) a Routing Label, containing the appropriate OPC and DPC;
(ii) a Message Type Code of 1 octet, uniquely defining the function and format of the message;
(iii) various parameters – parameters can be optional or mandatory for a particular type of message. Mandatory Parameters can be of fixed length or variable length. The formatting technique adopted for the SCCP is extremely flexible to enable future evolution to take place.

In Connection-Oriented services, a connection is first established using Connection Request and Connection Confirm Messages. SCCP Signalling Information is then transferred between nodes using N-Data Messages. Messages are correlated by using Local Reference Numbers allocated by the nodes involved in the connection. This use of Local Reference Numbers, rather than the number of a traffic circuit, confirms the nature of SCCP as being non-circuit related. The connection is cleared by using Released and Release Complete Messages.

In Connectionless Services, SCCP Signalling Information is transferred in N-Unitdata messages without establishing a connection.

The SCCP was defined internationally before widespread implementations of interim systems took place. It was therefore possible to capitalise on the benefits of economies of scale in terms of development costs when implementing the transfer of non-circuit-related information. Because the SCCP, in conjunction with the MTP, offers a Network Service, many higher-layer functions can use the SCCP, providing great flexibility for network operators.

5.9 References

1 ITU-T Recommendation Q.711: 'Functional Description of the SCCP' (ITU, Geneva)
2 ITU-T Recommendation Q.713: 'SCCP Formats and Codes' (ITU, Geneva)
3 ITU-T Recommendation Q.714: 'SCCP Procedures' (ITU, Geneva)

Chapter 6
CCSS7 Narrowband ISDN User Part

6.1 Introduction

The Message Transfer Part (MTP) of CCSS7 provides a comprehensive transport mechanism to carry Level 4 messages through the signalling network. However, the MTP cannot interpret the meaning of the messages being transferred. It is the User Parts (UPs) that define the meaning of the messages that are being transferred for circuit-related connections. The UPs determine the formats of messages and the sequence in which they are sent. The User Parts also interact with the call control function within a node to establish a controlling mechanism for calls.

The ISDN User Part (ISUP)[1] defines the messages and procedures for the control of switched services in integrated services digital networks (ISDNs). The ISUP covers both voice (e.g. telephony) and non-voice (e.g. circuit-switched data) applications.

For ISDN calls, there is a basic requirement to use digital transmission links and digital nodes, thus ensuring that there is a digital connection (e.g. 64 kbit/s) from one user to another. The ISUP is designed primarily to control the set-up and release of calls in this all-digital environment and the procedures and formats are optimised accordingly. Supplementary services are also defined using the advanced techniques available in the ISUP. The ISUP was developed from the outset for operation in both national and international networks. Hence, the adoption of ISUP will contribute to the reduction in signalling interworking costs between these networks.

The ISUP was designed for use in narrowband networks. A broadband version of ISUP is described in Chapter 15. To discriminate between the narrowband and broadband versions, a prefix is added to form N-ISUP for narrowband and B-ISUP for broadband applications. As this terminology was adopted after the specification of the narrowband version, many references to the narrowband version simply use the term ISUP. This Chapter continues with this terminology.

The principles of the ISUP formatting[2] are described in Section 6.2 and the procedures for providing calls[3] are described in Section 6.3. ISUP Supplementary Services[4] are described in Section 6.4.

6.2 ISUP Formats

6.2.1 Format principles

The format principles adopted for previous CCS systems were considered to be restrictive when applied in an ISDN environment. The aim when defining the formatting for ISUP was to provide more flexibility. By placing more emphasis on the use of variable and optional fields, within a framework defined by the specification, the ISUP is more responsive to changing requirements.

ISUP messages are carried in the Signalling Information Field (SIF) of Message Signal Units, the general format of which is explained in Chapter 4. The message consists of a Routing Label, a Circuit Identification Code, a Message Type Code and Parameters, as shown in Fig. 6.1.

The ISUP uses the circuit-related approach for communication identification, i.e. the number of the circuit is used in the message to identify information pertaining to that circuit. Hence, a Circuit Identification Code (CIC) is used in the ISUP to indicate the number of the traffic circuit between two nodes to which the message pertains.

The Message Type Code consists of a one-octet field and is mandatory for all messages. It describes the category to which the message belongs. The Message Type Code defines the function of each message and the nature of the parameters that can be included in the message.

Each message includes a number of parameters. In general, each parameter has three fields, namely (a) Parameter Name, (b) Length of Parameter and (c) Specific Information. This repetitive approach matches the trends in modern processing techniques and is similar to the concepts for the SCCP. However, some fields are not transported through the network to reduce the amount of unnecessary information carried. The parameters are divided into the Mandatory Fixed Part, the Mandatory Variable Part and the Optional Part.

Mandatory Fixed Parameters must be included in a given Message Type and are of fixed length. The position, length and order of these parameters are uniquely defined by the Message Type, so names and length indicators for these parameters are not included in the message.

Mandatory Variable Parameters must be included in a given Message Type and they are of variable length. A 'Pointer' is used to indicate the beginning of

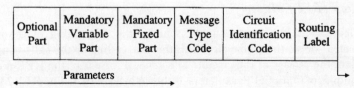

Fig. 6.1   Format of ISUP messages

92   Telecommunications Signalling

each parameter. A Pointer is an octet that can be used during processing of the SIF to find a particular piece of information: this avoids having to analyse the whole message to find one piece of information. The name of each parameter is implicit in the Message Type, so names are not included in the message for these parameters. A Pointer is included to indicate the beginning of the optional part of the SIF.

Optional Parameters may or may not occur in any particular Message Type. Each Optional Parameter includes the Name (one octet) and a Length Indicator (one octet) before the Parameter Contents. A defined octet is used to indicate the end of the Optional Part of the message. An illustration of the ISUP message format, giving the structure of the parameters, is shown in Fig. 6.2.

6.2.2   Examples of message formats

The CCSS7 specification of the ISUP gives a range of Message Types and parameters. Examples of some Message Types to illustrate the principles are given below:

(a)   Initial Address Message (IAM)

The Initial Address Message (IAM) is the first message to be sent during call set-up. It contains the address digits (e.g. digits dialled by the customer to route the call) and it results in a seizure of a circuit by each node. The general format of the IAM is shown in Table 6.1, with examples of Optional Parameters. The Message Type of the IAM is coded 00000001.

(b)   Address Complete Message (ACM)

The Address Complete Message (ACM) is sent by the destination node to indicate successful receipt of sufficient digits to route the call to the destination node. The general format of the ACM is shown in Table 6.2, with examples of Optional Parameters. The Message Type of the ACM is coded 00000110.

(c)   Answer (ANS) Message

The Answer (ANS) Message is sent by the destination node to indicate that the called customer has answered the call. The format of the Answer Message is shown in Table 6.3. The Message Type of the ANS message is coded 00001001.

(d)   Release (REL) Message

The Release (REL) Message can be sent by the originating or destination nodes to clear-down the traffic circuit. Either the calling or the called customer can initiate the release of the circuit. The general format of the REL message, with examples of Optional Parameters, is shown in Table 6.4. The REL Message has a Message Type coded 00001100.

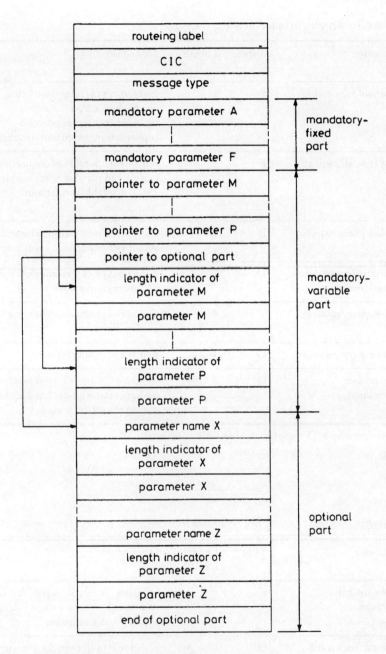

Fig. 6.2   ISUP message and parameter format
Source: ITU-T Recommendation Q.763

94 *Telecommunications Signalling*

Table 6.1   Format of initial address message

Parameter	Type	Length (octets)	Description
Nature of connection	F	1	gives status of connection being established, e.g. satellite included/not included, echo suppressor included/not included
Forward-call indicators	F	2	indicates capability of connection, e.g. end-to-end possible/not possible, ISUP available throughout connection
Calling-party category	F	1	states whether calling customer is subscriber or operator (plus language)
Transmission-medium requirements	F	1	requests type of transmission, e.g. 64 kbit/s connection
Called-party number	V	4–11	number of called customer (e.g. dialled digits)
Calling-party number	O	4–12	number of calling customer
User-to-user information	O	3–131	capacity can be provided to allow customers to send data to each other during the set-up sequence

F = fixed mandatory, V = variable mandatory, O = optional

Table 6.2   Format of address complete message

Parameter	Type	Length (octets)	Description
Backward-call indicators	F	2	indicates eventual capability of the connection, e.g. end-to-end signalling available/not available
Optional backward-call indicators	O	3	additional indicators for use with, e.g. supplementary services
User-to-user information	O	3–131	as described for IAM

Table 6.3 Format of answer message

Parameter	Type	Length (octets)	Description
Backward-call indicators	O	4	as described for ACM
User-to-user information	O	3–131	as described for IAM

Table 6.4 Format of release message

Parameter	Type	Length (octets)	Description
Cause indicators	V	≥ 3	gives reason for the release of the call, e.g. calling/called customer cleared-down
User-to-user information	O	3–131	as described for IAM

6.2.3 Format overhead

The ISUP makes extensive use of Optional Parameter Fields, thus increasing the flexibility available to network and service operators. However, such flexibility can increase the processing overhead within a node, because the node needs to determine the information that has been provided in a particular message. In addition, the size of some ISUP messages could add overhead if many optional fields are included within one message. The IAM can contain up to 29 Optional Parameters, including up to 131 octets of user-to-user information.

Provided that these features are treated carefully, the ISUP formatting technique is extremely flexible and will permit evolution to handle future requirements.

6.2.4 Message segmentation

In view of the possibility of long messages, there is potential to exceed the maximum message length of 272 octets that the MTP is capable of carrying. In

Table 6.5 Format of segmentation message

Parameter	Type	Length	Description
User-to-user information	O	3–131	as described for IAM
Generic Notification	O	3	gives information on, e.g., user status with supplementary services
Generic Number	O	5–13	gives number, numbering plan, source, etc.

this case, it is possible for an originating node to split the original message and send it in two parts. The first part is sent in a message of the original type and the rest of the message is sent in a Segmentation Message. The format of the Segmentation Message is illustrated in Table 6.5.

For example, if an IAM is too long because of the inclusion of user-to-user information, the other parameters are included in a shorter IAM and the user-to-user data is sent in a Segmentation Message. The Forward and Backward Call Indicator Parameters are used to indicate the segmentation of the message.

6.3 ISUP Procedures

6.3.1 Basic call set-up

The procedures for basic call set-up are illustrated in Fig. 6.3.

The information needed to route a call, e.g. the called party number, can be supplied to the originating node in enbloc form or in overlap form. In en bloc form, all the address digits required to identify the called party are provided in one message. In overlap form, the address digits are supplied in groups of one or more: in this case, routing takes place when sufficient address digits to route the call to the intermediate node have been received, with further digits being sent across the network in Subsequent Address Messages (SAMs).

Upon receipt of a request to set up a call from the calling party, the originating node analyses the routing information. The routing information can be stored within the originating node or at a remote database. Analysis of the called party number allows the originating node to determine to which node to route the call. In this example, it is recognised that the call must be routed to the intermediate node.

Examination of the information supplied by the calling party indicates the type of connection required and the signalling capabilities required. The type of connection is chosen from a range that includes 64 kbit/s and a variety of multi-

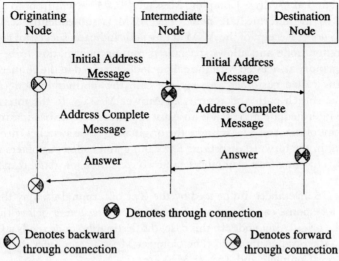

Fig. 6.3   ISUP basic call set-up

rate connections from 2×64 kbit/s to 1920 kbit/s. The signalling capabilities reflect the level of signalling required to meet the needs of the calling party: these include the choices of 'ISUP required', 'ISUP preferred' and 'ISUP not necessary'.

The analysed information is used to select an appropriate circuit and to formulate an Initial Address Message (IAM) with appropriate parameter settings, e.g. appropriate Transmission Medium Requirement Values. The IAM is sent to the intermediate node and the originating node switches through the corresponding circuit in the backward direction (to the calling party). Switching through in the backward direction only at this stage allows the calling party to hear tones provided by the network, but prevents the calling party from sending information on the speech path.

The intermediate node receives the IAM and analyses the appropriate parameters. By examination of the address digits, the intermediate node determines the routing to the destination node. By examination of the Connection Type and Signalling Capabilities Fields in the IAM, an appropriate circuit is selected.

Many parameters in the IAM are passed on transparently. However, parameters indicating the status of the connection can be modified by the intermediate node to reflect the history of the connection. The IAM is sent to the destination node and the circuit at the intermediate node is through-connected in both the forward and backward directions.

When the IAM is received at the destination node, it is analysed to determine the called party. The status of the called party's line is determined and a compatibility check is performed to allow implementation of, for example, supplementary services. The called customer is informed that a call is being

98 Telecommunications Signalling

established and an Address Complete Message (ACM) is sent from the destination node to the intermediate node. The ACM is subsequently sent to the originating node. Receipt of the ACM at any node indicates successful routing to the destination node and allows specialised routing information to be deleted from the memory associated with the call in nodes involved in the connection.

When the called customer answers the call, the destination node switches through the speech path and returns an Answer Message to the intermediate node. The intermediate node sends an Answer Message to the originating node. Upon receipt of the Answer Message, the originating node switches through the speech path in the forward direction. The calling and called customers are now connected, charging can commence and conversation or data transfer can occur.

In some circumstances, influenced by the type of terminal used by the called customer, a response can be received from the called customer before the ACM is sent to the originating node. In this case, the destination node sends a Connect Message to the originating node. The Connect Message combines the functions of the Address Complete and Answer Messages.

6.3.2 Basic call release

In the ISUP, either the calling or called party can initiate immediate release of the connection, i.e. the ISUP adopts the first-party release method of operation, thus catering for data applications. In Fig. 6.4, the calling customer requests disconnection from the originating node.

The originating node commences the clear-down of the connection and sends a Release Message to the intermediate node. The intermediate node sends a Release Message to the destination node and commences clear-down of the

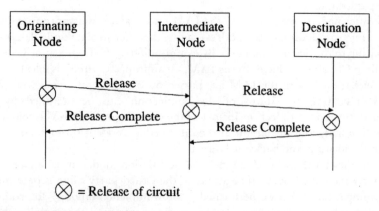

Fig. 6.4 *ISUP basic call release*

speech path. When the speech path has cleared-down and is ready for new traffic, the intermediate node returns a Release Complete Message to the originating node. Similarly, upon receipt of the Release Message, the destination node commences clear-down of the speech path. When the clear-down is complete, the destination node returns a Release Complete Message to the intermediate node.

One concept influencing the design of the release procedure is the need to ensure that either customer can release the connection as soon as possible. The ISUP transfers the Release Message immediately, without waiting for the clear-down to occur. This increases the speed of transmission of the Release Message through the network.

The original specification of the ISUP defined a three-message release sequence involving the Release, Released and Release Complete Messages. This was replaced with the procedure described above to achieve more commonality with the SCCP release procedures.

6.3.3 Additional features

ISDNs require a high degree of flexibility and a wide range of facilities to meet the needs of customers. Additional features are provided by ISUP to meet these needs and the following text gives a summary of common examples.

(a) End-to-end signalling

End-to-end signalling allows nodes to send and receive signalling information without the signalling information being analysed by intermediate nodes. End-to-end signalling is typically used between local nodes to transfer specialised information. In this context, the local nodes are termed 'End Points'. Two forms of end-to-end signalling are specified, namely Pass-Along and SCCP Methods.

The Pass-Along Method makes use of the routing information established for a call. When the ISUP establishes a call, the number dialled by the calling customer is translated into routing information for use in the signalling network. Routing information is held by each node in the connection for the duration of the call. A Message Type termed 'Pass Along' is defined and when a transit node receives a Pass-Along Message it uses the routing information already available to transfer the message to the next node. Analysis of the end-to-end information contained within the Pass-Along message is avoided at intermediate nodes.

The SCCP method uses the SCCP to transfer signalling information. Two forms of information transfer are specified, Connectionless Service and Connection-Oriented Service.

In the Connectionless Service, an ISUP Message (typically an IAM) being transferred from an originating node to a destination node includes a Call Reference. The inclusion of a Call Reference is an implicit indication to the destination node (Receiving End Point) that a Connectionless SCCP mode of information transfer is required. Upon receipt of a Call Reference at the

Receiving End Point, a corresponding Call Reference is returned to the originating node (Sending End Point) in an ACM. This exchange of Call References permits the transfer of Unitdata Messages using the SCCP. The Call References are used to correlate the Unitdata Messages with the call.

In the Connection-Oriented Service, a SCCP Connection Request (CR) Message is embedded within an ISUP Message. If end-to-end signalling is required at the same time as establishing an ISUP call, the CR Message is embedded within an IAM (other types of ISUP message can be used if a call already exists). Receipt of an IAM with an embedded CR Message at a destination node indicates that the originating node wishes to establish an end-to-end Connection. The CR Message is passed by the ISUP at the destination node to the SCCP. The SCCP at the destination node then responds directly to the SCCP at the originating node with a Connection Confirm (CC) Message. Information transfer then takes place with Data Messages using normal SCCP procedures.

The end-to-end techniques represent the ability to establish a logical connection between end-points (i.e. an efficient means of exchanging signalling information between end-points). This is a very flexible and powerful tool because it permits an increase in information transfer without undue impact on transit nodes.

(b) Call Progress

The Call Progress Message can be sent after the Address Complete Message towards the originating node to provide additional information. For example, the Mandatory Parameter 'Event Information' can be used to indicate to the originating node that a call is being diverted from its original destination to another. A wide range of Optional Parameters can also be included.

(c) Suspend/Resume

The Suspend Feature allows a call to be suspended for a period of time during the conversation/data phase. It can be initiated by an interworking node or an analogue called party. The feature can be used, for example, to allow a customer to change the terminal equipment that is being used, or to change locations within the customer's premises, without releasing the call. The Resume Procedure is initiated by a re-answer indication by the called party. On receipt of the Resume Indication, each node recognises that the original call needs to be re-enacted. Unintentionally long suspensions are avoided by one node in the network starting a timer upon receipt of a Suspend Indication.

(d) Fallback

The Fallback Procedure allows a calling party to specify both a preferred and a secondary connection type. For example, a calling party could prefer to use a 64 kbit/s connection, but recognises that this request cannot always be supplied and would accept an analogue connection. The procedure utilises User Service Parameters in the IAM to indicate the preferred and fallback connection types.

As an example, take the case when a 64 kbit/s connection is requested but the called party has an analogue line. When the IAM is received at the destination node, it is recognised that only an analogue connection can be made. In this case, the ACM is used to indicate the analogue nature of the connection by setting values within the Transmission Medium Parameter. Other backward messages are used if an intermediate node cannot select the preferred connection type.

(e) Propagation delay

If propagation delay is important to the usage of a connection, this procedure allows a calling party to determine the delay on the connection being established. The procedure is initiated by including a Propagation Delay Counter Parameter in the IAM. This requests each node in the connection to establish the delay on the next part of the connection and add this to the value in the parameter. In this way, the destination node can ascertain the total delay on the connection and inform the calling party by including values within a Call History Information Parameter in an appropriate backward message.

(f) Echo control

Some connections, e.g. on long international calls, can result in an echo effect. This can be disconcerting for the calling and called parties using speech and, in these cases, it is necessary to insert control devices to reduce the echo. In other cases, it is desirable to deactivate control devices that have already been inserted. The echo control procedure is used on a per call basis to convey information between nodes about the need for, and the ability to insert and deactivate, such echo control devices. The procedures involve using a Nature of Connection Parameter in the IAM and a Backward Call Parameter in backward messages like the ACM and Connect Messages.

(g) Blocking/Unblocking

A Blocking Procedure is used to remove a circuit from service, e.g. to respond to a fault condition or to enable circuit testing. The sending of a Blocking Message from one node to another causes the appropriate circuit to be made unavailable for all calls other than test calls. After fault rectification, or upon completion of testing, the circuit is re-instated by means of an Unblocking Message. Both Blocking and Unblocking Messages are acknowledged after the appropriate action has been performed.

In the case when a number of circuits need to be removed from service, a similar procedure applies using Group Blocking and Group Unblocking Messages.

(h) Abnormal conditions

The specification of a complex signalling system needs to cover abnormal conditions to ensure that appropriate action is taken by the network to overcome problems. ISUP covers many such conditions and examples are given below.

The Reset Procedure is used when the status of a circuit, or a number of circuits, becomes unclear to a node (e.g. due to memory mutilation within the node). In this case, procedures are defined for the node at each end of a circuit to reset the circuit to the idle condition. If a small number of circuits are affected, Reset Messages are used for each circuit. If a large number of circuits are affected, a Group Reset Message can be applied to reset all the circuits by sending one message.

Dual seizure is when two inter-connected nodes choose the same circuit for two different calls at approximately the same time. The risk of dual seizure is minimised by applying different algorithms for choosing speech paths at each end of the transmission link. In this way, the likelihood of two exchanges choosing the same speech circuit is reduced. To resolve the contention when a dual seizure does occur, each circuit is nominally controlled by one node, based on a series of rules, e.g. the type of connection established. The IAM sent by the controlling node for a particular circuit is processed normally by the non-controlling node. The IAM sent by the non-controlling node for the circuit is disregarded by the controlling node. Upon detection of dual seizure at the non-controlling node, an automatic repeat attempt is initiated for another circuit.

Rules are defined to cover the actions to be taken at each node upon receipt of an unexpected message. For example, if a node receives a Release Message relating to an idle circuit, the node will acknowledge the signal with a Release Complete Message.

Procedures are defined to cater for the reception of unreasonable messages and parameters. Unreasonable information can either be unexpected (e.g. a message is recognised but is received out of sequence) or unrecognised. Procedures for unrecognised information are typically invoked when the ISUP at one node is at a higher level of functionality than the ISUP at another node. For example, if one node is operating a new version of ISUP and another is operating the previous standard version of ISUP, the standard node could receive a message that it does not understand. In this case, the standard node can return a Confusion Message with a Cause Parameter indicating that unrecognised information has been received. The upgraded node can then send an alternative message, re-route the call to another node or inform the calling customer that the new feature is not yet available on the route requested.

The procedures are specified in terms of the type of node receiving the unreasonable information. Type A Nodes include those originating and terminating calls and those performing interworking functions. Type B Nodes are those that act as transit nodes without special interworking functions.

The procedures for handling unreasonable information will become increasingly important for the efficient operation of networks. The need for rapid changes to meet customer demands means that more versions of the ISUP will be implemented, resulting in a greater need for an efficient method of allowing versions to interwork.

6.4 ISUP Supplementary Services

ISUP defines a range of supplementary services that add a considerable variety of functions to the basic call establishment. The categories of supplementary service are:

- Number Identification;
- Call Offering;
- Call Completion;
- Multi-Party;
- Community of Interest;
- Charging;
- Additional Information Transfer.

The following text gives examples of each category of supplementary service to explain the principles involved.

6.4.1 Number identification

(a) Calling Line Identification Presentation (CLIP)

CLIP is a supplementary service in which the Calling Party's Number (Address) is provided to the called party before the call is answered. CLIP applies to all calls delivered to the called party (unless the calling party has restricted presentation, see Item (b)).

The Calling Party Address is provided either by the Calling Party Access Signalling System or by the originating node. The address is included in the IAM and passed through the network to the destination node. If the called party subscribes to this supplementary service, the Calling Party Number is included in the Set-Up Message. This allows, for example, the Calling Party Number to be displayed on the called party's terminal equipment.

(b) Calling Line Identification Restriction (CLIR)

CLIR allows the calling party to prevent the presentation of the Calling Party Address. This overrides the CLIP supplementary service. For CLIR, the Calling Party Number is still transmitted through the network in case the network or operator needs to identify the caller (e.g. to trace a call origin), but an indicator in the Calling Party Number Parameter in the IAM is set to prevent the number being presented to the called party.

6.4.2 Call Offering

The Diversion supplementary service is an example of Call Offering. Diversion permits a called party to re-route incoming calls to another party. Four cases are specified:

- Call Forwarding on Busy (CFB), in which the network detects that the called party is busy or the called party indicates its busy status;
- Call Forwarding on No Reply (CFNR), in which the call is diverted after a period of alerting;
- Call Forwarding Unconditional (CFU), in which all incoming calls are diverted;
- Call Deflection (CD), in which the called party requests a diversion after a call has been offered and before the call is answered.

CD is used on a per call basis, whereas the others apply to all calls that meet the appropriate characteristics.

(a) Call Forwarding on Busy (CFB)

Typical signalling flows for CFB are given in Fig. 6.5. In this case, Party B has requested that calls should be diverted to Party C when Party B is busy. Upon receipt of the IAM at Node B, a review of the status of the Called Party's Line (Party B) indicates that the called party is busy. Recognising that Party B subscribes to CFB, Node B determines the address to which to divert the call. This address has been registered previously by Party B.

Node B then initiates an IAM to Node C and also returns an ACM to Node A. The ACM includes information about the diversion in an appropriate

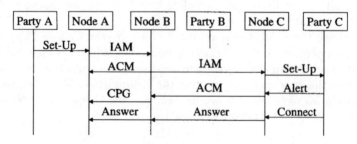

Fig. 6.5 Flows for Call Forwarding on Busy

        IAM = Initial Address Message
        ACM = Address Complete Message
        CPG = Call Progress

parameter, e.g. Redirection Information. The call is established to Party C as for a basic call, but with indicators in the Access System to show the history of the call. When Node B receives the ACM from Node C, Node B returns a Call Progress Message to Node A to inform Node A that Party C has been alerted. The Answer Message is returned from Node C to Node A as for a basic call.

(b) Call Forwarding on No Reply (CFNR)

Typical signalling flows for CFNR are given in Fig. 6.6. In this case, Party B has requested that calls should be diverted to Party C when Party B does not reply within a given time period.

The flows are similar to those for CFB, but in this case Party B receives a Set-Up Message and returns an Alerting Indication. Node B determines that Party B subscribes to CFNR and recognises that diversion might take place if there is no reply in the specified time period. The ACM returned from Node B to Node A therefore includes a Backward Call Indicator Parameter indicating that call diversion might occur. After the timer expires, the connection to Party B is disconnected and an IAM is sent to Node C. Node A is informed of the diversion in a Call Progress Message.

(c) Call Deflection (CD)

Typical signalling flows for CD are given in Fig. 6.7. In this case, Party B has not requested call diversion beforehand, but decides to request a diversion to Party C after a Set-Up Message has been received.

In this case, Party B receives a Set-Up Message and, deciding to divert the call, returns a Deflection Indication to Node B, including the address to which to

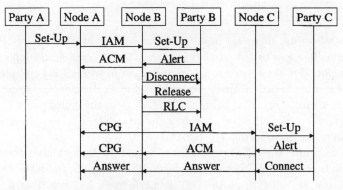

Fig. 6.6 Flows for Call Forwarding on No Reply

        IAM = Initial Address Message
        ACM = Address Complete Message
        RLC = Release Complete
        CPG = Call Progress

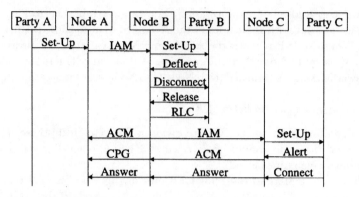

Fig. 6.7 Flows for Call Deflection

IAM = Initial Address Message RLC = Release Complete
ACM = Address Complete Message CPG = Call Progress

divert the call. The Deflection Indication causes Node B to abandon the set-up of the call to Party B and initiate a call to Party C. Node B returns an ACM to Node A with parameters set to indicate progress. The flow is then similar to CFB and CFNR.

(d) Call Forwarding Unconditional (CFU)

In this case, Party B has requested that all calls are diverted and the signalling flows are similar to those shown for CFB.

6.4.3 Call Completion

In basic call set-up, there are many circumstances that would normally result in a failed call, e.g. a called party being busy. Call Completion supplementary services respond to these circumstances and assist in successfully completing the call. An example is Call Waiting, in which a busy called party can be notified that another incoming call is being made. This gives the called party the option of ignoring the new call or of accepting the new call. The signalling flows for Call Waiting are illustrated in Fig. 6.8.

In this example, the ACM is sent from Node B to Node A after the Alerting Indication is received by Node B. Hence, Node B can indicate in the ACM that Call Waiting is being applied. This is performed by including a Generic Notification Indicator Parameter, with a value Call Waiting, in the ACM. If Party B accepts this new call, then a Connect Indication is sent to Node B, which returns an Answer Message to Node A. The call is then established as normal.

If Party B does not wish to accept the new call, a Disconnect Indication is sent to Node B and this initiates normal release procedures in the network.

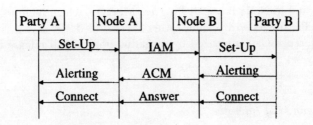

Fig. 6.8 Flows for Call Waiting

IAM = Initial Address Message
ACM = Address Complete Message

6.4.4 Multi-Party

Multi-Party supplementary services allow a number of users to communicate at the same time. The example chosen to illustrate the principles is Conference Calling, in which a conference bridge is used to permit simultaneous communication. Parties can be attached or disconnected from the conference without affecting other participants. Typical signalling flows are given in Fig. 6.9.

The conference begins from a call that is already established between Parties A and B. Party A initiates the Conference Call with a Facility Indication and Node A sends a Call Progress Message to Node B including a Generic Notification Parameter with a value Conference Established. Parties A and B enter a conference state.

Other parties can now be added to the conference by Node A sending a Call Progress Message to each new party (Party C in Fig. 6.9), which includes a Generic Notification Parameter with a value Conference Established. The existing participant (Party B) is informed of Party C joining the conference by

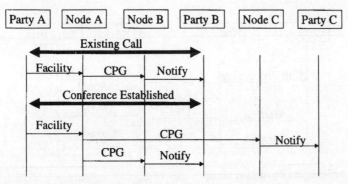

Fig. 6.9 Flows for Conference Calling

CPG = Call Progress

Node A sending to Node B a Call Progress Message, which includes a Generic Notification Parameter with a value Other Party Added.

The Conference Call is cleared by initiating normal call release procedures to each party.

6.4.5 Community of Interest

Community of Interest supplementary services refer to features that apply to groups of users. A typical example is Closed User Group (CUG), which enables parties to form a group or club. Within a CUG, members can communicate with each other, unless specific rules are applied to restrict inter-member communication. Similarly, members of a CUG can only make/receive calls external to the CUG if specific permission is established.

An Interlock Code is used to check the validity of membership of a CUG. A validation check is performed at call set-up to ensure that the Interlock Code is applicable to both the calling and called parties. The data for a party's CUG can be stored at the party's local node (decentralised operation) or at a remote point (centralised operation). Typical signalling flows for the decentralised CUG are given in Fig. 6.10.

Upon receipt of a call request from Party A, Node A performs a validation check on the validity of the call within the constraints of the CUG. If successful, an IAM is formed containing the Interlock Code for the CUG and a Call Indication determining whether or not external access is permitted.

When the IAM is received at Node B, a validation check is performed to determine the validity of the requested call. If the data stored at Node B indicates that Party B is a member of the CUG and the call meets the restraints imposed on Parties A and B, then the call proceeds as normal.

If the call fails the validation check at Node B, then release procedures are initiated with a cause indicator outlining the reason for failure.

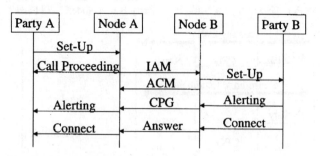

Fig. 6.10 Flows for Closed User Group

IAM = Initial Address Message
ACM = Address Complete Message

6.4.6 Charging

An example of a Charging supplementary service is the International Telecommunications Charge Card (ITCC). In the ITCC, the calling party can make use of various Telecommunications Facilities and the bill for usage is charged to the calling party's normal account.

In this service, the Card Issuer is the network/service operator that issues the charge card. The Card Acceptor is the network/service operator that provides Telecommunications Services to the card holder. In this terminology, the card holder is the calling party. The Card Issuer is responsible for collecting charges from the card holder and making appropriate payments to Card Acceptors for services rendered. Typical signalling flows are given in Fig. 6.11.

The ITCC is initiated by the calling party indicating to Node A that a Charge Card Call is required. The Card Acceptor initiates a validation procedure with the Card Issuer. The validation can be initiated by Node A or by the Card Acceptor's charge card system; Fig. 6.11 shows the former case. Node A retrieves from the calling party a Primary Account Number, a Personal Identification Number and the Called Party Number.

At this point, Node A sends a Transaction Capabilities Begin/Invoke Message to the Card Issuer's Service Data Point, which is where the calling party's profile and the validation algorithms are stored. The Service Data Point performs the validation exercise and returns an End/Return Result Message if successful or a Return Error Message if unsuccessful. If a successful validation is indicated, the call can proceed with the sending of an IAM in the normal manner. If unsuccessful, the calling party is denied access.

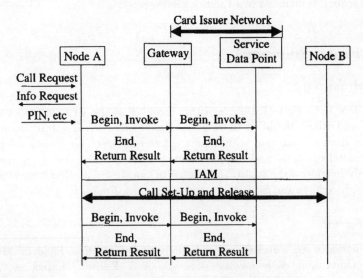

Fig. 6.11   Flows for International Telecommunications Charge Card

After the call has been released, Node A sends information on the call, e.g. start time and duration, to the Service Data Point. This is termed Disposition and the information is included in a Begin/Invoke Message. An End/Return Result Message indicates successful Disposition and completes the transaction.

6.4.7 Additional Information Transfer

Additional Information Transfer refers to user-to-user signalling (UUS). UUS allows ISDN calling and called parties to exchange a limited amount of data over the signalling channel.

Service 1 allows the calling and called parties to exchange data within ISUP Messages during call set-up and release. UUS is carried in the User-to-User Information Parameter in the Initial Address, Address Complete, Call Progress, Connect, Answer, Segmentation and Release Messages.

An explicit request for Service 1 is included in the IAM and a specific acceptance or rejection is carried in the other messages (except Segmentation). Service 1 can also be implicitly requested by including UUS in the IAM.

Service 2 allows the calling and called parties to exchange data in up to two User-to-User Information Messages in each direction. This occurs within the Call Set-Up Phase (i.e. between transmission of the ACM and the Answer Message). The request for Service 2 is included in the IAM.

Service 3 allows the calling and called parties to exchange data in User-to-User Information Messages during the Active (Conversation) Phase of the call (i.e. after the Answer Message). The request for Service 3 can be made during call set-up or during the Active Phase of the call. During call set-up, the request is included in the IAM in the User-to-User Indication Parameter. After call set-up, the request is included in a Facility Message.

6.5 Chapter Summary

6.5.1 Architecture

The ISDN User Part (ISUP) is within Level 4 of the structure adopted for CCSS7 and utilises the MTP to transfer messages between signalling points. The ISUP is designed for use within narrowband ISDNs and covers telephony, circuit-switched data and supplementary services. The ISUP is designed primarily to control the set-up and release of calls in an all-digital environment and the procedures and formats are optimised accordingly.

6.5.2 Format

ISUP messages are carried in the Signalling Information Field of Message Signal Units and each message consists of a Routing Label, a Circuit Identification Code, a Message Type and Parameters.

The Routing Label is described in Chapter 4. The Circuit Identification Code (CIC) is used to indicate the number of the circuit between two nodes to which the message pertains. The Message Type Code describes the category to which the message belongs and defines the function of each message.

Each message includes a number of parameters, which are divided into the Mandatory Fixed Part, the Mandatory Variable Part and the Optional Part. In general, each parameter has three fields: Parameter Name, Length of Parameter and Specific Information. However, to reduce the amount of unnecessary information carried, some fields are not transported through the network. If messages become too long to be carried by the MTP, a Segmentation Capability is specified that permits information to be split at the originating node and then reassembled at the destination node.

6.5.3 Procedures

A basic call set-up sequence is initiated by the originating node sending an Initial Address Message (IAM) to the next node in the connection (forward direction). The IAM contains all the information relating to the characteristics of the required speech path. In Enbloc Operation, all the address digits required to identify the called party are contained in the IAM. In Overlap Operation, the IAM contains only sufficient address digits to route the call to the next node in the connection and further information is provided in Subsequent Address Messages. The IAM is forwarded to the destination node over appropriate signalling links.

The destination node returns an Address Complete Message (ACM) to confirm receipt of the Called Party's Number. While returning the ACM to the originating exchange, the destination node alerts the called party about the incoming call. The destination node also returns appropriate tones over the allocated speech path to the calling party. When the called party answers, the tones are stopped and an Answer Message is returned to the originating node, stimulating through connection in the forward direction. It is usual for the originating exchange to commence charging upon receipt of the Answer Message.

The Basic Call Release Procedures use Release and Release Complete Messages. The Release Message is passed as quickly as possible through the network and causes the circuit to be disconnected. The Release Complete Message confirms that disconnection is complete and that the circuit is available for use again. Both the calling and called parties can initiate the release procedures.

6.5.4 Features

The ISUP provides a range of additional features to enhance the services offered to customers. These include end-to-end signalling (avoiding processing at transit nodes), Suspend/Resume (allowing a called party to suspend a call temporarily), Propagation Delay (to determine the Propagation Delay on a particular connection) and Echo Control. Facilities are also provided to allow different versions of ISUP to interwork, e.g. using a Confusion Procedure.

6.5.5 *Supplementary services*

The ISUP provides a wide range of supplementary services categorised as Number Identification, Call Offering, Call Completion, Multi-Party, Community of Interest, Charging and Additional Information Transfer. Examples of the supplementary services are given in the main text. The ISUP provides a user-to-user signalling capability in which customers can exchange data over the signalling channel without it being analysed by the network.

6.6 References

1 ITU-T Recommendation Q.761 'Functional Description of ISUP' (ITU, Geneva)
2 ITU-T Recommendation Q.763 'ISUP Formats and Codes' (ITU, Geneva)
3 ITU-T Recommendation Q.764 'ISUP Signalling Procedures' (ITU, Geneva)
4 ITU-T Recommendation Q.730 'ISUP Supplementary Services' (ITU, Geneva)

Chapter 7
Transaction Capabilities

7.1 Introduction

Networks are evolving to meet the need for more uninhibited information flow. Part of this evolution is a trend towards the use of more non-circuit-related signalling. Transaction Capabilities (TC)[1] is a protocol that provides a non-circuit-related capability to transfer information between nodes. TC uses the Network Service (Layers 1-3) defined by the OSI Model (explained in Chapter 3), e.g. as provided by the SCCP and MTP.

7.2 Users

TC is a general protocol that can be used to support a broad range of users, including:

(*a*) public land mobile networks;
(*b*) intelligent networks;
(*c*) supplementary services requiring non-circuit-related signalling;
(*d*) operations, maintenance and administration.

   (*a*) In mobile communications, the Mobile Application Part (MAP) provides a range of functions to route calls to and from the mobile customer. For example, the location of the customer is regularly fed into Location Registers. These provide the network with routing information for the mobile customer's incoming calls. The process of updating the registers uses TC to carry the location information. MAP is explained in Chapter 8.

   (*b*) In intelligent networks, specialised information is stored in separate nodes (databases). The Intelligent Network Application Part (INAP) uses TC to perform non-circuit-related information transfer between exchanges and databases, e.g. to provide additional routing information. INAP is explained in Chapter 9.

   (*c*) Some advanced supplementary services require non-circuit-related information transfer to complete a call. For example, a procedure can be adopted in which a local exchange can gain information about a called number, before setting up a physical speech path. This information can be used to influence the actions taken by the originating exchange.

114 *Telecommunications Signalling*

(*d*) Modern networks require a comprehensive operations, maintenance and administration infrastructure to ensure reliable service. These functions can require the transfer of large amounts of data between nodes or they can simply require the transfer of an instruction and a corresponding response. The Operations, Maintenance and Administration Part (OMAP) of CCSS7 uses TC to monitor and control the signalling network. OMAP is explained in Chapter 10.

7.3 Network Service

In real-time situations, information transfer is required quickly to complete a function. For example, consider a local exchange needing to gain access to a network database for specialised routing information during the establishment of a call. The time taken to achieve the information transfer adds to the post-dialling delay encountered by the calling customer. Real-time usage generally involves the transfer of a small amount of data, e.g. a telephone number and its translation.

In off-line situations, the length of time taken to achieve an information transfer is less critical. For example, if there is a need to transfer bulk information from one node to another, the critical factor is to ensure accurate delivery of the information. Off-line situations generally involve the transfer of large amounts of data.

Chapter 3 describes the services offered by the Network Layer of the OSI Model as being either Connectionless or Connection-Oriented. To make early progress, the ITU has concentrated on specifying TC based on the Connectionless Network Service. Hence, a Connection-Oriented Service is not available at the OSI Layer 3/4 boundary at this stage. However, a Connection-Oriented Service is offered by the Transaction Sub-Layer (see Section 7.5.5).

7.4 Portability

An objective of TC is that it is comprehensively portable. This means that TC should be a general protocol that is independent of (*a*) the type of user and (*b*) the means of achieving the Network Service. User-Independence means that TC is flexible enough to handle a broad range of non-circuit-related applications, including those not yet specified.

Independence from the Network Service means that TC can use not only the SCCP and MTP to transfer messages, but any transfer mechanism meeting the requirements of the OSI Layer 3/4 interface. For example, if a network uses a Packet-Data Transfer Mechanism[2] as part of its existing operations and maintenance infrastructure, TC can be used to control appropriate information flow without modifying the transfer mechanism. This demonstrates the powerful nature of the OSI-Model approach and it is a step in the direction of allowing network operators to 'mix and match' protocols according to local circumstances.

7.5 Architecture

7.5.1 OSI layers

TC is specified in accordance with the OSI Model. TC consists of an Intermediate Service Part (ISP) and a Transaction Capabilities Application Part (TCAP), as illustrated in Fig. 7.1. TC, together with its users, transcends Layers 4 to 7 of the OSI Model.

7.5.2 Intermediate Service Part (ISP)

The ISP covers the functions of Layers 4 to 6 of the OSI Model, thus handling the Transport, Session and Presentation aspects of non-circuit-related communication. The ISP is required when TC uses the Connection-Oriented Network Service. As explained earlier, the use of TC in this manner is not yet defined. Hence, details of how Layers 4 to 6 are applied to TC are not yet specified. For the Connectionless Network Service, the functions of ISP are not required and Layers 4 to 6 can be considered to be transparent.

Some recent texts use the terms TC and TCAP interchangeably: this assumes that the ISP will always remain transparent. This book continues to discriminate between the terms.

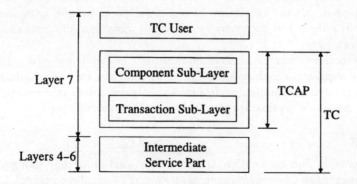

Fig. 7.1   Achitecture of Transaction Capabilities

        TC = Transaction Capabilities
        TACP = Transaction Capabilities Application Part

7.5.3 TCAP

TCAP and the TC user perform the functions of Layer 7 of the OSI Model. User-Independence is achieved by including all application-dependent functions within the TC user. For example, in the case of an intelligent network database, TCAP is responsible for transferring information between exchanges and the database. The TC user, INAP, performs the Application-Specific Functions. TCAP is a form of the Remote Operations Service Element (ROSE) explained in Chapter 3.

TCAP is divided into the Component Sub-Layer and the Transaction Sub-Layer. The Component Sub-Layer deals with 'Operations'. An Operation is (*a*) a request for an action to be performed by another node and (*b*) a response to such a request. Operations are carried in Information Elements termed Components. The format of Components is defined in an application-independent manner

The Transaction Sub-Layer is responsible for exchanging messages containing Components between two nodes. The Component Sub-Layer can be regarded as the user of the Transaction Sub-Layer.

7.5.4 Component Sub-Layer

(a) Operations

The Component Sub-Layer in one node communicates with the corresponding sub-layer in another node by sending and receiving 'Components'. A Component is the means by which TC conveys Operations. For example, consider an exchange that needs the assistance of translation facilities in a remote database. The exchange sends a Component to the database requesting that a translation be performed. A parameter in the component gives the original number. Upon completion of the translation by the database, a component is returned to the exchange, with the translated number included within a parameter.

Several Operations can be active simultaneously, but only one reply can be returned in response to each request. The reply indicates success or failure in completing the action. Components in a message are delivered to the remote TC User in the same order as they are provided by the Originating User.

(b) Classes

The Component Sub-Layer Operations are divided into four categories, termed Classes. Each Class corresponds to an expected form of response.

In Class 1, both success and failure to conduct an action are reported back to the node initiating the request. An example of this type of Operation is when an exchange requests a remote database to perform a routing translation on a telephone number. It is required in this case for the database to report back to the exchange either the successful completion of the Operation (with the result being a translated number) or the failure to complete the Operation (with a reason for the failure).

In Class 2, only failures to complete the Operation are reported. This category can be used when, for example, there is a need to conduct a routine test and a reply is only necessary when there is a fault preventing the completion of the test.

Class 3 Operations are used when it is necessary to report only successful results. This can be used in the case when a fault is suspected and the likely outcome is the failure of the operation. It is assumed that the operation fails unless a successful result is reported back.

If neither success nor failure needs to be reported, then Class 4 can be used. An example of its use is when a node wishes to send a warning of an event to several other nodes and a response or acknowledgement is not required.

(c) Dialogue

A Dialogue is an exchange of information between two TC users. The Component Sub-Layer can perform a role of co-ordinating several Dialogues on behalf of the TC user.

7.5.5 Transaction Sub-Layer

The Transaction Sub-Layer provides the capability to exchange Components between nodes by means of Dialogues. Two services are offered, namely an Unstructured Dialogue and a Structured Dialogue.

In an Unstructured Dialogue, there is no explicit establishment and ending of an Association between two Transaction Sub-Layers. The only facility offered to the TC user is the capability to send Components for which replies are not expected (Class 4 Operations). This can be likened to the Connectionless Service described for the SCCP.

In a Structured Dialogue, an explicit Association is formed between two nodes. The exchange of information between TC users takes the form of starting the Dialogue, exchanging Components and ending the Dialogue. This can be likened to the Connection-Oriented Service described for the SCCP.

7.6 Primitives

The primitives used by TC are in line with the OSI Model explained in Chapter 3.

7.6.1 Component Sub-Layer

Examples of the primitives exchanged between TC users and the Component Sub-Layer for Component Handling are given in Table 7.1.

Table 7.1   Component Handling Primitives

Primitive name	Description
TC-Invoke	initiates the commencement of an Operation (action)
TC-Result-Last*	gives the result of a successful Operation or the last part of a segmented result
TC-Result-Not-Last*	gives part of a segmented result with the expectation that more will follow
TC-User-Error	indicates that a request to perform an Operation failed
TC-User-Reject	indicates that a request to perform an Operation was abnormal

*Primitives exchanged when result is too long or available at different times

If a result is too long to be carried, or parts of the result become available at different times, the result is Segmented and returned in Return-Result-Not-Last Primitives. The final part is included in a Return-Result-Last Primitive.

The interface between the TC User and the Component Sub-Layer also carries the primitives to establish dialogues, which are passed on to the Transaction Sub-Layer.

7.6.2 Transaction Sub-Layer

Examples of the primitives exchanged between the Component Sub-Layer and the Transaction Sub-Layer are given in Table 7.2.

Table 7.2   Transaction Sub-Layer Primitives

Primitive name	Description
TR-Uni	requests or indicates an Unstructured Dialogue
TR-Begin	begins a dialogue
TR-Continue	continues a dialogue
TR-End	ends a dialogue
TR-User-Abort	aborts a dialogue

7.7 General TCAP Message Format

7.7.1 Information Elements

TCAP Messages consist of 'Information Elements'. Each Information Element has a standard structure[3] consisting of three fields; Tag, Length and Contents.

Transaction Capabilities 119

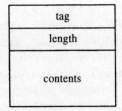

Fig. 7.2 Primitive Information Element (Source: ITU Recommendation Q.773)

The Tag Field distinguishes between one type of Information Element and another and governs the meaning of the Contents Field. The Length Field gives the length of the Contents Field.

The Contents Field contains the substance of the Information Element, i.e. the primary information that the element conveys. The contents can consist of a single value or one or more Information Elements. If the contents consist of a single value, the Information Element is known as a Primitive Information Element. The format of a Primitive Information Element is shown in Fig. 7.2.

If the contents consist of one or more Information Elements, then the encompassing Information Element is called a 'Constructor'. The format of a Constructor Information Element is shown in Fig. 7.3.

The recursive approach used in Constructor Information Elements allows the formatting of TCAP to be extremely flexible while remaining Application-Independent. Each user can make use of Information Elements to build simple or complex messages meeting the needs of that user.

7.7.2 *TCAP message structure*

A TCAP message consists of a Constructor Information Element. The message contains a Transaction Portion, containing Information Elements pertinent to the Transaction Sub-Layer, and a Component Portion, containing Information Elements pertinent to the Component Sub-Layer. Optionally, a Dialogue

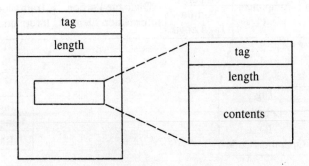

Fig. 7.3 Constructor Information Element (Source: ITU Recommendation Q.773)

Portion can also be included. The format of the message is illustrated in Fig. 7.4, with examples of Information Elements to highlight the principles.

The message commences with a Message Type Tag, which identifies the message and indicates whether the message refers to an Unstructured Dialogue or a Structured Dialogue. For an Unstructured Dialogue, the message is identified as Unidirectional. For a Structured Dialogue, the message is identified as being of the Begin, Continue, End or Abort Type. The Message Length Field indicates the length of the Contents Field.

The Contents Field consists of a series of Information Elements. The first Information Element belongs to the Transaction Portion. In the example in Fig. 7.4, the Transaction Information Element shown is a Transaction Identifier, with the contents of the Information Element identifying the transaction that is taking place (ID). Because the contents consist of a single value, this is a Primitive Information Element.

The second Information Element in the example is a Dialogue Information Element, which is an Optional Field. This portion is used to transfer information between TC users that is not in the form of Components, e.g. user-to-user Information.

The third Information Element in Fig. 7.4 is a Constructor Information Element, consisting of the Component Portion Tag and Length, followed by two Components. Each Component is itself a Constructor Information Element and the example shows the contents of Component 1 as being a Component Identifier (ID) and Operation Code. In this case, the Operation Code is the action that is requested of the remote node.

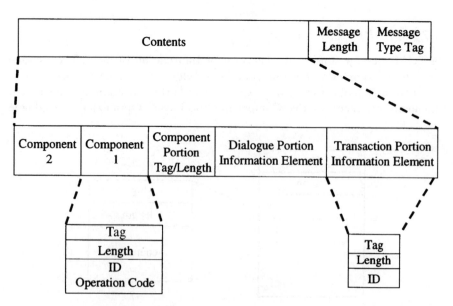

Fig. 7.4   Principle of TCAP message format

The recursive approach used by TCAP, in which the Contents Field of one Information Element contains the Tag/Length/Contents of other Information Elements, is an important difference between the formatting technique of TCAP and that of the ISUP. Whereas the ISUP technique is flexible, it is not Application-Independent. The recursive approach can increase the processing overhead of a message: for example, in simple messages, some of the information that is implicit in the Message Type has to be provided explicitly to conform with the overall message structure. However, the technique is extremely flexible and this outweighs the disadvantages of the recursive approach for non-circuit-related applications.

7.8 Transaction Sub-Layer Format

7.8.1 Unstructured Dialogue

In the case of an Unstructured Dialogue, the Transaction Sub-Layer does not establish an explicit Association between two nodes. The Message Type used in an Unstructured Dialogue is termed Unidirectional. This reflects the concept that an answer is not normally expected. The format of a Unidirectional Message is illustrated in Table 7.3.

Because an Association is not formed at the Transaction Sub-Layer, it is inappropriate to provide a Transaction Identity. Thus, the message consists of information associated with Components only (and Dialogue if used).

7.8.2 Structured Dialogue

In the case of Structured Dialogue, the Transaction Sub-Layer establishes an explicit Association between two nodes. This is achieved using three Message Types, namely Begin, Continue and End.

Table 7.3 Unidirectional Message

Field	Form	Mandatory/optional
Message Type Message Length	Constructor	M
Dialogue Portion	Constructor	O
Component Portion Tag Component Portion Length	Constructor	M
Components	Constructor	M

M = Mandatory, O = Optional (Source: ITU-T Recommendation Q.773)

Table 7.4 Begin Message

Field	Form	Mandatory/optional
Message Type Message Length	Constructor	M
Originating Transaction ID Tag Transaction ID Length Transaction ID	Primitive	M
Dialogue Portion	Constructor	O
Component Portion Tag Component Portion Length	Constructor	O
Components	Constructor	O

M = Mandatory, O = Optional (Source: ITU-T Recommendation Q.773)

(a) Begin

The Begin Message is illustrated in Table 7.4. The Dialogue commences by including an Originating Transaction Identity (OTID), consisting of a Tag, Length and Contents. A Transaction Identity is a number generated by a node to identify to which Dialogue a message refers. The OTID is a number selected by the node generating the Begin Message. As the contents relate to a single value, the Transaction ID is a Primitive Information Element.

(b) Continue

The Continue Message Type is illustrated in Table 7.5. One or more Continue Messages can be exchanged between nodes as part of the Structured Dialogue. The Continue Message includes both an OTID and a Destination Transaction Identity (DTID) Information Element. Thus, both the originating and destination nodes can associate a Continue Message with a particular Dialogue. It is the inclusion of Transaction Identities that permits the maintenance of the Association and thus allows a Structured Dialogue to take place. The Originating Identity always refers to the node generating the particular message.

(c) End

When either node involved in a dialogue has completed the transfer of information relevant to that transaction, the Dialogue is ended by the transmission of an End Message. The format of the End Message is illustrated in Table 7.6.

Table 7.5 Continue Message

Field	Form	Mandatory/optional
Message Type Message Length	Constructor	M
Originating Transaction ID Tag Transaction ID Length Transaction ID	Primitive	M
Destination Transaction ID Tag Transaction ID Length Transaction ID	Primitive	M
Dialogue Portion	Constructor	O
Component Portion Tag Component Portion Length	Constructor	O
Components	Constructor	O

M = Mandatory, O = Optional (Source: ITU-T Recommendation Q.773)

Table 7.6 End message

Field	Form	Mandatory/optional
Message Type Message Length	Constructor	M
Destination Transaction ID Tag Transaction ID Length Transaction ID	Primitive	M
Dialogue Portion	Constructor	O
Component Portion Tag Component Portion Length	Constructor	O
Components	Constructor	O

M = Mandatory, O = Optional (Source: ITU-T Recommendation Q.773)

(d) Abort

An Abort Message is also defined. This is used when the Transaction Sub-Layer or TC user cannot process a message due to an error. For example, if an unrecognised message type is received, an Abort Message is returned with an explanation of the fault.

7.9 Component Sub-Layer Format

The Component Portion contains one or more Components, each Component being a Constructor Information Element. The Component Portion uses three basic types of Component to initiate actions and return the results of actions, namely Invoke, Return Result (Last) and Return Result (Not Last). Two other Components, Return Error and Reject, are used to indicate that a request to carry out an action has not been completed or that aspects of a message received (e.g. a component) are unrecognised.

7.9.1 Invoke

The Invoke Component is used to request a node to perform an action. For example, one node might request another node to perform a routing translation: an Invoke Component is used to initiate the action of translation. An example of the format of the Invoke Component is shown in Table 7.7.

The Component Type Tag indicates that the Component is an Invoke. The Invoke Identity (ID) is a Reference Number that identifies the Invoke Component. This allows the originating node to correlate any response with the initial request to perform an action. If it is desired to associate two Invoke Components together, then a Linked ID is also used.

Table 7.7 Invoke Component

Field	Form	Mandatory/optional
Component Type Tag Component Length	Constructor	M
Invoke ID Tag Invoke ID Length Invoke ID	Primitive	M
Linked ID Tag Linked ID Length Linked ID	Primitive	O
Operation Code Tag Operation Code Length Operation Code	Primitive	M
Parameter Tag Parameter Length Parameters	Primitive/ Constructor	O

M = Mandatory, O = Optional

The Operation Code Information Element indicates the precise Operation (or action) that is to be performed (e.g. Perform a Translation). The Component Portion does not interpret or analyse the meaning of the Operation Code. The code is received from the TC user, inserted into the appropriate location in the Component Portion, transmitted to the destination node and delivered to the destination TC user.

Parameters may be included in the message if appropriate, e.g. if the Operation requested is to perform a routing translation upon a telephone number, the number can be included as a parameter following the Operation Code Information Element.

7.9.2 Return Result

The Return Result (Last) Component is used to report the successful completion of an Operation. The Return Result (Not Last) Component is used to indicate that only part of the results of an Operation is being supplied, with the implication that the remainder of the results will be supplied separately. When the final part of the results is supplied, a Return Result (Last) Component is used.

The Return Result (Last) and Return Result (Not Last) Components have the same general format, similar to that for the Invoke Component shown in Table 7.7. The Invoke Identity Information Element is included to correlate the result of the Operation with the original invocation. The Return Results Component can include Optional Parameters. For example, if the original invocation requested a translation of a telephone number, the Parameters Field contains the translated number. If parameters are present, then the Operation Code Field is also required.

7.10 Component Sub-Layer Procedures

The Component Sub-layer Procedures[4] provide a TC user with the capability of invoking Operations at a remote node and receiving replies. The Component Sub-Layer can also receive Dialogue Control Information and TC User

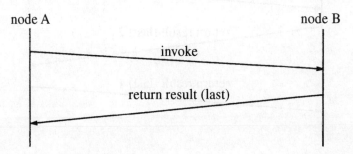

Fig. 7.5 Example of basic Component Flow

Information and generate an appropriate Dialogue Portion. Component Sub-Layer Procedures are very flexible to ensure that they can handle the requirements of a range of Applications. Examples of Component Flows are given in the following Figures to illustrate the principles.

7.10.1 Single invocation

Fig. 7.5 shows the basic Component Flow in which Node A sends an Invoke Component to Node B and Node B returns the results to Node A.

Suppose that Node A is a local exchange that needs information from Node B to assist in deciding whether or not to set up a call. In this case, Node A sends an Invoke Component to Node B seeking relevant information. For this example, the Invoke Component has a Parameter Information Element containing the Called Customer's Number. After checking the status of the called customer, Node B returns a Return Result (Last) Component including the required information. This information allows Node A to determine whether or not to establish a traffic path through the network.

7.10.2 Multiple invocation

In Fig. 7.6, Node A sends an Invoke Component (1) to Node B, but Node B needs more information before processing the component. In this case, Node B initiates its own Invoke Component (2), soliciting a response from Node A in a Return Result (Last) Component (2) . Upon analysis of the result, Node B responds to the original invocation with a Return Result Last Component (1).

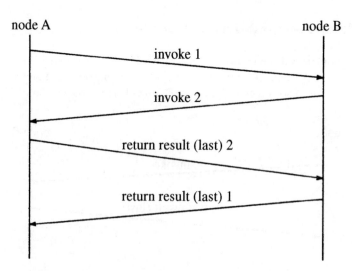

Fig. 7.6 Example of Multiple Invocation

An example of this sequence is when Node A is an exchange seeking the translation of a telephone number into routing information from a database (Node B). In the example, the database needs more information from Exchange A, e.g. the Calling Customer's Number might be required to provide appropriate routing information. Once this information is provided to the database, the original invocation can be processed and the routing information is supplied to Exchange A in a parameter within a Return Result (Last) Component.

7.10.3 Segmentation

In Fig. 7.7, the information resulting from processing an Invoke Component at Node B is too long to be sent in a single Return Result Component. Hence, the information is segmented into the parameters of three Components, two of the Components being of the type Return Result (Not Last) and the final Component being of the type Return Result (Last).

7.10.4 Abnormal conditions

The ITU-T specification of TCAP includes a wide range of procedures to deal with abnormal conditions. For example, if an Invoke Component is received with a syntax error, a Reject Component is returned indicating the reason for the failure. Details of these abnormal conditions are left for the specification.

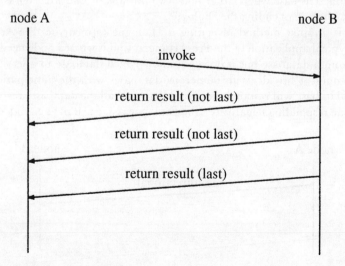

Fig. 7.7 Example of Segmented Result

7.11 Transaction Sub-Layer Procedures

7.11.1 Unstructured Dialogue

In the Unstructured Dialogue Case, the Transaction Sub-Layer of a node sends a Unidirectional Message to another node, as illustrated in Fig. 7.8. Upon receipt of a Unidirectional Message, the Transaction Sub-Layer in Node B passes the information to the TC user at Node B without analysis.

7.11.2 Structured Dialogue

An example of the Transaction Sub-Layer Procedures in a Structured Dialogue is given in Fig. 7.9.

Node A initiates the commencement of a Structured Dialogue by sending a Begin Message. The OTID selected by Node A is Value X and is included in the Begin Message. Node B analyses the Begin Message and agrees to establish a Dialogue. Node B returns a Continue Message to confirm this decision. Node B selects the OTID of Value Y for inclusion in the Continue Message. The DTID Field contains the Identity X (corresponding to the ID selected by Node A).

Upon receipt of the Continue Message from Node B, Node A analyses the information and sends a Continue Message to Node B. In this case, the OTID is X and the DTID is Y. After reception and analysis of the Continue Message from Node A, Node B determines that the Dialogue can now be terminated and returns an End Message. There is no OTID in the End Message and the DTID is X.

In the example above, Node B initiated the end of the Dialogue, but it would be just as appropriate for Node A to perform this function, depending upon the Application. The case when either node can initiate an End Message is termed the 'Basic Method' of ending the Dialogue.

There is another method of ending a Dialogue termed the 'Pre-Arranged End'. A typical application of the Pre-Arranged End is when a node needs information from a database but it does not know which database to query. In this case, a request is broadcast to numerous databases with the anticipation that only one database will respond positively. To avoid all the databases responding (all but one responding negatively), the Dialogue is deemed to have ended unless

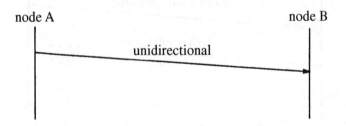

Fig. 7.8 Unstructured Dialogue

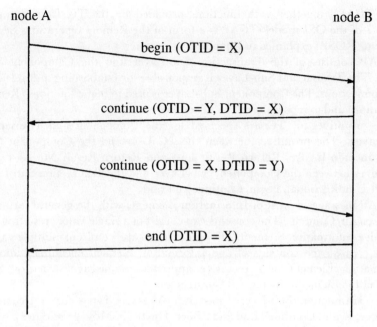

Fig. 7.9 *Structured Dialogue*

a positive response is received. The Dialogue between the node and the database responding positively then continues as described above.

7.12 Chapter Summary

Transaction Capabilities (TC) is a protocol that provides non-circuit-related transfer of information between nodes. TC is used in conjunction with an appropriate Network Layer Service (e.g. as provided by the MTP and SCCP). TC conforms to the OSI 7-Layer Model and it is therefore a general protocol that can be employed by a wide range of users, e.g.

- Operations, Maintenance and Administration Part (OMAP);
- Intelligent Network Application Part (INAP);
- Mobile Application Part (MAP).

TC is designed to be Application-Independent. It is currently specified as a Connectionless Network Service. A Connection-Oriented Network Service will be defined at a later stage.

TC lies within Layers 4 to 7 of the OSI 7-Layer Model. Layers 4 to 6 are known as the Intermediate Service Part and they are transparent for Connectionless Network Services. The Transaction Capabilities Application

Part (TCAP), together with functions provided by the TC User, comprise Layer 7 of the OSI model. TCAP is a form of the Remote Operations Service Element (ROSE) explained in Chapter 3.

TCAP consists of the Transaction Sub-Layer and the Component Sub-Layer. The Transaction Sub-Layer is responsible for establishing and releasing TC Connections. The Component Sub-Layer is used to request actions (Remote Operation) and to report the results of actions.

The primitives for TC are specified for the Component and Transaction Sub-Layers. The primitives between the TC User and the Component Sub-Layer include Invoke, Return Result Last and Return Result Not Last. The primitives between the Component Sub-Layer and the Transaction Sub-Layer include Unidirectional, Begin, Continue and End.

TCAP messages consist of Information Elements with the general format of Tag/Length/Contents. The contents can consist of a single value, resulting in a Primitive Information Element. The contents can also consist of multiple values. In these cases, the contents of one Information Element encompass another Information Element in a recursive approach to message structure. Such Information Elements are termed Constructors.

The Transaction Sub-Layer uses four Message Types for a structured Dialogue: Begin, Continue, End and Abort. The Begin Message initiates a transaction. Continue Messages are used during the transaction and an End Message releases the transaction. The Abort Message is used to handle abnormal conditions.

The Component Sub-Layer defines several components. The Invoke Component is used to request an action to be performed by another node. The response can be successful (in which case, a Return Result Component is returned) or unsuccessful (in which case, a Return Error Component is returned). The Return Result Component can be qualified as Last or Not-Last. A Reject Component is also used in failure conditions.

7.13 References

1 ITU-T Recommendations Q.771-775: 'Transaction Capabilities' (ITU, Geneva)
2 ITU-T Recommendations X.25 and X.75: 'Packet Mode Data' (ITU, Geneva)
3 ITU-T Recommendation Q.773: 'TCAP Formats' (ITU, Geneva)
4 ITU-T Recommendation Q.774: 'TCAP Procedures' (ITU, Geneva)

Chapter 8
Mobile Communications

8.1 Introduction

8.1.1 Evolution

One of the most dramatic changes in the telecommunications industry over the last few years has been the growth in mobile communications. In the 1980s, mobile communications systems were based on analogue technologies. Several types of system were adopted, based on differing standards, but all were restricted in the variety of services that could be offered and it was difficult to raise the quality of service to meet increasing customer expectations. Mobile stations (handsets) were clumsy and the opportunities for miniaturisation were limited.

In the 1990s, a major transformation took place in which digital technologies were introduced. These technologies overcame many of the limitations of their analogue predecessors. It is now possible to offer a wide range of services at a suitable quality level. Digital technologies applied to mobile stations have allowed wholesale reduction in size and evolving battery technologies have assisted this trend.

The aim of mobile communications systems is to offer services to customers wherever the customers are located. This requires the provision of radio spectrum over the geographic area to be covered. Terrestrial mobile systems aim to achieve this target by providing a network of radio stations. Such systems are termed 'public land mobile networks' (PLMN)[1]. Coverage for PLMNs is typically 95% to 98% of the population in mature networks. The areas that remain uncovered are in terrain that is difficult to cover economically, e.g. in valleys, and many of these areas are unlikely to receive service.

Satellite technologies are also evolving and systems using multiple satellites are coming into service. Satellite systems can achieve 100% coverage for those customers who need to be in contact at all times. However, they are expensive and will not replace the terrestrial systems. The two forms of technology will complement each other by adopting mobile stations that can operate in both the terrestrial and satellite modes.

The early analogue mobile communications systems were based on national boundaries and there were a limited number of service providers. The systems used a range of technologies, thus reducing the benefits of standardisation in terms of economies of scale for development, compatibility, etc. With the emergence of digital technologies, it was hoped to establish a worldwide standard that would permit global roaming by customers. This aim was in line with the increasingly international nature of business, the holiday industry, etc. However, it has not been fully achieved, and several standards have been adopted across the world.

8.1.2 Technologies

The dominant technology is the Global System for Mobile Communications (GSM) and this is being adopted throughout Europe, with strong advocates in Africa and Asia/Pacific. The Digital Communications System (DCS) 1800 uses the same principles as GSM, but a higher frequency is used over the radio path. Other continents have adopted differing standards, e.g. North America is adopting a mixture of GSM, GSM-based personal communication services (PCS) 1900, code division multiple access (CDMA) and time division multiple access (TDMA) standards.

A third generation mobile system, the Universal Mobile Telecommunications System (UMTS), is being developed. The opportunity for a worldwide application of standards is again tantalisingly possible, this time more focused on data and IP communications.

In the absence of current worldwide support for a single standard, some of the limitations on roaming can be overcome by the use of multimode handsets. These can operate in, for example, the GSM and DCS bandwidths and can use both systems.

8.1.3 Signalling implications

In a fixed telecommunications system, the locations of the calling and called customers are derived from the numbering scheme adopted in the network. In mobile communications, the position is different because a customer can roam across the coverage area at will. When making outgoing calls, the approximate location of the customer is known from the identity of the radio system used by the mobile station. However, for incoming calls there is no direct translation between the number of the mobile station and the location of the customer. Procedures are therefore needed to determine the location of a called customer on a real-time basis to allow incoming calls to be delivered. It is this requirement that places the heaviest demand on network resources for mobile calls.

So, what is the impact on signalling? The complexities of roaming, customer location, billing etc. place additional demands upon signalling systems. For example, a customer may establish a call in one area and move to another area during the call: the network needs to transfer enough information to re-route the call without loss of service or degradation of quality. To locate a called customer requires the transfer of large amounts of information between databases. Signalling systems play the key role in meeting these needs.

The major impact of the diversity of technologies being adopted in different countries is on the use of the radio spectrum and this is beyond the scope of this book. On the other hand, all the mobile technologies need a secure, flexible and economic means of transferring signalling information. Many common principles apply in the network, regardless of the standard adopted for the radio spectrum. The next generation of mobile system is based on packet data techniques, with the signalling based on CCSS7 and Internet Protocol technologies (explained in Chapter 18). This chapter outlines the principles of signalling in mobile networks by describing the Mobile Application Part (MAP) of CCSS7.

Section 8.2 briefly outlines a typical network structure for a PLMN. Sections 8.3 and 8.4 describe the services that are used by MAP: in this context, the services are of the type described for the OSI Model explained in Chapter 3. Section 8.5 outlines the salient procedures for basic calls, the Short Message Service and supplementary services.

8.2 Network Structure

8.2.1 Numbering

Each customer is allocated a number from the national numbering scheme upon subscription to the mobile communications service. This is termed the 'mobile station ISDN number' (MSISDN). This number is the only identity used by the customer and it is the number dialled by a calling customer to establish a call.

However, the mobile station itself is also allocated other numbers. The International Mobile Subscriber Identity (IMSI) is used over the radio path and through the network. The IMSI is used for all signalling throughout the PLMN and it consists of a Mobile Country Code, Mobile Network Code and a Mobile Station Identification Number. The International Mobile Equipment Identity (IMEI) identifies the mobile station as a piece of equipment. It consists of a Type Approval Code, a Final Assembly Code, a Mobile Station Identification Number and a serial number.

8.2.2 Network elements

PLMNs contain a number of network elements and a typical network[2] is illustrated in Fig. 8.1.

The Mobile Switching Centre (MSC) performs the switching functions for a designated area. The MSC is like a normal Public Switched Telephone Network (PSTN) exchange, but it needs to take account of the mobile nature of customers. It therefore needs to implement procedures that track the location of a customer (Location Registration) and allow for customers moving from one area to another during a call (handover). In mobile systems, the amount of

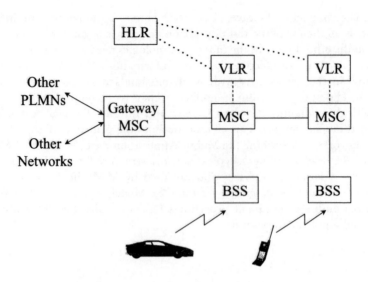

Fig. 8.1 Typical network structure

        PLMN = Public Land Mobile Network
        HLR = Home Location Register
        MSC = Mobile Switching Centre
        VLR = Visitor Location Register
        BSS = Base Station System

radio spectrum available is tightly constrained and a role of the MSC is to take account of such limitations when performing the switching function.

The Gateway MSC is a normal MSC with the additional functions necessary to interconnect a PLMN with other networks, e.g. a PSTN.

The Mobile Station is the terminal (handset) used by the customer. It contains a removable Subscriber Identity Module (SIM) Card that stores an identification number and details of the customer subscription.

The Home Location Register (HLR) is a database that stores customers' profile information, particularly their location. The HLR stores the IMSI and MSISDN of each customer. The information stored also includes the services that a customer can receive, e.g. subscription to a supplementary service. A PLMN can contain more than one HLR, if needed, to overcome capacity constraints.

When a customer moves from one area to another (i.e. roams), a Visitor Location Register (VLR) takes responsibility for tracking the customer. The VLR stores the location area of each visiting mobile station together with each IMSI and MSISDN.

The Base Station System (BSS) is responsible for all the radio functions and includes Base Station Controllers and Base Transceiver Stations.

Mobile Communications 135

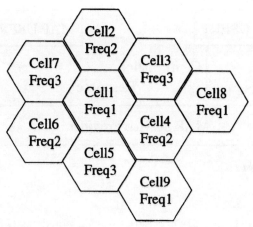

Fig. 8.2 Cellular coverage

8.2.3 Cellular structure

The frequency spectrum for use in a PLMN is very limited and yet wide radio coverage is needed to permit customers to roam. The spectrum limitation in a cellular system is overcome by providing radio coverage using a mesh of cells, as illustrated in Fig. 8.2. Each cell is served by a base station (radio station).

Each cell uses a number of frequencies, depending on the expected load. These frequencies can then be re-used in other cells, provided that adjacent cells do not use the same frequencies (this avoids interference between cells). In Fig. 8.2, only one frequency is shown per cell to illustrate the re-use principle. Cell 1 uses Frequency F 1. This frequency is not used in the adjacent Cells 2-7, thus avoiding interference, but it is used in Cell 8. The re-use of frequencies achieves high efficiency in the allocation of spectrum.

The Location Area is defined as the area in which a mobile station may move without updating the Location Register. A Location Area can contain one or more cells. A paging message is sent to all cells within a Location Area. Each Location Area is identified by a Location Area Identity. The area that is served by an MSC contains a number of Location Areas.

8.3 MAP Services

8.3.1 MAP service model

MAP is based on a model of a service provider supplying services to users, as illustrated in Fig. 8.3. The services are of the type explained in Chapter 3 for the OSI Model, in which tiers are deemed to offer a service to a higher tier. The transfer of information between tiers is by primitives. MAP uses the primitives in the OSI Model, namely Request, Confirmation, Response and Indication. The transfer of information between MAP Users is termed a MAP Dialogue.

136 *Telecommunications Signalling*

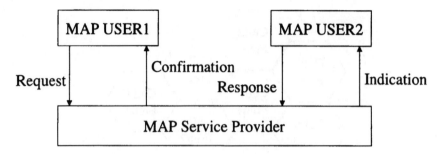

Fig. 8.3   MAP service model

Services initiated by MAP Users are classified as Unconfirmed or Confirmed. Unconfirmed Services are when a User sends information and does not expect validation of delivery. Confirmed Services provide such validation. Services can also be initiated by the Service Provider.

MAP services are also classified as Common and Specific. Common Services are available to all MAP Users, whereas Specific Services are restricted in their use.

8.3.2 MAP relationship with TC

Transaction Capabilities (TC) is described in Chapter 7. TC provides a non-circuit-related information transfer capability and spans Layers 4 to 6 and part of Layer 7 of the OSI Model. Layers 1 to 3 of the OSI Model are provided by the SCCP and MTP. The MAP services are users of TC, as illustrated in Fig. 8.4.

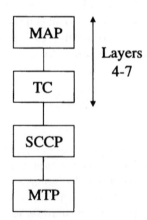

Fig. 8.4   MAP services as users of TC

8.4 MAP Service Descriptions

8.4.1 Common services

The common services are used to perform basic functions within the OSI Layer 7 (Application Layer). These include establishing and releasing Communication Dialogues between MAP users. They are also used to gain access to functions at lower layers in the OSI Model and to report abnormal conditions. Examples of common services are described in Table 8.1. The table also shows typical parameters that would be transferred by the Request, Indication, Response and Confirmation Primitives.

It is possible for different versions of MAP to be in operation. The Application Context Parameter is used to identify the version of MAP at each user. If the versions are different, the version with the lowest capability is adopted for the dialogue.

The Release Method Parameter indicates how the Dialogue was ended. This can be a normal release initiated by one of the users, or it could be a Pre-Arranged End that was agreed beforehand by the users.

The Reason Parameters in the Abort Services include resource limitations, procedure errors, etc.

The Problem Diagnostic Parameter indicates abnormal events or User rejection of a Dialogue.

8.4.2 Location services

The location services are used to pass information within the PLMN on the location of the customer. Examples of location services are given in Table 8.2.

Table 8.1 Examples of common services

MAP service	Description	C/U	Typical parameters Req/Ind	Res/Con
Open	establishes a MAP Dialogue between two Users	C	Application Context Destination Address	Result
Close	releases a MAP Dialogue	U	Release Method	-
U-Abort	allows a User to abort the Dialogue	U	User reason	-
P-Abort	allows the Provider to abort the Dialogue	U	Provider reason Source	-
Notice	notifies a User about protocol problems	U	Problem Diagnostic	-

C/U = confirmed/unconfirmed service

138 *Telecommunications Signalling*

Table 8.2   Examples of location services

MAP service	Description	C/U	Typical parameters Req/Ind	Res/Con
Cancel Location	HLR deletes customer record from VLR	C	Invoke ID IMSI	Invoke ID
Send ID	new VLR seeks information from previous VLR	C	Invoke ID TMSI	Invoke ID Information
Update Location	VLR updates location information in the HLR	C	Invoke ID IMSI MSC address VLR number	Invoke ID
Update Location Area	MSC provides VLR with updated location of customer	C	Invoke ID Target location area ID IMSI Previous location area Location update type	Invoke ID

C/U = confirmed/unconfirmed service

The Invoke ID Parameter is a Reference Number that identifies the Dialogue to which a primitive refers. The user can thus co-ordinate the various primitives during the Dialogue. The Target Location Area ID Parameter identifies the Location Area into which the customer is roaming. The TMSI is a Temporary Mobile Station Number, used while the mobile station is roaming.

Location updating can be triggered by several factors, e.g. the mobile station moving into a new area or the mobile station confirming its presence on a periodic basis. The Location Update Type Parameter defines the type of trigger used.

8.4.3 Handover services

When a customer roams from one MSC area to another, a handover procedure takes place. The first MSC (MSC 1) retains control of the call and information is passed between the MSCs accordingly. Examples of handover services are given in Table 8.3. In the table, MSC 2 is the MSC to which the handover occurs.

The BSS-APDU Parameter is an Application Protocol Data Unit that refers to detailed protocol definition on the MSC/BSS Interface. The Handover Number is a temporary identity used until the Handover Procedure is completed. After completion, it is re-allocated to another Handover Occurrence.

Mobile Communications 139

Table 8.3 Examples of handover services

MAP service	Description	C/U	Typical parameters Req/Ind	Res/Con
Allocate Handover Number	MSC2 requests handover number from VLR2	C	Invoke ID	Invoke ID
Forward Access Signalling	MSC1 passes information to MSC2	U	Invoke ID BSS-APDU	
Prepare Handover	MSC1 prepares to handover to MSC2	C	Invoke ID Request handover number	Invoke ID handover number
Process Access Signalling	MSC2 passes information back to MSC1	U	Invoke ID BSS-APDU	Invoke ID BSS-APDU
Send End Signal	(a) MSC2 informs MSC1 that radio path established (b) MSC1 releases call	C	Invoke ID BSS-APDU	Invoke ID
Send Handover Report	VLR transfers handover number to MSC2 to be used by MSC1	C	Invoke ID Handover number	Invoke ID

C/U = confirmed/unconfirmed service

140 *Telecommunications Signalling*

Table 8.4 Examples of call handling services

MAP service	Description	C/U	Typical parameters Req/Ind	Res/Con
Complete Call	VLR requests MSC to set up call	C	Invoke ID	Invoke ID
Send Info for Incoming Call	MSC requests information from VLR	C	Invoke ID MSRN	Invoke ID
Send Info for Outgoing Call	MSC requests information from VLR	C	Invoke ID Called number	Invoke ID

C/U = confirmed/unconfirmed service

Table 8.5 Examples of SMS parameters

MAP service	Description	C/U	Typical parameters Req/Ind	Res/Con
Forward Short Message	transfers short messages between MSCs	C	Invoke ID	Invoke ID
Send Info/SMS	MSC seeks subscriber information	C	Invoke ID	Invoke ID
Send Routing Info For SM	Gateway MSC seeks routing information	C	Invoke ID MSISDN	Invoke ID IMSI MSC No.

C/U = confirmed/unconfirmed service

8.4.4 Call handling services

The call handling services relate to basic call control (e.g. setting up and releasing calls to and from the mobile station) and examples are given in Table 8.4.

8.4.5 Short Message Service (SMS)

The Short Message Service (SMS) allows an alphanumeric message to be delivered to or from the mobile station. Examples of the parameters used are given in Table 8.5.

8.4.6 Supplementary services

Supplementary services are extremely complex in nature. The interaction of supplementary services further complicates the position, e.g. establishing the procedures when a Call Forwarding supplementary service is active at the same time as Call Waiting. This aspect needs full evaluation before implementation. The complexity increases again when the evolution of supplementary services is considered: it must be possible to evolve one supplementary service without major impact on others.

The aim is to keep MAP as independent as possible from the implementation of supplementary services, thus establishing a common protocol that can evolve effectively. In this way, numerous supplementary services can be implemented by means of a limited number of MAP services and parameters. This is achieved by using MAP services and parameters to route control information to specific processes that deal with individual supplementary services. Thus, the processes handle the wide deviation in the nature of supplementary services.

Typical services and parameters are given in Table 8.6.

The SS Code Parameter indicates the supplementary service that the customer is registering. The Forwarding Information Parameter indicates a successful outcome of a Call Forwarding supplementary service.

SS Data is returned to indicate successful operation of the Call Waiting supplementary service. The CLI Restriction Info Parameter is returned by the responding User when the request relates to the Calling Line Identity Restricted supplementary service.

8.4.7 Miscellaneous mobility services

Table 8.7 gives some miscellaneous services that are used in the description of procedures in Section 8.5. The CM Service Type Parameter defines the category of the call, e.g. a normal originating call, an emergency call or a supplementary service request.

Table 8.6 Examples of supplementary service parameters

MAP service	Description	C/U	Typical parameters Req/Ind	Res/Con
Activate	VLR & HLR activate an SS	C	Invoke ID SS code	Invoke ID Forward info Call barring info SS data
De-Activate	VLR & HLR de-activate an SS	C	As activate	As activate
Erase	VLR & HLR erase data related to SS	C	Invoke ID SS code	Invoke ID Forwarding information
Interrogate	VLR & HLR retrieve data related to SS	C	Invoke ID SS code	Invoke ID SS status CLI restriction info
Invoke	MSC checks subscription to SS after call set up	C	Invoke ID SS code	Invoke ID
Register	VLR & HLR register data related to SS	C	Invoke ID SS code Forwarded to number Forward info	Invoke ID Forwarding information
Register Password	VLR & HLR register new password	C	Invoke ID SS code	Invoke ID SS code New password

C/U = confirmed/unconfirmed service; SS = supplementary service

Table 8.7 Examples of miscellaneous mobility services

MAP service	Description	C/U	Typical parameters Req/Ind	Res/Con
Insert Sub. Data	HLR updates a VLR	C	Invoke ID	Invoke ID SS
Process Access Request	MSC initiates processing for mobile station access	C	Invoke ID CM service type Connection status Current location ID	Invoke ID
Page	VLR initiates paging for mobile station in specific location area	C	Invoke ID IMSI	Invoke ID
Station Search For Mobile	VLR initiates paging for mobile station in all location areas	C	Invoke ID IMSI	Invoke ID Location area ID

C/U = confirmed/unconfirmed service; SS = supplementary service

8.5 MAP Procedures

This section describes typical procedures for the key functions in a PLMN. For simplicity, the opening and closing of dialogues are not shown on the flow diagrams. Focus is placed on the specific services adopted.

8.5.1 Mobile station location updating

The aim of a mobile communications system is that customers are free to move around at will and still receive service. This ability to move around is termed 'Roaming'. Roaming requires that the network is aware of the location of the mobile station whenever it is the active state. This is achieved by the mobile station informing the network of its position. This triggers a Location Updating (Registration) Procedure[3].

The carrier frequency in each cell transmits the Location Area Identity. This frequency is monitored continuously by the mobile station. When the identity changes, the mobile station informs the MSC by using a Location Update Request Service.

(a) Location updating within VLR area

Fig. 8.5 illustrates the procedures adopted when the mobile station remains within the area served by the VLR. Upon receipt of the Update Request, the MSC opens a Dialogue with the VLR and sends an Update Location Area Indication. In this case, the VLR recognises that the mobile station is still within its coverage area. The VLR updates the Location Area Information and returns an acknowledgement to the MSC. A confirmation is passed on to the mobile station and the Dialogue is closed.

To prevent too many interactions with the HLR, the HLR is only informed of a change of location when the mobile station moves to a new VLR area. Hence, in this case, the HLR is not informed of any change.

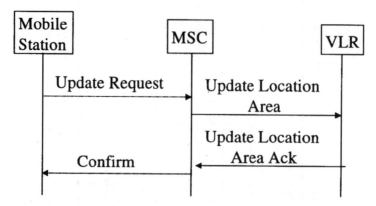

Fig. 8.5   Location updating within VLR area

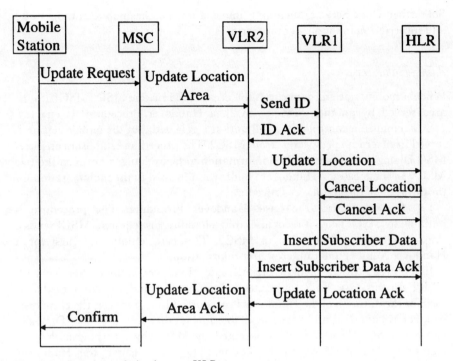

Fig. 8.6   Location updating between VLRs

(b)   Location updating between VLRs

When a mobile station moves from one VLR area to another, the procedures are more complex because the VLRs have to communicate and the HLR has to be informed. Fig. 8.6 shows the procedures in this case.

Again, the mobile station triggers the procedure by recognising a change in Location Area Identity on the carrier frequency. Upon receipt of the Indication from the mobile station, the MSC establishes a Dialogue with VLR2 and sends an Update Location Area Indication. VLR2 recognises that the mobile station is new to its area by analysing either the previous Location Area Identity or the IMSI. VLR2 establishes a Dialogue with VLR1 and transfers a Send Identification Indication to seek identity and authentication information. This is provided by VLR1 to VLR2 by means of a Send Identification Acknowledgement Indication.

By analysing the information received from VLR1, VLR2 identifies the mobile station's HLR. VLR2 establishes a dialogue with the HLR and sends an Update Location Indication showing that the mobile station is now within the area served by VLR2.

The HLR is now in a position to update the two VLRs. The HLR instigates the release of information held by VLR1 by using the Cancel Location Service. The HLR also provides appropriate information to VLR2 by using the Insert

Subscriber Data Service, including information on the mobile station status and subscription to supplementary services.

8.5.2 Handover

When a mobile station roams from the area served by one MSC (MSC1) to the area served by another MSC (MSC2), a Handover Procedure[4] is enacted. MSC1 remains in control of the call until it is released, but the mobile station is served from a radio perspective from MSC2. The procedure establishes an inter-MSC Dialogue and permits BSS information to be exchanged between the two MSCs. In some cases, a Handover Number is allocated to the mobile station for the duration of the Handover Procedure.

Fig. 8.7 summarises the basic Handover Procedure. The procedure is initiated by MSC1 when it recognises that a handover is required. MSC1 sends a Prepare Handover Indication to MSC2. This can contain a request for a Handover Number and radio resource information.

If required, MSC2 invokes the Allocate Handover Number Service with VLR2. In this case, VLR2 allocates a Temporary Handover Number that is used for the duration of the Handover Procedure. After handover, the Handover Number is cancelled and made available for another instance. The Dialogue between MSCs 1 and 2 is then accepted by MSC2 by returning a Prepare Handover Acknowledgement Indication. This can include the Handover Number and information about radio resources. Further information, e.g.

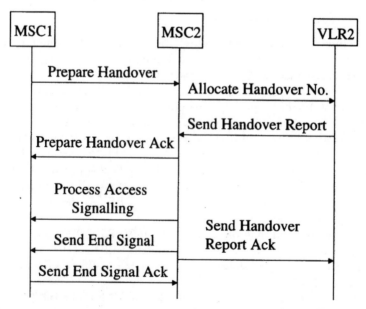

Fig. 8.7 Basic Handover Procedure

concerning radio resources, can be passed from MSC2 to MSC1 using a Process Access Signalling Indication.

When the handover is completed and the mobile station is connected to resources controlled by MSC2, MSC2 sends a Send End Signal Indication to MSC1. MSC2 also sends a Send Handover Report Acknowledgement Indication to the VLR to release the Handover Number.

When the call has finished, MSC1 sends a Send End Signal Acknowledgement Indication to MSC2 to end the Dialogue and release the call.

8.5.3 Outgoing Call Set-Up

The procedures for a customer making an outgoing call can take two forms. In the first form, the radio channel is allocated at an early stage. In the second form, the radio channel is allocated when the called customer answers (Off-Air Call Set-Up). There is little impact upon the network procedures and these are summarised in Fig. 8.8 for the early allocation.

When a customer wishes to make an outgoing call, the mobile station sends a Service Request Indication to the MSC. The MSC checks the authority of the mobile station to make a call by sending a Process Access Request Indication to the VLR. The VLR checks identities, authenticates the mobile station and determines if roaming is allowed. Confirmation of authority to proceed is returned to the MSC by sending a Process Access Request Acknowledgement. This is passed to the mobile station by the MSC as a Service Acceptance Indication.

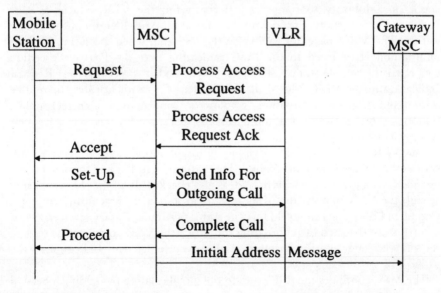

Fig. 8.8   Outgoing Call Set-Up

Upon receipt of the Service Acceptance Indication, the mobile station sends a Set-Up Indication that includes the call control information, e.g. Called Party Address and Bearer Capability (required connection characteristics). The MSC sends a Send Info For Outgoing Call Indication to the VLR to determine parameters applicable to establish the call.

The VLR checks for the activation status of supplementary services and verifies the Bearer Capability Request for the correct subscription. The status of Call Barring supplementary services (e.g. all outgoing calls barred) is also checked. If all the checks are successful, the VLR informs the MSC that it can establish the call by sending a Complete Call Indication including the status of supplementary services, etc.

The MSC then proceeds to establish the call for the mobile station by sending an ISUP Initial Address Message. If the MSC is also the Gateway MSC, the message is sent to the PSTN. Otherwise, the message is sent to the Gateway MSC. The call set-up then proceeds according to the ISUP Procedures.

8.5.4 Incoming Call Set-Up

An incoming call to a mobile station is initiated by the receipt of an ISUP Initial Address Message at the Gateway MSC. The Gateway MSC determines the appropriate MSC to which the call should be routed and the Initial Address Message is passed to the appropriate MSC. The procedures for setting up an incoming call to a mobile station are summarised in Fig. 8.9.

The MSC sends a Send Info For Incoming Call Indication to the VLR to check the parameters needed to set up the call to the mobile station. The VLR checks for authority to roam, etc. If successful, the VLR returns a Page Indication that includes the Identity and Location Area Identity of the mobile station. The MSC instigates paging by the BSS and, when successful, Layer 3 information is returned to the MSC indicating that the Paging Procedure contacted the mobile station. The MSC then sends a Process Access Request Indication to the VLR. The VLR checks identities, authenticates the mobile station and determines if roaming is allowed. Confirmation of correct mobile station authorities is returned to the MSC by sending a Process Access Request Acknowledgement.

The VLR checks for the activation status of supplementary services and verifies the Bearer Capability Values. The status of Call Barring supplementary services (e.g. all incoming calls barred) is also checked. If all the checks are successful, the VLR informs the MSC that it can establish the call by sending a Complete Call Indication including the status of supplementary services, etc. If no errors are detected or reported by the MSC, the VLR completes its role by returning a Send Info For Incoming Call Acknowledgement Indication to the MSC.

The MSC continues the call set-up to the mobile station by sending a Set-Up Indication. The MSC ensures correct channel assignment and alerts the mobile

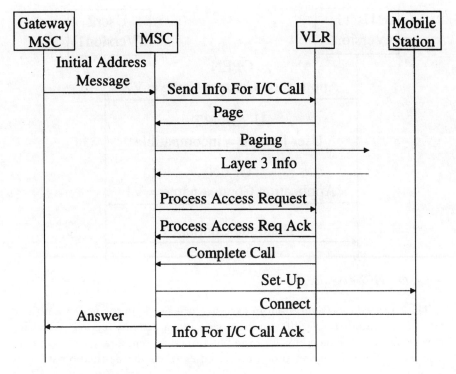

Fig. 8.9 *Incoming Call Set-Up*

station. Upon answer, the MSC returns an ISUP Answer Message to the Gateway MSC.

8.5.5 Version compatibility

As specifications and technologies evolve, different versions of signalling systems will be implemented in networks. It is essential that these versions can interwork effectively and MAP uses a negotiation procedure to achieve compatibility, as illustrated in Fig. 8.10.

When MAP Service User 1 initiates a Dialogue, an Application Context Name Parameter is supplied, indicating the level of communication capability supported by User 1. In this case, User 1 is capable of supporting Version 2 of MAP. This information is transferred to the receiving user in an Open Indication.

If User 2 is also capable of supporting Version 2 of MAP, the Dialogue continues as normal. However, if User 2 is capable of supporting only Version 1 of MAP as shown in the figure, a U-Abort Indication is returned to User 1 with the User Reason Parameter indicating a version incompatibility. User 1 then has to open a new Dialogue based on Version 1 capabilities.

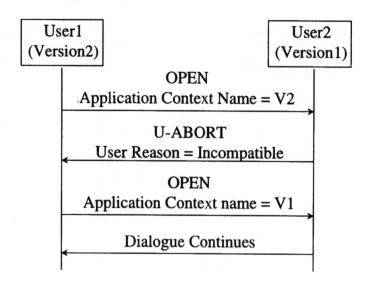

Fig. 8.10   Version negotiation

These procedures could result in excessive processing in the network if there were large numbers of different versions over a lengthy period. Excessive overheads can be avoided by, for example, providing temporary look-up tables that avoid the negotiation procedure in cases when the destination status is known.

8.5.6 Short Message Service

The Short Message Service provides the ability to transfer information to or from the mobile station in alphanumeric form. Fig. 8.11 illustrates the procedure for a short message incoming to the mobile station.

When a short message is received at the gateway MSC, routing information is requested from the HLR by sending a Send Routing Info For Short Message Indication. The HLR checks the appropriate parameters, data and consistency with supplementary service status. If the Short Message Service is allowed, the IMSI and MSC numbers are returned to the Gateway MSC in an Acknowledgement Indication.

The Gateway MSC establishes a dialogue with MSC1 and sends a Forward Short Message Indication which includes the short message to be delivered to the mobile station. The IMSI is included within the Open Dialogue Indication or within the Forward Short Message Indication. MSC1 seeks authority to proceed from the VLR by sending a Send Info For SMS Indication. The VLR checks that all the mobile station information is correct. If there are no errors, the paging procedure is initiated. If the Location Area Identity is known, a Page Indication is sent to MSC1. This indicates paging within the Location Area. If not known, a Search For Mobile Subscriber Indication is sent, indicating paging on a wider scale.

Mobile Communications 151

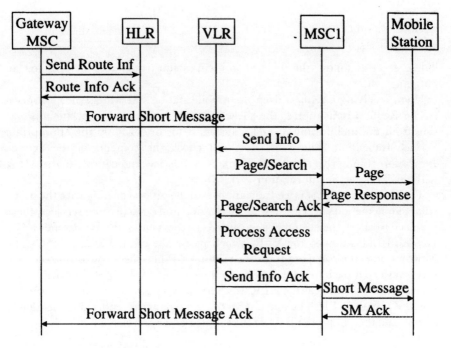

Fig. 8.11 Incoming Short Message Service

The VLR awaits confirmation that the mobile station has been contacted. If the Search Procedure was used, MSC1 sends a Search For Mobile Subscriber Acknowledgement Indication followed by a Process Access Request Indication. If the Page Indication was used, then only the Process Access Request Indication is used. In either case, the VLR returns a Send Info For SMS Acknowledgement Indication to MSC1. MSC1 then initiates the forwarding of the message to the mobile station. Upon confirmation of delivery to the mobile station, MSC1 returns a Forward Short Message Acknowledgement Indication to the Gateway MSC.

8.5.7 Supplementary services

To cater for the complexity of supplementary services, a Process Co-ordinator is provided within each network node. The Process Co-ordinator opens and validates the Dialogue and then selects the appropriate processes to handle the requested supplementary service.

The procedures for supplementary services adopt a transparent communication between the mobile station and the HLR. The services include:

- activation;
- de-activation;

- erase;
- interrogation;
- register.

These services follow the general guidelines illustrated in Fig. 8.12 for an outgoing call.

Upon receiving a request from the mobile station, the MSC sends a Process Access Request Indication to the VLR. The MSC collects appropriate information from the mobile station and transfers it to the VLR by the appropriate Service Indication. The MSC does not check the contents of the Service Indication. Similarly, the VLR initiates a Service Indication to the HLR without checking the information.

The HLR maps the Service Indication onto the process dealing with the particular supplementary service. For the activation and de-activation services, there is a need to check the customer's password. In these cases, the HLR sends a Get Password Indication to the VLR, which passes the request to the MSC. The MSC requests the password from the customer. When the password is received, it is transferred back to the HLR without checking by the MSC or VLR. After checking the password at the HLR, the Service Acknowledgement is returned to the MSC via the VLR. The outcome of the interchange is reported to the mobile station and the call proceeds using the supplementary service.

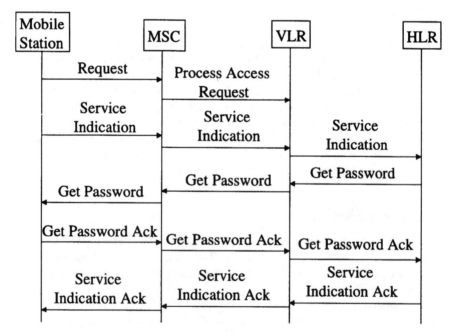

Fig. 8.12 Supplementary services

8.6 Chapter Summary

8.6.1 Introduction

Major growth is occurring in mobile communications. Several technical systems are implemented and planned, and all require a secure, flexible and economic way of transferring signalling information in the network. This Chapter outlines the principles by describing the Mobile Application Part (MAP) of CCSS7.

In fixed communications, the locations of the calling and called customers are derived from the numbering scheme adopted in the network. However, for calls to a mobile station, there is no direct translation between the number of the mobile station and the location of the user. Procedures therefore need to be adopted to determine the location of a called customer on a real-time basis. Roaming, user location, billing etc. place heavy demands on network (signalling) resources.

8.6.2 Numbering

Several numbers are used for addressing and identification purposes. Each user is allocated a number, the mobile station ISDN number (MSISDN), from the national numbering scheme. This is the number dialled by a calling customer to establish a call. The International Mobile Subscriber Identity (IMSI) is used over the radio path and for all signalling throughout the Public Land Mobile Network (PLMN). The International Mobile Equipment Identity (IMEI) identifies the mobile station as a piece of equipment. The TMSI is a Temporary Mobile Station Number, used while the mobile station is roaming.

8.6.3 Network elements

PLMNs consist of a number of network elements:

- The Mobile Switching Centre (MSC) performs the switching functions for a designated area, including Handover and Location Registration.
- The Gateway MSC interconnects a PLMN with other networks.
- The Mobile Station is the terminal (handset) used by the customer.
- The Home Location Register (HLR) is a database that stores customers' profile information, particularly their location.
- The Visitor Location Register (VLR) takes responsibility for tracking a customer. The VLR stores the location area of each visiting mobile station together with its IMSI and MSISDN.
- The Base Station System (BSS) is responsible for all the radio functions and includes Base Station Controllers and Base Transceiver Stations.

8.6.4 Services

MAP uses Transaction Capabilities (TC) to transfer messages. Primitives defined in the OSI Model are used to communicate with TC. MAP provides a range of services that can be Confirmed (validation of delivery) or Unconfirmed. The services are classified as Common (available to all MAP Users) or Specific (restricted in use). Services include:

- Common Services, which perform basic functions within OSI Layer 7, e.g. establishing and releasing Communication Dialogues between MAP Users;
- Location Services, which pass information within the PLMN about the location of a customer;
- Handover Services, to facilitate a customer roaming from one area to another;
- Call Handling Services, which setup and release calls to and from a mobile station;
- Short Message Service (SMS), which allows an alphanumeric message to be delivered to or from a mobile station;
- Miscellaneous mobility services, e.g. Insert Subscriber Data, in which a HLR updates a VLR.

In addition, a range of supplementary services is facilitated. However, many supplementary services are complex, especially when two or more are used in conjunction with each other. MAP avoids undue complexity by using MAP Services and Parameters to route control information to specific processes that deal with individual supplementary services. This allows MAP to remain a common protocol that can evolve effectively.

8.6.5 Procedures

A range of procedures is described in the main text.

(a) Location updating

In location updating, a mobile station recognises a change in Location Area and informs the MSC. The MSC opens a dialogue with the VLR, provides the updated information and the transaction is confirmed. More complex procedures apply for location updating between VLRs, in which the VLRs and HLR communicate.

(b) Handover

When a mobile station roams from the area served by one MSC to the area served by another MSC, a handover procedure is enacted. If necessary, the new MSC can request the allocation of a temporary handover number from an associated VLR. The original MSC remains in control of the call, but the new MSC serves the call from a radio perspective.

(c) Outgoing call

To establish an outgoing call, the mobile station sends a request to the MSC. The MSC checks with the VLR to determine the authority of the user to make a call. If successful, the mobile station supplies the called number and bearer requirements. This is again checked with the VLR and, if successful, the MSC initiates a call to the Gateway MSC.

(d) Incoming call

The Gateway MSC passes an incoming call to the appropriate MSC. The MSC communicates with the VLR to establish the location of the mobile station. The VLR performs checks, provides the location information and instructs the MSC to page the mobile station via the BSS. If successful, the mobile station returns Layer 3 information, which is checked by the VLR and the MSC is given the go-ahead to connect the call.

(e) Short Message Service

Short Message Service procedures are similar to those for an incoming call.

8.7 References

1. ITU-T Recommendation Q.1001: 'General Aspects of PLMNs' (ITU, Geneva)
2. ITU-T Recommendation Q.1002: 'PLMN Network Functions' (ITU, Geneva)
3. ITU-T Recommendation Q.1003: 'Location Registration' (ITU, Geneva)
4. ITU-T Recommendation Q1005: 'Handover' (ITU, Geneva)

Chapter 9
Intelligent Network Application Part

9.1 Introduction

9.1.1 Distributed intelligence

In the past, the intelligence within networks needed to set up and release calls was typically distributed across the local and transit nodes. Each node had routing tables and information on supplementary services to facilitate the control of calls. This architecture provides a secure foundation for communications, but it suffers from a lack of flexibility. If a new service needs to be introduced, or an existing service needs to be modified, all nodes providing call control functions have to be upgraded. In a large network, such upgrades can take many months to complete. As software is being modified in a large number of nodes, the risk of major faults occurring rises significantly.

The distributed call control architecture also means that individual nodes have limited information on how to deal with calls. They are therefore restricted in their ability to provide advanced supplementary services. To illustrate this point, consider a customer who owns two sites in different geographic locations. Each site has its own telephone number and calls are routed accordingly. However, if one site has reached its threshold for receiving calls, the customer would like the overflow calls to be routed to the second site. This is difficult to achieve cost effectively in a distributed call control network because the nodes in the network are unaware of the status of the sites. Similarly, if the customer wants all calls to be routed to one site outside working hours, the changes in routing data would be difficult and expensive to implement.

In a competitive environment, such limitations place network operators at a serious disadvantage and customers are left dissatisfied by the apparent lack of responsiveness.

9.1.2 Centralised intelligence

A significant step towards providing flexibility in networks is the concept of the Intelligent Network (IN). In an IN, aspects of call control are centralised in a single functional node, which takes the role of providing call control for a defined set of services.

The approach of centralising call control intelligence in an IN overcomes the limitations of traditional networks. In this case, the introduction of a new service, or changing an existing service, requires the modification of only a single node. This can be performed very quickly and with minimum risk to the rest of the network. In addition, the central node has available a wide range of information to provide advanced supplementary services. In the example of the customer who wants flexible routing arrangements, the central node can monitor the number of calls routed to each site. When a threshold is reached for the number of calls delivered to the first site, the overflow calls can be automatically routed to the second site. Outside working hours, the central node can be used to route all calls automatically to one of the sites.

The IN approach is applicable to all forms of network, including PSTN, ISDN, PLMN, IP and broadband networks.

9.1.3 IN structure

IN has a number of Functional Entities (FEs) and is illustrated in Fig. 9.1. Nodes A and B are local or transit nodes.

The Call Control Functions (CCF) of Nodes A and B deal with standard (non-IN) calls. For IN, Nodes A and B have the additional capability of recognising that a particular call needs extra information before it can be routed. For example, if a call from a User to Node A includes a request for an advanced

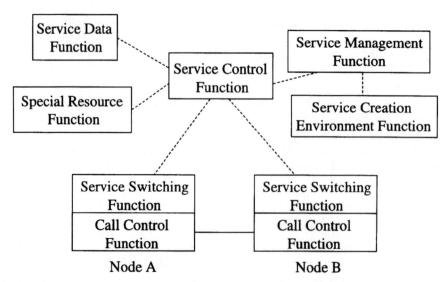

Fig. 9.1 IN structure

service within the dialled number, Node A recognises the request and holds the call until further routing information is sought. This is achieved by triggering a signalling communication. The ability to trigger a request for the information is known as the Service Switching Function (SSF).

The Service Control Function (SCF) contains the service logic and processing capability for advanced services. Thus, the SCF performs the call control for advanced services and issues instructions to the other FEs to facilitate call completion.

The Specialised Resource Function (SRF) provides a range of resources during a call. These resources include the ability to establish a speech path to the calling user to provide two key features. The first feature is to play an announcement or send a tone to a user, giving information on the progress of a call. The second feature relates to the collection of extra information from the calling user, e.g. by receiving tones from the user and converting them into digital format for sending to the SCF. The SCF is also supported by a Service Data Function (SDF) that stores specialised information.

The Service Management Function (SMF) provides management capabilities, e.g. service provision and service deployment. The Service Creation Environment Function (SCEF) facilitates flexible and speedy derivation and modification of services. The SCEF permits the definition, development and testing of services for input to the SMF.

Physical Entities (PEs) in this context are physical implementations of the FEs and these are termed 'Points'. Corresponding terminology is adopted. The Service Control Point (SCP) and Service Switching Point (SSP) refer to the physical nodes performing the corresponding functions.

9.1.4 Modular approach

The key target attribute of the IN approach is flexibility. To achieve maximum benefit, a modular approach is adopted in terms of service definition and network capabilities. The aim is to achieve independence between the network and the services that it supports. This allows services to be introduced without changes being necessary within the network. It also permits evolution of the network without affecting the services in operation.

The modular approach to the IN provides a range of attributes:

- Service providers are able to implement services that are independent of the underlying network infrastructure.
- Multi-vendor operation is facilitated, with differing vendors providing FEs.
- Rapid service creation is possible using the SCEF.
- Efficient service deployment is possible through the capabilities of the SMF.

The modular approach consists of building network and service capabilities from basic units, as outlined below.

(a) Services

Services are standalone commercial offerings. Services are built from a set of basic units termed 'Service Features' (SFs). These provide modularity and flexibility in design and implementation by allowing service designers to mix and match SFs to form new services.

(b) Network functions

The network capabilities to support services are defined independently of the services to provide maximum evolutionary potential for the network infrastructure. Service Independent Building Blocks (SIBs) are re-usable modular network functions that are available to support service provision. Network designers can mix and match SIBs to build network capabilities to support service provision.

(c) Capability Set

A Capability Set (CS) defines IN attributes in terms of network interfaces and service definitions. ITU-T has defined a first set of attributes termed Capability Set 1 (CS-1)[1]. The aim is to evolve the specification of Capability Sets through CS-2, CS-3, etc. towards a full-functionality IN. The capability sets are intended to support all forms of service. However, as part of the evolutionary process, two types of service have been identified.

Type A Services relate to Single Ended, Single Point of Control. Single Ended means that a service feature relates to only one user in a call. Single Point of Control indicates that a particular aspect of a call is influenced by only one Service Control Function at a given time. Thus, the control of a call is restricted to one SSF and one SCF at a given time. These constraints are applied to reduce the operational and control complexity. CS-1 is defined to meet the needs of Type A services.

In Type B services, the constraints for Type A are removed and calls can adopt any combination of user and control function features. Several IN users can be associated within a call and users can join and/or leave a call. Type B services are not yet covered by the Capability Set Approach.

9.1.5 Service invocation and control

Consider a user connected to Node A in Fig. 9.1. To request a service, the user typically initiates the sending of a dialled number to Node A. The Call Control Function (CCF) in Node A is programmed to recognise that a CS-1 Service Request has taken place. The CCF suspends call processing and passes information on the status of the call to the SSF. The SSF interprets the Service Request, prepares a standard query message and passes it to the SCF.

The SCF decodes the query, interprets the requirements from a processing perspective and implements the appropriate processing application. The processing application can involve other FEs, e.g. reference to an SDF or to an SRF. For example, if the SCF needs to gain more information from the calling user before the applications can be completed, the SRF needs to be involved. The SCF does not have a physical link with users. It therefore requests the SSF to connect the user to the SRF. The SCF also instructs the SRF on which procedures to implement, e.g. collect tones from the user and translate them into digital format. When the procedures are completed, the SRF formulates a report and returns it to the SCF.

The SCF completes the appropriate processing and formulates a response to the SSF. The SSF interprets the response and issues specific instructions to the CCF to complete the call. The signalling procedures for a similar example are explained in Section 9.4.3.

9.1.6 Signalling

Signalling amongst the SCF, SSF, SDF and SRF FEs is by means of the Intelligent Network Application Part (INAP). INAP is designed to be a User of Transaction Capabilities (TC). TC can be supported by a range of systems, but the most common implementations use the SCCP and MTP of CCSS7. INAP is currently defined to cover CS-1 functions and will evolve to handle CS-2, CS-3, etc., as these are specified.

9.2 IN Architecture

9.2.1 Conceptual Model

The IN Architecture[2] is based on the IN Conceptual Model. This is a tool for describing the capabilities and characteristics of an IN. The model structures IN into four planes, namely the Service Plane, Global Functional Plane, Distributed Functional Plane and Physical Plane.

(a) Service Plane

The aim of the Service Plane is to realise a wide range of Service Features (SFs). In the way CS-1 was derived, the SFs were not applied to the Service Plane and there is no information provided on the general data structures on which services operate. Hence, for CS-1, the Service Plane has little relevance. The Service Plane will be more applicable in future Capability Sets.

(b) Global Functional Plane

The Global Functional Plane covers the activation of services, monitoring functions and addressing. The plane contains the SIBs that reflect the elements of network functionality needed to provide flexible solutions.

(c) Distributed Functional Plane

The Distributed Functional Plane contains the FEs illustrated in Fig. 9.1. In particular, the information flows between FEs are covered in this plane; these act as an input to the design of the INAP.

(d) Physical Plane

In the Physical Plane, the model identifies the different physical entities and protocols that can exist in an IN. The protocols reflect the information flows outlined in the Distributed Functional Plane.

9.2.2 Implementation components

IN CS-1 contains several basic components that act as a foundation for implementation. A brief description is given here to help put INAP into context. The components are Service Independent Building Blocks (SIBs), Service Logic, Processing Models, Information Flows and Application Service Elements (ASEs).

(a) Service Independent Building Blocks

SIBs are re-usable blocks of software that represent network capabilities. Each SIB comprises a set of actions that is used to model service entities. Examples of SIBs are:

- Algorithm, which applies a mathematical algorithm to data;
- Charge, which determines if there are charging arrangements additional to those carried out in basic call processing;
- Compare, which performs a comparison between a parameter and a reference value;
- Queue, which provides sequencing of calls to be completed to a called user;
- Translate, which determines output information from input information;
- Verify, which confirms that information meets syntax guidelines.

(b) Service Logic

Service Logic is a set of routines and rules for developing and implementing services. Global Service Logic describes the order in which SIBs can be sequenced to provide services. Distributed Service Logic uses actions and information flows to execute services.

(c) Processing Models

Processing Models are tools used to describe the distribution of functions amongst FEs. Call modelling provides a high level concept of call and connection processing in the SSF and SCF. It is both service and vendor independent.

Service Logic Processing provides a concept of SCF activities needed to support service logic execution.

(d) Information Flows

Information Flows relate to the transfer of information between FEs. The concept of resources contained within one FE that another FE can manipulate is covered by defining Logical Objects. In CS-1, information flows between two FEs consist of either a Client/Server Response Pair or a Client Request alone.

(e) Application Service Elements

An Application Service Element (ASE) defines a function or a set of functions to facilitate communication between two Applications. ASEs can be general or specific. General ASEs can be used by various Application Protocols. Specific ASEs are used by one Application. As an example, a general ASE that INAP uses is Transaction Capabilities (TC). Specific ASEs for INAP have also been defined and examples are described in Section 9.3. ASEs are explained further in Chapter 3.

An Application Context consists of a combination of ASEs and the relationship between the ASEs. CS-1 of INAP supports a wide range of Application Contexts, but none is standardised at this stage.

9.2.3 INAP architecture

The INAP architecture is based on the OSI layer 7 functions described in Chapter 3. Readers unfamiliar with these functions are recommended to review Chapter 3 to facilitate a deeper understanding of the INAP architecture.

INAP is a user of Transaction Capabilities (TC), as described in Chapter 7. TC provides a non-circuit-related information transfer capability and spans Layers 4 to 6 and part of Layer 7 of the OSI Model. Layers 1 to 3 of the OSI Model are provided by other protocols, e.g. the SCCP and MTP of CCSS7. The architecture is illustrated in Fig. 9.2.

INAP contains a range of Specific ASEs. The Specific ASEs comprise groups of Operations, where an Operation is a request to perform an action. The Operations are similar to those in TC.

A Physical Entity (PE) can either have single interactions with other PEs or multiple interactions. For single interactions, coordination of ASEs is performed by a Single Association Control Function (SACF), as illustrated in Fig. 9.3. The Single Association Object (SAO) represents the capabilities of the SACF and a set of ASEs to be used for a single transaction between two FEs.

For multiple interactions between PEs, a Multiple Association Control Function (MACF) is used to coordinate a number of SAOs, each of which interacts with an SAO in another PE. This is illustrated in Fig. 9.4.

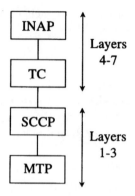

Fig. 9.2   INAP as a TC User

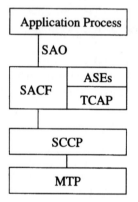

Fig. 9.3   INAP architecture, single interaction

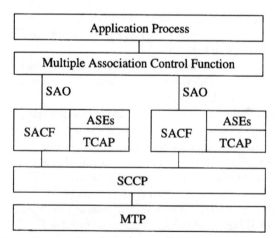

Fig. 9.4   INAP architecture, multiple interactions

9.3 INAP Formats

INAP defines a set of Operations[3] that cause actions to take place. The Operations are grouped into INAP ASEs. Table 9.1 gives examples of the ASE types adopted for CS-1. The Detection Point (DP) is the juncture in the SSF call processing at which an IN call is triggered.

Each Operation contains parameters that carry information about the nature of the task to be implemented. The parameters in INAP are also called Data Types. Operations related to a particular call contain a Call ID Parameter to indicate to which call the Operation pertains. Examples of Operations and Parameters are given in Table 9.2.

9.4 INAP Procedures

9.4.1 General

A key goal of INAP is to provide a flexible foundation for evolution, in which the protocol can handle new services without major modification. Although

Table 9.1 Typical ASE types

ASE type	Function	Typical Operations
SCF Activation	SSF requests SCF service	Initial DP
Connection Establishment	SCF initiates call connection from SSF to SRF	Connect to Resource
Connect	SCF initiates routing at SSF	Connect
Traffic Management	SCF implements traffic control measures	Call Gap
Specialised Resource Control	activates special resources	Play announcement Prompt/Collect User Information Specialised Resource Report
Basic Call Processing	communicates progress on call set up and release	Called Party Busy Answer
Charging	initiates charging	Apply Charging Apply Charging Report
Service Management	SCF requests a service to be handled in a particular manner	Activate Service Filtering

DP = Detection Point

Table 9.2 Typical Operations

Operation	Typical Parameter	Purpose
Initial DP	Call ID Service key Dialled number	SSF requests service from SCF
Connect	Call ID Called user number	SCF requests SSF to route a call to a given number
Connect to Resource	Call ID Routing address	SCF requests SSF to connect to SRF
Establish Temporary Connection	Call ID	SCF requests SSF to establish connection to another SSF or SRF
Play Announcement	Call ID Information to send	SCF requests SRF to send information to calling user
Prompt and Collect User Info	Call ID Information to send	SCF requests SRF to interact with calling user
Specialised Resource Report	Call ID Info	Response to Collect User Info
Call Gap	Gap criteria Gap indicators	SCF requests SSF to reduce specific service attempts

DP = Detection Point, ID = Identity

additional ASE types and operations might be necessary in the future, the versions defined for CS-1 should provide the vast majority of functions that are needed. With this background, the aim is to avoid defined procedures for INAP. This allows a flexible use of INAP Operations and Parameters without unduly restraining network operators and service providers in service implementation.

However, to help understand the concepts within INAP, the following examples of procedures are given. These do not imply specific implementations, but they describe the principles of using INAP to provide services in a flexible and speedy manner. This applies particularly for the cases involving the SRF, in which several implementations and types of procedure can be adopted. Dotted lines in the Figures indicate messages that are not part of the INAP protocol, but they are shown to clarify the relationship between INAP and other systems, e.g. ISUP.

9.4.2 *Translation*

Consider a service in which a special address is needed by the network in order to reach the called user. The digits dialled by the calling user refer to a generic address (e.g. a published telephone number) and the network needs to translate the dialled number into the special address. This scenario applies to many forms of Freephone services in which the called user pays for the call. The procedure is illustrated in Figure 9.5.

The local exchange to which the calling customer is connected receives the call request (e.g. a Set-Up Message in DSS1) and analyses the dialled digits. In this case, the local exchange has trigger devices that enable it to recognise that additional routing information is required. The local exchange therefore has IN Service Switching Functions (SSF).

At this stage, the local exchange formulates an Initial Detection Point Operation and sends it to the Service Control Function (SCF), including the Service Key and Dialled Number as parameters. The Service Key indicates the form of service required.

The SCF analyses the Service Key and translates the Dialled Number into the special number required to reach the called user. This information is returned to the SSF in a Connect Operation. The local exchange then continues the call set up in a normal manner by sending an Initial Address Message (IAM) to the next exchange.

9.4.3 Translation with announcement

In this case, an announcement or tone can be relayed to the calling user while the call is being established, e.g. to inform the calling user that the set-up time could be significant as a complex translation is being sought. Communication with the SRF can be (i) from the SSF, (ii) from the SCF and transferred transparently via the SSF or (iii) directly between the SCF and SRF. The procedure is illustrated in Fig. 9.6 for the direct communication. The set-up to the local exchange and SCF is the same as in Section 9.4.2.

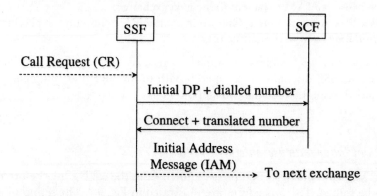

Fig. 9.5 Simple Translation

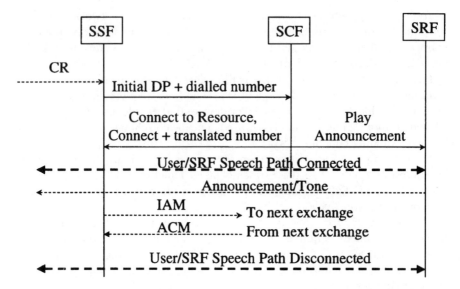

Fig. 9.6 Translation with announcement

Note that the 'Connect to Resource Operation' could also be 'Establish Temporary Connection'

When the SCF analyses the called number, it recognises the need to give information to the calling user (e.g. the set up time is likely to be significant). The SCF:

- returns a Connect to Resource Operation to the SSF, which causes the SSF to establish a speech circuit from the user to the Specialised Resource Function (SRF) – the manner in which this is achieved varies according to the type of SRF and is not described in detail here;
- returns a Connect Operation, including the translated number, to the SSF, which causes an IAM to be sent to the next exchange;
- sends a Play Announcement Operation to the SRF to initiate the playing of the announcement to the calling user.

Upon receipt of the Address Complete Message (ACM) from the next exchange, the SSF disconnects the speech path to the SRF and completes the provision of the call between the two users.

9.4.4 Translation with user interaction

In this case, additional information is required from the calling user before the call can be routed. The procedures are illustrated in Fig. 9.7. The set-up to the local exchange and SCF is the same as described previously.

Intelligent Network Application Part

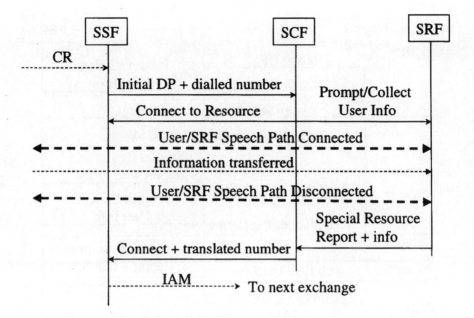

Fig. 9.7 *Translation with user interaction*

Note that 'Connect to Resource Operation' could also be 'Establish Temporary Connection'

When the SCF analyses the called number, it recognises that additional information is required from the calling user before the call can be routed. The SCF therefore sends a Connect to Resource Operation to the SSF and a Prompt/Collect User Information Operation to the SRF. These Operations establish a speech circuit from the user to the Specialised Resource Function (SRF) and permit the calling user to provide the additional information (e.g. by dialling additional digits). The additional information is passed back to the SCF in a Specialised Resource Report. The SCF can then provide the necessary translation for the Dialled Number and the translation is passed back to the SSF to facilitate completion of the call.

9.4.5 Credit card calling

This service provides an authentication procedure to allow use of a credit card payphone. The procedure is illustrated in Fig. 9.8.

The Call Request to the SSF is initiated when the calling user picks up the telephone. The Call Request does not include the Called User Number because authentication is required before service can be provided. Upon analysis of the Call Request, the SSF sends an Initial DP Operation, with the Service Key Parameter appropriately set, to the SCF. The SCF recognises that authentication procedures are needed and initiates a speech path to be established between

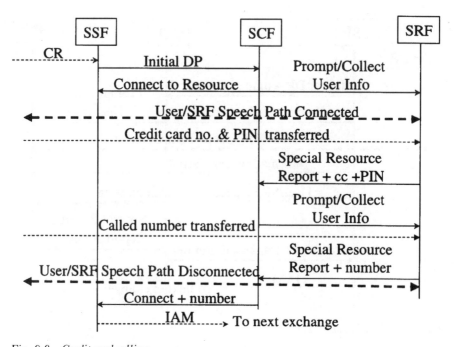

Fig. 9.8 Credit card calling

Note that 'Connect to Resource Operation' could also be 'Establish Temporary Connection'

the user and the SRF by sending a Connect to Resource Operation to the SSF. The SCF also sends a Prompt/Collect User Information Operation to the SRF.

The SRF requests the credit card number and personal identification number (PIN) from the user and these are passed back to the SCF for analysis in a Specialised Resource Report Operation. If authentication is successful, the SCF sends a second Prompt/Collect User Information Operation to the SRF to initiate the collection of the Called Number. The SCF checks the called number for validity (e.g. to ensure that international calls are allowed) and passes the number to the SSF in a Connect Operation. This initiates the sending of an IAM towards the called user.

9.4.6 Call gapping

This procedure reduces the number of call attempts sent to the SCF for a particular service or called number. This caters for high loads for a limited period of time. The procedure is illustrated in Fig. 9.9 for the case when a translation is also required.

The receipt of Call Request 1 at the SSF stimulates the sending of an Initial DP Operation, including a Dialled Number Parameter, to the SCF. The SCF

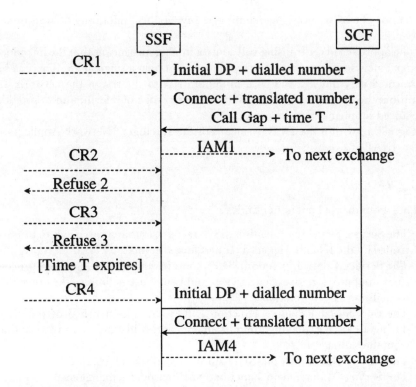

Fig. 9.9   Call gapping

realises that an internal threshold has been reached in terms of its capability to handle calls to this address. The SCF returns a Connect Operation for Call 1, including the appropriate Translated Number and the SSF completes the call to the called user. The SCF also sends a Call Gap Operation, with parameters that define the nature of the gapping to be applied, e.g. the length of time (T) before another call to the specific destination is allowed. Call Requests 2 and 3 are refused by the SSF. Upon expiry of time T, the SSF recognises that it can send more requests pertaining to the address to the SCF. Upon reception of Call Request 4, an Initial DP Operation is transferred and the procedures revert to a normal translation.

9.5   Chapter Summary

9.5.1   Introduction

The intelligent network (IN) is a significant step in providing flexibility in networks. Aspects of call control are centralised in a single functional node, the Service Control Pont (SCP). The SCP takes the role of providing call control for a defined set of services, overcoming many of the limitations of traditional networks.

The approach of centralising call control intelligence means that the introduction of a new service requires the modification of only a single node. This can be performed very quickly and with minimum risk to the rest of the network. In addition, the central node has available a wide range of information to provide advanced supplementary services.

Signalling within the IN is by means of the Intelligent Network Application Part (INAP).

9.5.2 IN structure

IN has a number of Functional Entities:

- The Service Switching Function (SSF) recognises that a call needs to be controlled by the IN and triggers a request for assistance.
- The Service Control Function (SCF) contains the service logic and processing capability for advanced services and issues instructions to the other FEs to facilitate call completion.
- The Specialised Resource Function (SRF) provides a range of resources during a call, e.g. playing announcements and collecting extra information from the calling user.
- The Service Data Function (SDF) stores specialised information.
- The Service Management Function (SMF) provides management capabilities, e.g. to assist in service provision and service deployment.
- The Service Creation Environment Function (SCEF) facilitates flexible and speedy derivation and modification of services.

Corresponding Physical Entities (PEs) are termed 'Points' (Service Switching Point, etc.).

9.5.3 Modularity

To achieve maximum flexibility, a modular approach is adopted for service definition and network capabilities. Services are standalone commercial offerings and are built from a set of Service Features (SFs). The network capabilities to support services are defined independently of the services to provide maximum evolutionary potential for the network infrastructure. Service Independent Building Blocks (SIBs) represent re-usable modular network functions that are available to support service provision.

The modular approach to the IN provides:

- services that are independent of the underlying network infrastructure;
- facilitation of multi-vendor operation;
- rapid service creation using the SCEF;
- efficient service deployment through the capabilities of the SMF.

9.5.4 Services

A Capability Set (CS) defines IN attributes in terms of network interfaces and service definitions. The ITU-T has defined a first set of attributes termed Capability Set 1 (CS-1). The aim is to evolve the specification of Capability Sets through CS-2, CS-3, etc. towards a full-functionality IN.

Type A Services relate to Single Ended, Single Point of Control. CS-1 is defined to meet the needs of Type A services. Type B services provide calls that can adopt any combination of user and control function features. Several IN users can be associated within a call and users can join and/or leave a call. Type B services are not yet covered by the Capability Set Approach.

9.5.5 INAP architecture

INAP makes use of Transaction Capabilities (TC), the SCCP and MTP for transferring information. INAP and the Transaction Capabilities Application Part (TCAP) form Layer 7 of the OSI Model. INAP is structured according to Layer 7 of the OSI Model (explained in Chapter 3). It contains a range of Specific ASEs, comprising groups of Operations (requests to perform actions). A Single Association Control Function (SACF) coordinates the ASEs for a Single Association Object. A Multiple Association Control Function (MACF) performs a corresponding role for multiple Association Objects.

9.5.6 Format

Operations are grouped into a number of Specific ASEs, including:

- SCF Activation, in which an SSF requests a service from the SCF;
- Connect, in which the SCF initiates routing at an SSF;
- Specialised Resource Control, in which special resources are activated;
- Service Management, in which the SCF requests a service to be handled in a particular manner.

9.5.7 Procedures

A key goal of INAP is to provide a flexible foundation for evolution, in which the protocol can handle new services without major modification. With this background, the aim is to avoid defined procedures for INAP. This allows a flexible use of INAP Operations and Parameters without unduly restraining service implementation.

However, to help understand the concepts within INAP, the following examples of procedures are given in the text:

- translation;
- translation with announcement;
- translation with user interaction;
- credit card calling;
- call gapping.

9.6 References

1 ITU-T: Recommendation Q.1211: 'Introduction to IN Capability Set 1' (ITU, Geneva)
2 ITU-T: Recommendation Q.1218: 'Interface Recommendation for IN Capability Set 1' ITU, Geneva)
3 ITU-T: Recommendation Q.1214: 'Distributed Functional Plane for IN CS-1' (ITU, Geneva)

Chapter 10
Management Aspects of CCSS7

10.1 Introduction

Networks are extremely complex and an effective management system is needed to optimise the functions that can be performed. It is essential that operators can plan, provide and maintain services and networks, while reducing operational costs.

Network management includes gathering information, analysing the consequences of the information and taking appropriate actions. It includes:

- Performance management, i.e. gathering statistics to enable the behaviour of the network to be evaluated. Performance management includes the collection of measurements, modifying equipment capacities and adjusting routing tables.
- Fault management, i.e. detecting, isolating and correcting abnormal operation of the network. Fault management includes handling alarm conditions, activating measurements and taking corrective actions.
- Configuration management, i.e. controlling resources within the network, allowing changes to the structure of the network, operating data, etc. in a flexible manner. Configuration management includes composing and verifying routing tables, installing link sets and initialising timers.

CCSS7 is a key part of telecommunications networks. More than this, it forms a signalling network in its own right. It is therefore essential that CCSS7 should have a comprehensive management capability. This Chapter briefly reviews the management aspects of CCSS7.

10.2 CCSS7 Management Functions

10.2.1 Scope

CCSS7 has three types of management capability, as illustrated in Fig. 10.1. The names of the types are not ITU-T standard and are used here to facilitate explanation.

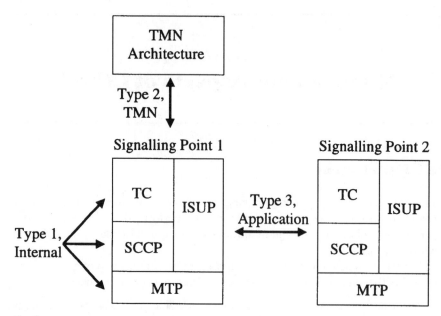

Fig. 10.1 CCSS7 management types

(a) Internal Type

The first type (Internal Type) relates to the management functions specified as part of the signalling system itself. For example, the MTP Signalling Link and Signalling Route Management Functions are specified as part of the MTP. This type of capability is described in each Chapter of this book (e.g. Chapter 4 for the MTP).

(b) Telecommunications Management Network (TMN) Type

The second type of capability (TMN Type) describes how CCSS7 is used as part of the overall approach to network and service management, including the collection of measurements. This overall approach is termed the Telecommunications Management Network (TMN) system and is explained in Section 10.3.

(c) Application Type

The third type of capability (Application Type) refers to the exchange of management information between Signalling Points (SPs), e.g. between SPs1 and 2 in Fig. 10.1. In this type of management, CCSS7 is used to verify and validate routing information, e.g. routing tables, Circuit Identification Codes (CICs), etc. The functions performed and the means of communication are described in Section 10.4.

(d) Operations, Maintenance and Administration Part

The second (TMN) and third (Application) types of management functions are specified within the Operations, Maintenance and Administration Part (OMAP) of CCSS7. The first type is covered as part of the signalling system itself.

10.2.2 OMAP

The OMAP has three main requirements to fulfil:

- provide an interface to the TMN architecture;
- work with other parts of TMN to form an integrated approach to network management;
- extend the functions specified in the CCSS7 parts (e.g. the MTP) to provide a more comprehensive management capability.

10.3 TMN Type of Management Functions

10.3.1 Principles of TMN

The ITU has specified a Network Management System called the Telecommunications Management Network (TMN)[1]. The basic concept underlying TMN is the provision of a formal structure (architecture) to achieve the transfer of management information between Operations Systems (Support Systems) and telecommunications networks. Operations Systems include Network Management, Service Management, Customer Care and Billing.

TMN is based on a model incorporating Business Management, Service Management, Network Management and Network Element Management. A Network Element refers to a functional object within the network, e.g. a Signalling Point (SP) or a signal transfer point (STP). Network Management refers to capabilities that transcend more than one functional object and to the correlation of information throughout the network.

Fig. 10.2 illustrates the relationship between TMN and a telecommunications network. The telecommunications network consists of a wide range of Network Elements such as exchanges, transmission equipment, signalling terminals, etc.

The TMN provides management functions and offers communications between Operations Systems and with the various parts of the telecommunications network. The TMN is conceptually separate from the telecommunications network that it manages. The TMN has interfaces with the network at various points to permit an exchange of information and allow the TMN to control the operation of the network.

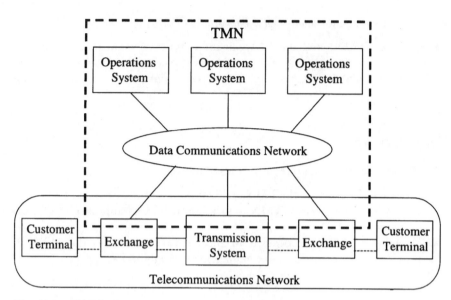

Fig. 10.2   TMN/telecommunications network relationship

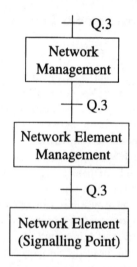

Fig. 10.3   TMN interfaces

10.3.2 Architecture

OMAP relates primarily to Network Element Management and Network Management. The structure is illustrated in Fig. 10.3. TMN defines reference points between the levels of management, denoted by the letter 'q'. When OMAP is applied, the q reference points are known as Q.3 Interfaces. Q.3 is a regularly used term and refers to the interfaces defined by the TMN architecture for OMAP.

10.3.3 Managed objects

The most basic element of the network that is managed is termed a 'managed object'[2]. Typical managed objects are a Signalling Termination in an exchange and a Signalling Link Set. Some managed objects, e.g. a Signalling Termination, are located within one Network Element and are therefore managed wholly by the Network Element Management Functions. OMAP can take a 'nodal' view of these items as they affect only one SP.

Other managed objects involve more than one Network Element, e.g. a Signalling Link Set involves the management functions of two SPs. In these cases, the managed objects are managed by the Network Management Functions. In these cases, OMAP takes a 'network' view of the item, i.e. how the item fits within the network.

A simplified managed object model is illustrated in Fig. 10.4.

In SP A, the Functional Elements (MTP, SCCP, etc.) are mapped into the managed objects. These managed objects then have a relationship with the objects in the Operations System (in this case, the Operations System is a Network Management System). The overall management function within the network is known as the Management Information Service (MIS). The User of the MIS within the Operations System manages a corresponding Agent Function within SP A.

For managed objects wholly within SP A, the model for SP A applies. For other managed objects that extend between SPs A and B, e.g. when testing between the nodes, the object is spread between SPs A and B. However, in non-failure conditions, SP A is responsible for interacting with the Operations System. The necessary co-ordination between SPs A and B is an internal function of the object and is defined as part of OMAP. An example of this type of managed object is the MTP Routing Verification Test (MRVT) described below.

10.3.4 Managed object specification

Management Information Models are being derived by the ITU-T for CCSS7. These consist of defining the managed objects within the Network Element Management and Network Management Functions. The managed objects are being categorised into classes.

The work involves the formal specification of the objects, including their attributes, notifications and management information. Attributes refer to the properties of the objects and usually include quantitative values. Notifications relate to information emitted by a managed object when an event has occurred within the object.

Much work remains to be performed within the ITU-T, but the specification for the MTP Network Management and Network Element Management is in progress. Table 10.1 gives examples of the managed object classes for MTP Network Management (NM) and Table 10.2 gives examples for MTP Network Element Management (NEM).

180 *Telecommunications Signalling*

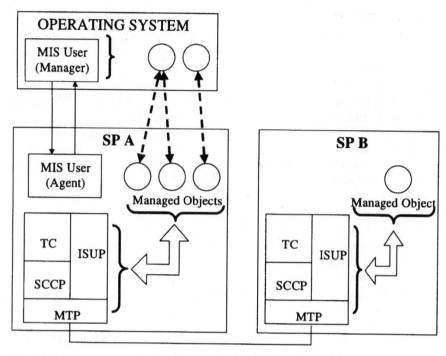

Fig. 10.4 Managed object model (Source: ITU-T Recommendation Q.750)

Table 10.1 Examples of managed object classes – MTP NM

Object	Role
CIC Table	assign or modify CIC values at two SPs
Signalling link	reconfigure a signalling link (e.g. allocate different signalling terminals) modify attributes (e.g. congestion thresholds)
Signalling Link Set	change the state of a Signalling Link Set (e.g. repair status) modify attributes
Signalling Point	start a test programme and ascertain the state of an SP
Signalling Route	change the state of a Signalling Route
Signalling Route Set	define, modify or change the state of an existing Route Set

Table 10.2   Examples of managed object classes – MTP NEM

Object	Description
Managed Switching Element	an exchange, comprising all switching functions, call control and signalling functions
MTP Level 2	Level 2 Functions, i.e. signalling link
MTP Level 3	Level 3 Functions, i.e. MTP signalling network
MTP Signalling Point	covers End Point (e.g. Local Exchange), STP and Combined End Point/STP
Link Set Termination	covers one end of a Link Set. An Operations System can manage a complete Link Set by managing both ends
Link Termination	covers one end of a link
Route Set Network Element Part	covers the Network Element Part of a Route Set, gathering information within a node about another node in the network
Signalling Termination	physical interface terminating a signalling link at a node
STP Screening Table	provides information to deal with unauthorised CCSS7 Messages

10.3.5   Measurements

To manage a CCSS7 network effectively, it is necessary to monitor and measure the performance, utilisation and availability of the resources within the network. Monitoring and measurement is passive in the sense that carrying out these functions does not alter the attributes of the network. The ITU-T specification for monitoring and measurements[3] lists a comprehensive range. To illustrate the principles, this Section gives some examples. Table 10.3 outlines the main classes and sub-classes of measurements that are specified for CCSS7. Table 10.4 gives examples of measurements for some of the sub-classes.

10.4   Application Type of Management Functions

10.4.1   OMAP Model

The Application type of CCSS7 management functions provide an audit capability for CCSS7 data and the use of the data within the network. This requires

Table 10.3 Measurement classes

Element	Measurement class	Sub-class
MTP	Fault and Configuration	Detection of Link Failure Surveillance of Network Status Detection of Routing Table Errors
MTP	Performance	Link, Link Set, etc. Utilisation Component Reliability
SCCP	Fault Management	Routing Failures Unavailability Segmentation Faults
SCCP	Performance	Utilisation Quality of Service
ISUP	Fault and Configuration	ISUP Availability Protocol Errors Performance
TC	Fault Management	Protocol Errors User Generated Problems
TC	Performance	Messages Sent/Received Components Sent/Received Number of Transactions

communication between SPs. Fig. 10.5 illustrates the OMAP Model for the purpose of providing the requisite communication and performing the appropriate functions. The OMAP Model uses the OSI architecture explained in Chapter 3. Readers unfamiliar with this architecture are advised to review Chapter 3.

In the OSI Model, Layer 7 contains an Application Process and an Application Entity. A key component of an Application Entity is an Application Service Element (ASE). An ASE defines a group of functions and can be regarded as a sub-programme within the Application Entity. For OMAP, the ASE is termed the Operations and Maintenance ASE (OMASE).

The Management Process makes use of the OMASE User and OMASE to perform the CCSS7 management functions. The Management Process provides the mapping between the SP Management Functions and the OMASE User. The logic to perform the management functions is located in the OMASE User. OMASE provides the communications functions needed between the two SPs.

OMASE uses the TCAP, SCCP and MTP stack to transfer management messages.

Table 10.4 Examples of measurements

Sub-class	Measurements
MTP, Detection of Link Failure	Number of link failures
MTP, Detection of Routing Table Errors	Duration of unavailability Adjacent SP inaccessible Duration of adjacent SP inaccessible Number of MSUs discarded
SCCP, Utilisation	SCCP traffic received SCCP traffic sent Messages handled
ISUP, Protocol Errors	Inability to release a circuit Abnormal release condition Release due to unrecognised information Missing Blocking Acknowledgement
TC, Number of Transactions	Number of new transactions Mean number of open transactions Mean duration of transactions Peak number of transactions

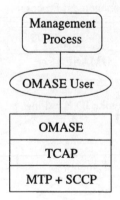

Fig. 10.5 OMAP Model for CCSS7 interface

10.4.2 OMASE User

The OMASE User[4] provides the logic for the Application type of management functions. The OMASE User consists of three sets of functions, namely the MTP Routing Verification Test (MRVT), the SCCP Routing Verification Test (SRVT) and the Circuit Validation Test (CVT) functions.

The MRVT determines the consistency of data in the MTP routing tables in the network. The test is based on an initiating SP sending messages on all routes to a destination SP. The test is stopped either by the destination SP or by any SP detecting a fault on the route. The identity of each SP transferring the test messages is noted to ensure that a full analysis of the route can be made.

The SRVT is very similar to the MRVT. For SRVT, the aim is to verify SCCP routing data and therefore the procedures focus on testing the Global Title Translation Service.

The CVT is used to check the continuity of a circuit and that the data associated with the circuit is consistent in two exchanges.

The structure of the OMASE User is illustrated in Fig. 10.6. Primitives from the OMASE (Indication and Confirmation) need to be distributed to the appropriate set of functions. This is performed by the Object ID Function, which distributes the primitives on the basis of the Object Identifier (e.g. the MTP Routing Tables for MRVT and the SCCP Routing Tables for the SRVT). All primitives contain the Invoke and Dialogue Identities and the Called and Calling Party Addresses.

For Request Primitives, the ID Function assigns unique Dialogue and Invoke Identities across the MRVT, SRVT and CVT Functions.

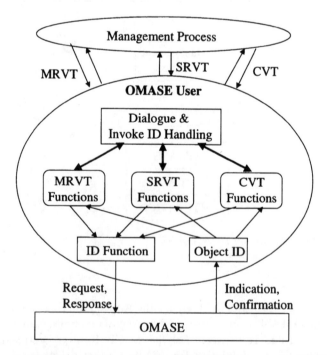

Fig. 10.6   OMASE User structure (Source: ITU-T Recommendation Q.753)

10.4.3 OMASE primitives

The OMASE[5] defines the management information used by the MRVT, SRVT and CVT Functions. The OMASE provides the inter-nodal communications for these functions in messages within the CCSS7 network, making use of Transaction Capabilities (TC).

The OMASE Services are derived from those defined in the Common Management Information Protocol[6] (CMIP). CMIP is a protocol, defined by the International Standards Organisation and by the ITU, which is used to exchange management information between Application Layer Entities.

The primitives used by OMASE include OM-Action and OM-Event-Report. The use of the OM-Action Primitive is illustrated in Fig. 10.7. The principles of the primitives are as described in Chapter 3.

10.4.4 OMASE services

The services offered by OMASE include the OM-Action Service and the OM-Event-Report Service. An example of the parameters used in the OM-Action service is illustrated in Table 10.5. The parameters used in the OM-Event-Report Service are similar to those in Table 10.5 but, for example, an Event Type Parameter is used instead of an Action Information Parameter.

10.4.5 OMASE parameters

The OM-Action Primitive includes an Action Information Parameter. To invoke the MRVT, this parameter is coded to indicate Test Route Action. An example of the contents of Test Route Action is given in Table 10.6.

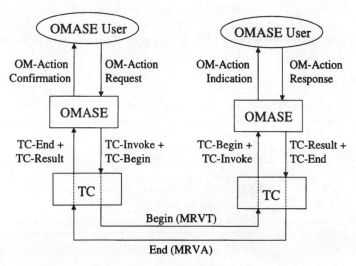

Fig. 10.7   OMASE primitives

Table 10.5   Example of OM-Action Parameters

Parameter	Description
Calling Party Address	initiating exchange Point Code
Called Party Address	responding exchange Point Code
Dialogue ID	mapped by TCAP into Transaction ID
Invoke ID	TCAP Invoke ID
Object Class	type of object being addressed, e.g. MTP Routing Tables
Object Instance	Point Code of the test destination
Action Information	specifies a particular action to be performed

Table 10.6   Example of Test Route Action

Element	Description
Initiating SP	Point Code of the SP requesting the MRVT/SRVT
Trace Requested	requests that a trace of all routes used to reach the destination should be reported to the Initiating Exchange
Threshold	the maximum number of SPs to deal with the test, thus identifying over-length routes
Point Codes Traversed	each SP adds its Point Code as it is traversed, thus providing a record of the route

The OM-Event-Report Primitive includes an Event Type Parameter. To invoke a trace of a route, this parameter is coded to indicate Route Trace. Route Trace initiates a report on the MTP or SCCP routing status. It can be invoked explicitly, by including the appropriate element in the Event Type Parameter, or upon detection of a fault by an SP. An example of the contents of Route Trace is given in Table 10.7.

Table 10.7   Example of Route Trace Event

Element	Description
Success	on successful completion, the trace of the Point Codes is included
Detected Loop	indicates that the Point Codes of the SPs involved in a route have formed a loop
Excessive Length	when the threshold of SPs is exceeded, the route is considered to be excessive
Point Codes Traversed	each SP adds its Point Code as it is traversed, thus providing a record of the route

10.4.6 MTP Routing Verification Test (MRVT)

The MRVT is used to determine that routing data in the MTP network is consistent. The test detects:

- MTP routing loops;
- excessive route lengths;
- unknown destinations, e.g. non-existent destinations, missing routing entries, etc.;
- unidirectional signalling links, i.e. the inability to signal in both directions.

The procedure is based on sending test messages on all routes to a destination. The messages record the identities of each STP encountered. The test can be initiated by any SP and is typically initiated when routing data has been added or modified. If an error is detected at an SP, the test is terminated and the results are reported back to the initiating node.

10.4.7 MRVT messages

Three messages are used for MRVT. The first message is the MRVT Message. This is sent from an SP to an adjacent SP and contains, for example:

- the point code of the destination;
- the point code of the initiator;
- the threshold for the maximum number of SPs within a route;
- a list of the SPs traversed (this field is filled in as the message progresses through the network).

The second message is the MTP Routing Verification Acknowledgement (MRVA) Message. This is sent from an SP receiving an MRVT Message to the SP sending the MRVT Message. The use of the MRVA Message inherently indicates successful completion of the test at the sending SP, unless additional information is included, e.g. a reason for failure. Such reasons include a detected loop, unknown DPC, etc.

The third message is the MTP Routing Verification Result (MRVR) Message, which provides additional information. For example, if the test is successful, the MRVR contains the list of point codes traversed. If the test is a failure, the MRVR contains detailed information on the reason for the failure.

10.4.8 MRVT procedures

The procedures for MRVT are illustrated in Fig. 10.8. In the example, SP1 is the Initiating SP and SP4 is the Destination SP. SPs2 and 3 are STPs on routes between SPs1 and 4. The following terminology is adopted: M = MRVT Message, A = MRVA Message and R = MRVR Message. The numbers following the message name indicate the SPs that have dealt with the Message.

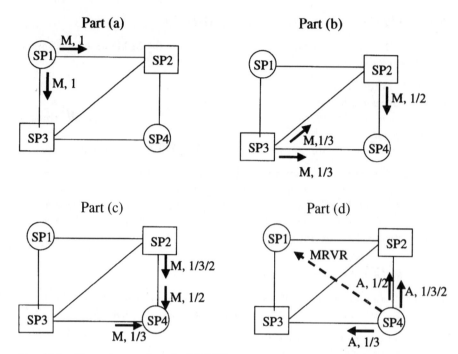

Fig. 10.8 Example procedures for MRVT

In Part (a) of the example, SP1 initiates the MRVT procedure by sending an MRVT Message on all routes to SP4, i.e. to SPs2 and 3.

Part (b): Upon receipt of the MRVT Message at SP2, an MRVT Message is sent to SP4 indicating the identities of the SPs that have dealt with the message (i.e. SPs1 and 2). At the same time, SP3 sends an MRVT Message to SPs2 and 4.

Part (c): SP4 receives the MRVT Messages from SPs2 and 3. In addition, SP2 sends a second MRVT (M, 1/3/2 from SP3) to SP4.

Part (d): Upon receipt of the second MRVT from SP2, SP4 now deduces that it has received MRVT Messages from all anticipated sources. SP4 therefore acknowledges receipt of the MRVT Messages to each adjacent SP. SP4 also formulates a report on the procedure and returns it in an MRVR Message to SP1 by any appropriate route.

10.4.9 SCCP Routing Verification Test (SRVT)

The messages and procedures for SRVT are very similar to the MRVT.

10.4.10 Circuit Validation Test (CVT)

In CCS systems, the signalling is transferred onto a separate path from the speech or data traffic. The correct performance of the signalling system is therefore not a guarantee that the traffic circuits are performing satisfactorily. The Circuit Validation Test (CVT) ensures that exchanges at each end of a connection have sufficient and consistent data to place a call on a specific circuit. The test also checks that the data at each exchange refers to the same circuit. The test is typically applied when a circuit is first brought into service or when changes to the transmission systems have occurred.

The test consists of checks made at both the near-end (i.e. the Initiating Exchange) and far-end (i.e. Responding Exchange). The test is started by management action requesting the Initiating Exchange to gain access to the circuit to be tested. The circuit is identified by a code that is mutually agreed by the two exchanges.

The near-end check ensures that data exists to (a) enable a transceiver to be connected to the circuit and (b) derive a Circuit Identification Code (CIC) and Routing Label for CCSS7. The transceiver can be a tone generator (for analogue lines) or a digital pattern generator (for digital lines).

If the near-end check succeeds, the Initiating Exchange sends a CVT Initiation (CVI) Message to the Responding Exchange. A transceiver is also attached to the appropriate circuit. The Initiating Exchange then waits for the tone or bit pattern to be received on the return path of the circuit.

At the Responding Exchange, checks are made on the status of the CIC and also that a loop (or transceiver) can be connected to the circuit. The Responding Exchange then applies the loop to the circuit, thus connecting the send and return paths (or transceiver). After a time-out period, the loop (or transceiver) is removed and the Responding Exchange sends a CVT Response Message to the Initiating Exchange, including the identity of the circuit from the responding perspective.

The Initiating Exchange compares the identity codes for the circuit. If the codes match, then the continuity of the circuit and the identification data have been validated.

10.5 Chapter Summary

CCSS7 has a comprehensive Network Management Function to enable optimum operation. Network Management includes Performance Management, Fault Management and Configuration Management.

Three types of management functions are provided by CCSS7. The Internal Type is specified as part of the signalling system itself. The TMN Type fits within the Telecommunications Management Network (TMN) Framework defined by ITU-T. The Application Type refers to the exchange of information using CCSS7 to verify network data. The Operations, Maintenance and Administration Part (OMAP) specifies the management functions of CCSS7 (except the Internal Type, which are specified within the signalling system).

190 *Telecommunications Signalling*

The TMN architecture is based on Network Management (NM) and Network Element Management (NEM). The interfaces between the levels are defined as Q.3 Interfaces.

A Managed Object Model is used to specify both NM and NEM functions. Managed Objects can be viewed from within a node (thus being managed by NEM) or between two or more nodes (thus being managed by NM). Examples are given for MTP Managed Object Classes for NM and NEM.

It is essential to take measurements to facilitate effective management. Examples of measurements are given for each of the Functional Elements.

The Application Type of Management Functions use an OMAP Model comprising a Management Process, an Operations and Maintenance ASE (OMASE) User and OMASE itself. The OMASE User provides the logic for management functions. The OMASE provides the communications functions.

The OMASE User provides the logic for the MTP Routing Verification Test (MRVT), the SCCP Routing Verification Test (SRVT) and the Circuit Validation Test (CVT). The primitives and services are given in the text.

The MRVT is used to determine that routing data in the MTP network is consistent. The procedures involve launching an MRVT Message on all routes and awaiting a report from the destination node or a node detecting failure. The SRVT is very similar. The CVT involves checking the data for a circuit and performing a continuity check.

10.6 References

1. ITU-T Recommendation Q.750: 'Overview of CCSS7 Management' (ITU, Geneva)
2. ITU-T Recommendation Q.751.1: 'Network Element Management Information for the MTP' (ITU, Geneva)
3. ITU-T Recommendation Q.752: 'Monitoring and Measurements for CCSS7' (ITU, Geneva)
4. ITU-T Recommendation Q.753: 'CCSS7 Management Functions' (ITU, Geneva)
5. ITU-T Recommendation Q.754: 'CCSS7 Application Service Elements' (ITU, Geneva)
6. ITU-T Recommendation X.711: 'Common Management Information Protocol Specification' (ITU, Geneva)

Chapter 11
DSS1 Physical and Data Link Layers

11.1 Introduction

In response to increasingly demanding customer requirements, network operators throughout the world are implementing integrated services digital networks (ISDNs). These provide user-to-user connections using digital transmission links and nodes. ISDNs need responsive and efficient signalling systems, both within the network and for users to gain access to the network. In the past, international standards organisations concentrated on the specification of internodal signalling systems. However, the benefits of ISDNs can only be realised by users if the access signalling systems are similarly responsive. Flexible information transfer is required between users and the network, and directly between users. The Digital Subscriber Signalling System No.1 (DSS1)[1] is defined to meet the demanding requirement of providing flexible access signalling for users.

DSS1 is responsible for transferring information between users and local nodes in both circuit-switched and packet-data applications. Chapter 3 describes the architecture of DSS1 (three layers). This Chapter describes the functions of the Physical and Data Link Layers of DSS1. Chapter 12 describes the functions of the Network Layer.

11.2 Physical Layer

Layer 1 describes the physical, electrical and functional characteristics of the interface between the user and the local node (network). Access from a user to a local node in an ISDN is by one of two connection types: basic access and primary rate (multi-line) access. The basic access is shown in Fig. 11.1.

The transmission link between the customer (user) and the local node provides an information bit rate of 144 kbit/s. The bit rate is structured to allow two traffic channels at 64 kbit/s (B channels) and a signalling channel at 16 kbit/s (D channel). The D channel is dedicated to controlling the two B channels. However, DSS1 is still a Common Channel Signalling System because signalling capacity is allocated dynamically according to need. The signalling channel can handle numerous terminals at the user's premises by dynamically allocating signalling capacity as required.

192   *Telecommunications Signalling*

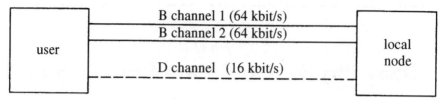

Fig. 11.1   Basic access

The primary (multi-line) rate access is shown in Fig. 11.2. This access technique uses a transmission link in which a 64 kbit/s signalling channel (D channel) controls up to 30 traffic channels (B channels).

The characteristics of Layer 1 functions allow the ability to connect a number of customer terminals to a line, each terminal having access to the B channels for traffic and the D channel for signalling. They also allow point-to-point and broadcast working. Point-to-point working is when the network node interacts with a specific terminal. Broadcast working is when the node broadcasts information to a range of terminals.

11.3 Data Link Layer Function

The Data Link Layer (Layer 2) is responsible for transferring messages between a user and a local node. The messages carry information (generated by Layer 3) in an information field. The Layer-2 functions do not understand the meaning of the information field, but it is the job of Layer 2 to deliver the information with a minimum of loss or corruption.

The functions of Layer 2 can be described in terms of the formats used and procedures adopted[2]. The formats and procedures are based on a High Level Data-Link Control (HDLC) Protocol[3] defined by the International Standards Organisation.

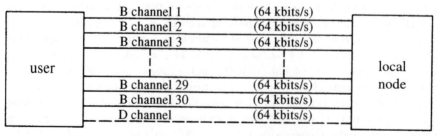

Fig. 11.2   Primary rate access

11.4 Data Link Layer Formats

The exchange of information by the Data Link Layer takes place in messages, termed 'frames'. These are similar to the Signal Units described for the CCSS7 MTP in Chapter 4. The frames are designated as being either 'commands' (requesting an action to be performed) or 'responses' (reporting the result of a command). The general format of a frame is given in Fig. 11.3.

11.4.1 Flag

Each frame starts and ends with a flag, which is a unique code that delimits the frame. The flag is coded 01111110. Imitation of the flag by any other field in the frame is avoided by banning any consecutive sequence of more than five 1's. This is achieved by a Data Link Layer function that inserts a zero after any sequence of five consecutive 1's, except in the flag, before transmitting a frame. Upon receipt of a frame, any zero that is detected following five consecutive 1's is discarded.

11.4.2 Address field

The address field consists of two octets. The command/response (C/R) bit of the address field indicates whether the frame is designated as a command or a response. If the frame is designated as a command, the address field identifies the receiver (i.e. the local node or user equipment receiving the frame). If the frame is designated as a response, the address field identifies the sender (i.e. the local node or user equipment sending the frame).

The extension bits illustrate a technique for expanding the length of a field in a flexible manner. The extension bit in the first octet is set to value 0 to

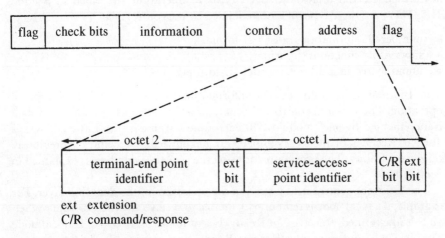

ext extension
C/R command/response

Fig. 11.3 Format of a frame

indicate that another octet follows. The extension bit in the second octet is set to Value 1 to indicate that the second octet is the last of the address field. If, in the future, it is necessary to expand the size of the address field, the extension bit in the second octet can be changed to Value 0 to indicate that a further octet follows. The third octet, in this case, would have an extension bit of Value 1 to indicate that it is the last octet. The increase in size is achieved without affecting the rest of the frame.

11.4.3 Identifiers

The Service Access Point Identifier (SAPI) and Terminal Endpoint Identifier (TEI) are used to identify the connection and terminal to which a frame pertains. The Service Access Point is used to describe the location at which the Data Link Layer provides services to Layer 3. Thus, the SAPI indicates which entity at Layer 3 needs to analyse the contents of the information field. For example, the SAPI indicates that the contents of the information field are relevant to the call control procedures for circuit-switched calls or to the call control procedures for packet data.

The TEI indicates the terminal equipment to which the message refers. The TEI code 1111111 ($=127$) indicates a broadcast call to all terminals associated with a Service Access Point. The remaining values ($0-126$) are used to identify terminals. The range is split between those terminals that use a TEI selected by the network (automatic TEI assignment) and those terminals that use a TEI chosen by the user (non-automatic TEI assignment).

11.4.4 Control field

The control field indicates the type of frame being transmitted, reflecting the acknowledged and unacknowledged types of operation, described in Section 11.5. There are three types of format defined by the control field:

- numbered information transfer (I-format);
- supervisory functions (S-format);
- unnumbered information transfer (U-format).

The I-format is used to transfer information in the acknowledged type of operation. The I-format control field contains a 'Send Sequence Number', which is incremented by one (modulo 128) each time a frame is transmitted. There is also a 'Receive Sequence Number', to acknowledge receipt of previously received I-frames. The procedure for using the sequence numbers is explained in Section 11.5.

The S-format control field is used in the operation of the data link layer. For example, if a local node is temporarily unable to receive any more I-frames, then an S-frame termed 'Receive Not Ready' is sent to the user. When the local node can receive I-frames again, a 'Receive-Ready' S-frame is used. The S-frame can also be used to acknowledge an I-frame. The S-frame contains a receive sequence number but not a send sequence number.

The U-format is used to transfer information in the unacknowledged type of operation and to transfer some administrative instructions. The U-format control field does not contain sequence numbers. A summary of the important commands and responses is given in Table 11.1.

11.4.5 *Information field*

The information field contains the information generated by one Layer 3 to be sent to another Layer 3. The information field can be omitted if the frame is not

Table 11.1 Examples of commands and responses

Format	Commands	Responses	Description
Information transfer (I)	Information	-	used in the acknowledged type of operation to transfer sequentially numbered frames containing information fields provided by Layer 3
Supervisory (S)	Receive ready	Receive ready	used to indicate that a data-link layer is ready to receive an I-frame or acknowledge previously received I-frames
	Reject	Reject	used to request retransmission of the I-frame
Unnumbered (U)	Unnumbered information (UI)		used in the unacknowledged type of operation to transfer information fields provided by Layer 3
	Set-asynchronous balanced-mode extended (SABME)		used to set up an acknowledged type of operation
	Disconnect		used to terminate an acknowledged type of operation
	Unnumbered acknowledgement		used to acknowledge and report acceptance of mode-setting commands, eg. SABME, disconnect

relevant to a specific call (e.g. in supervisory frames). If the frame is pertinent to the operation of the Data Link Layer, and Layer 3 is not involved in the formation of the frame, then the relevant information is included in the control field.

11.4.6 Check bits

The check-bits field, or more formally the 'Frame Check Sequence' Field, is generated by the Data Link Layer transmitting the frame. The bits are generated by applying a complex polynomial algorithm to the rest of the frame. The algorithm is the same as that described for the MTP in Chapter 4. The reverse process is applied by the Data Link Layer receiving the frame. If the results of the reverse process align with the check bits, then the frame is deemed to have been transmitted without errors. If the results do not align, then an error has occurred during the transmission of the frame.

11.5 Data Link Layer Procedures

11.5.1 Unacknowledged information transfer

The procedures for the unacknowledged transfer of information are illustrated in Fig. 11.4. Consider the case when the Layer-3 functions at the local node need to transfer information to the Layer-3 functions at the user's premises. The Layer 3 at the local node requests the transfer of information by passing a Unitdata Request Primitive to Layer 2. Layer 2 formulates an Un-numbered Information (UI) Command Frame, enclosing the information to be transferred within the information field. The frame is transmitted, via Layer 1, to Layer 2 at the user.

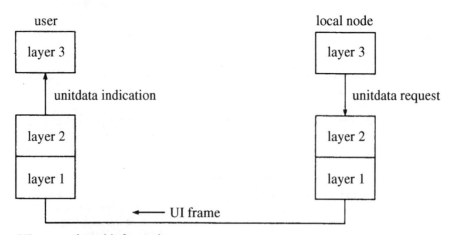

UI unnumbered information

Fig. 11.4 Unacknowledged information transfer

DSS1 Physical and Data Link Layers   197

If it is intended to broadcast the frame to all terminals, the TEI in the address field is coded 127. If a specific terminal is being addressed, i.e. point-to-point working is required, then the TEI is given the code (within the range of 0–126) appropriate to that terminal. Upon receipt of the UI Command Frame at Layer 2 of the user, the information contained within the information field is delivered to Layer 3 using a Unitdata Indication Primitive. In unacknowledged information transfer, there is no Data Link Layer error-recovery procedure. Hence, it is left to the Layer 3 functions to determine the logical recovery from the loss or corruption of a frame.

11.5.2 Terminal Endpoint Identifier procedure

Procedures are specified for the assignment, checking and removal of TEIs. All messages used for TEI management procedures are carried in the information field of UI command frames. The assignment procedure allows a user's automatic TEI equipment to request a local node to assign a TEI value that can be used in subsequent communications. Fig. 11.5 illustrates the principles involved.

The user Layer 2 formulates a command U-frame. The information field of the U-frame contains a message consisting of a message type field, a reference number and an action indicator. The message type field in this case is called 'Identity Request', reflecting the request for the local node to supply a TEI. The reference number is randomly generated for each request for a TEI. The number is used to discriminate between simultaneous requests for a TEI assignment. The action indicator is used to request the assignment of any appropriate TEI value.

The frame is transmitted to Layer 2 at the local node, where it is analysed and a suitable TEI value is selected. Layer 2 at the local node formulates a command U-frame. The information field of the U-frame contains a message

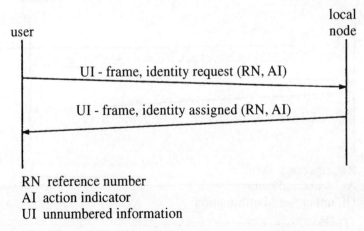

Fig. 11.5   Terminal Endpoint Identifier procedure

consisting of a message type (in this case called 'Identity Assigned'), the reference number provided by the user and the assigned TEI value in the action indicator field. The reference number is used to associate the response with the corresponding request. Several procedures are defined in case the assignment of a TEI value fails, but these are not described here.

11.5.3 Terminal Endpoint Identifier Check procedure

The check procedure allows the local node to check, for example, whether or not a TEI value is in use. The procedure is illustrated in Fig. 11.6.

Layer 2 at the local node formulates a command U-frame consisting of a message type (named 'Identity Check Request') and an action indicator including the TEI value to be checked. The frame is transmitted to the user Layer 2.

If any user equipment recognises the TEI value as its own, it responds by transmitting a U-frame consisting of a message type (named 'Identity Check Response') and an action indicator field including the TEI value. A reference number is also included in the response for other, more detailed, purposes. Again, procedures are specified to overcome failures in the logical sequence described. The check procedure can, optionally, be initiated by the user generating an 'Identity Verify' message type.

11.5.4 Terminal Endpoint Identifier Removal procedure

If the local node determines that a TEI value should be deleted, the 'Removal' procedure is invoked. The local node Layer 2 formulates a frame consisting of a message type (named 'Identity Remove') and an action indicator field including the TEI value to be removed. The frame is sent twice to reduce the risk of loss.

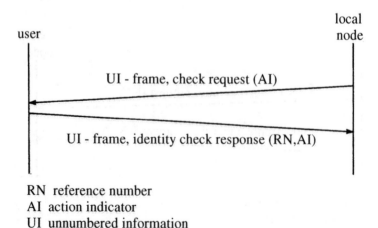

Fig. 11.6 Terminal Endpoint Identifier Check procedure

Actions at the user Layer 2 depend upon local circumstances. However, a typical action is to remove the TEI value and initiate an assignment procedure to request an alternative TEI value.

11.5.5 Acknowledged information transfer

The procedures for acknowledged information transfer are illustrated in Fig. 11.7.

(a) Connection set up

Consider the case of Layer 3 at the user needing to exchange information with Layer 3 at the local node. The procedures are initiated by the user Layer 3 sending an 'Establish Request' Primitive to user Layer 2. The user Layer 2

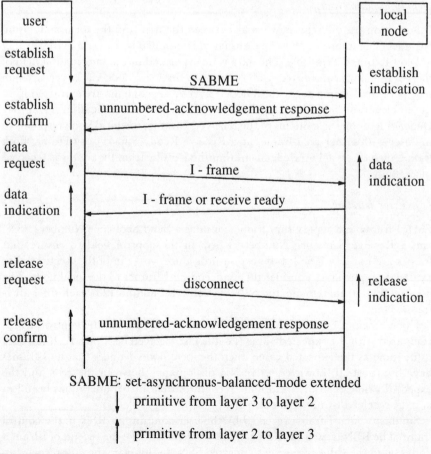

Fig. 11.7 Acknowledged information transfer

formulates a 'Set Asynchronous Balanced Mode Extended (SABME)' command frame, indicating the need to establish a Layer 2 acknowledged information transfer with the local node. The SABME command frame is transmitted to the local node via Layer 1.

Upon receipt of the SABME command at the Layer 2 of the local node, the frame is analysed and the conditions required to establish an acknowledged information transfer are checked (e.g. to ensure that all appropriate equipment is available). Provided that the appropriate conditions are met, the local node Layer 2 sends an 'Establish Indication' Primitive to Layer 3 to indicate that an acknowledged information transfer is being established. The local node also returns an Unnumbered Acknowledgement Response to the user Layer 2. Upon receipt of the Unnumbered Acknowledgement Response at the user Layer 2, an 'Establish Confirm' Primitive is passed to the user Layer 3 to indicate that acknowledged information transfer can begin.

(b) Information transfer

Information transfer can now occur between the user and the local node using I-frames. The information to be transferred is supplied by Layer 3 in the form of a 'Data Request' Primitive. The data is incorporated in the information field of the I-frame and transmitted from the user to the local node via Layer 1. Upon receipt of the I-frame at the local node Layer 2, the contents are extracted from the information field and passed to Layer 3 in a 'Data Indication' Primitive. Depending upon the contents of the received I-frame, the local node responds to the user with either an I-frame, or a Receive Ready Supervisory Frame. Both frames include an acknowledgement that the I-frame from the user was received successfully.

(c) Error detection

Each I-frame and supervisory frame contains a Send Sequence Number (SSN) and a Receive Sequence Number (RSN) in the control field to ensure that frames are not lost. The loss-detection procedure works in both directions. For explanation purposes, consider the user sending I-frames to the local node. The user equipment increments the SSN by Value 1 (modulo 128) each time an I-frame is sent.

Upon receipt of an I-frame at the local node, the SSN in the control field is compared with the expected value (i.e. the last received SSN plus 1). If the SSN is the same as the expected value, then the local node deduces that no I-frames have been lost. If, however, there is a discrepancy between the SSN and the expected value, then the local node recognises that an I-frame(s) has been lost and the received I-frame is discarded.

Similarly, each I-frame received by the user contains an RSN in the control frame. The RSN is used by the local node to acknowledge the receipt of I-frames and is set to the value of the next-expected SSN. In this way, the user receives an acknowledgement that the I-frames which have been sent have been successfully received by the local node. If the local node discards a frame (e.g. because of lack of correlation of SSN and expected value), the RSN is not incremented and hence the frame is not acknowledged. After a short time, the user equipment recognises that a particular frame has not been acknowledged and the frame is re-transmitted.

(d) Connection release

When the exchange of I-frames is completed, the release procedures apply that are shown in Fig. 11.7. In the Figure, the user Layer 3 sends a 'Release Request' Primitive to Layer 2. The user Layer 2 formulates a Disconnect Frame which is transmitted, via Layer 1, to the local-node Layer 2. Upon receipt of the Disconnect Frame at the local node Layer 2, a 'Release Indication' Primitive is passed to Layer 3 and an Unnumbered Acknowledgement Frame is returned to the user. Upon receipt of the Unnumbered Acknowledgement Frame by the user Layer 2, a 'Release Confirm' Primitive is passed to Layer 3 to complete the release procedure.

11.6 Chapter Summary

It is essential to provide a modern access signalling capability to take full advantage of the provision of ISDNs. DSS1 has been defined to provide such a capability. DSS1 is defined in terms of a 3-layer model.

Layer 1 of DSS1 defines the physical, electrical and functional characteristics of the transmission link between the user and the local node. Two forms of access are defined. The basic access is formed of two B channels (each of 64 kbit/s) and a D channel of 16 kbit/s. The D channel is the signalling channel. The primary access is formed of 30 B channels and a D channel of 64 kbit/s.

Layer 2 of DSS1 is responsible for transferring information between the user and the local node. Information is transferred in frames. Frames consist of a number of fields. Flags delimit the frame. The address field identifies the receiver of a command frame or the sender of a response frame. The control field indicates the type of frame being transmitted (I-format, S-format or U-format). The information field contains the information supplied by Layer 3 to be transferred. The check bits are used to detect errors during transmission of the frame.

Two forms of transfer are defined. In unacknowledged information transfer, information is exchanged in unnumbered information frames without sequencing. In this case, there is no error-correction method defined at Layer 2. In acknowledged information transfer, a Layer 2 connection is established by using a Set Asynchronous Balanced Mode Extended (SABME) Frame and information is exchanged using sequenced information frames. The connection is cleared by using a disconnect frame.

11.7 References

1 ITU-T Recommendations Q.920, 921, 930 and 931: 'Digital Subscriber Signalling System No. 1' (ITU, Geneva)
2 ITU-T Recommendation Q.921: 'Data-Link Layer Specification' (ITU, Geneva)
3 ISO Standard 4335: 'High Level Data-Link Control Procedures'

Chapter 12
DSS1 Network Layer

12.1 Introduction

The DSS1 Network Layer (Layer 3) is responsible for establishing, maintaining and clearing connections between users and the narrowband ISDN (e.g. between users and local nodes). The procedures in DSS1 control (*a*) circuit-switched connections, (*b*) user-to-user signalling connections and (*c*) packet-switched connections. The specifications of DSS1 Layer 3 adopt the term 'user' to describe a calling/called party[1]. This Chapter follows this terminology.

It is Layer 3 of DSS1 that formulates and analyses the messages carried by Layers 1 and 2. In this role, the DSS1 Network Layer interacts with the user and translates the user requirements into messages to be sent to an ISDN local node. At the local node, the DSS1 Network Layer interacts with the call control functions to stimulate actions meeting the user requirements. As outlined in Chapter 3, the DSS1 Layer 3 does not correspond to Layer 3 of the OSI Protocol Reference Model.

DSS1 Layer 3 can be described in terms of its formats and the procedures (defining the logical sequence of events) in meeting user requirements. Section 12.2 describes the principles of the format technique and Section 12.3 gives examples of messages to illustrate the principles. The basic procedures for establishing circuit-switched calls are described in Section 12.4 and the procedures for clearing such calls are described in Section 12.5.

DSS1 Layer 3 includes features additional to those necessary for establishing and clearing calls and these features are outlined in Section 12.6. DSS1 Layer 3 is required to provide access to packet-data facilities and the means of achieving this requirement are described in Section 12.7. Section 12.8 covers the invocation and transfer of user-to-user signalling over the access signalling channel. Three generic procedures for the control of supplementary services are outlined in Section 12.10.

12.2 Format Principles

The general format for DSS1 Layer 3 messages is shown in Fig. 12.1. Each message contains a protocol discriminator, a call reference and a message type. Other information elements are defined according to the type of message.

other information elements	message type	call reference value	length of call reference	protocol discriminator

first bit transmitted →

Fig. 12.1   Format of Layer 3 messages

12.2.1 Protocol Discriminator

The first part of every message is the Protocol Discriminator, consisting of one octet. The purpose of this field is to distinguish between DSS1 messages for call control and any other type of message that can be transported over the signalling channel. For example, it is possible to transmit other forms of data over an access signalling channel and it is essential that a local node can distinguish between DSS1 messages and such data. The Protocol Discriminator is coded as 00010000.

12.2.2 Call reference

The call reference is a number that is used to identify the call to which a message refers. The call reference is assigned during set-up from a pool of numbers and remains fixed for the duration of the call. At the end of the call, the call reference is returned to the pool of numbers in readiness to be used again. The call reference is valid for the particular user-network interface and does not have significance within the network.

The format of the Call Reference Information Element is illustrated in Fig. 12.2. The first four bits of the first octet indicate the length of the call reference value. The remaining bits of the first octet are spare. For basic access, the length of the call reference value is at least one octet and for primary-rate access, the length is at least two octets.

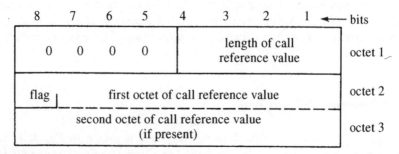

Fig. 12.2   Call Reference Information Element

The call reference is assigned either by the user equipment or by the local node. If the user originates a call, then the user assigns the call reference from its pool of numbers. Similarly, if the local nodes originates a call (in response to a request from the network), then the local node selects the call reference. With this approach, it is possible that both the user and the local node select the same call reference value for different calls. To ensure that it is possible to discriminate between the two calls in this case, a 'flag' is included as the last bit of the first octet of the call reference value. The flag indicates whether the user or local node originated a particular call reference value. Note that the flag in this context is not the same as the flag used to delimit frames.

12.2.3 Message type

The message type is used to identify the function of the message being sent. The message type field consists of one octet, with the last bit being reserved for use in the future to extend the length of the field. There are five categories of message type:

(a) Call establishment messages, which deal with the set-up of calls. For example, the Set-up Message is sent by a calling user to the local node to initiate a call.
(b) Call information phase messages, which are used when a call has been established. For example, a User Information Message can be sent during the conversation/data phase of a call to transfer user-to-user information.
(c) Call clearing messages, which deal with the clear-down of calls. For example, the Disconnect Message is sent by the user to the local node to clear a connection.
(d) Miscellaneous messages, e.g. an Information Message can be sent by the user or the local node to supply information additional to that provided in other basic messages.
(e) Nationally specified messages, which are not standardised internationally. The Message Type code 00000000 is used to denote that the next field is a message type that is defined by a network operator.

Examples of key messages are given in Table 12.1 (excluding acknowledgements).

12.2.4 Other information elements

Two categories of information element are defined, depending on whether they are one octet in length (Single Octet Information Elements) or more than one octet in length (Variable Length Information Elements). There are two types of Single Octet Information Element. Type 1 is illustrated in Fig. 12.3.

Bit 8 (Value 1) indicates that the element belongs to the single-octet category and Bits 5-7 are used to define the name of the information element in the form of an Information Element Identifier. Bits 1-4 provide the contents of the information element.

Table 12.1   Message types

Message	Category	Description
Set-up	Establishment	initiates call establishment
Alerting	Establishment	call being offered to user
Call Proceeding	Establishment	requested call establishment has been initiated
Connect	Establishment	call acceptance by called user
Progress	Establishment	information on interworking or in-band activity
Suspend	Information	removes call from active use
Resume	Information	reverses Suspend Message
User information	Information	allows additional information to be exchanged
Disconnect	Clearing	initiates clearing
Release	Clearing	preparing to release call reference
Release complete	Clearing	connection and call reference released
Notify	Miscellaneous	gives information pertaining to call, e.g. user suspended

Type 2 is illustrated in Fig. 12.4. Again, Bit 8 (Value 1) is used to show that the information element belongs to the single-octet category. However, the remainder of the octet is used entirely as the Information Element Identifier.

The format of the Variable Length Information Element is shown in Fig. 12.5. In this case, Bit 8 of the first octet is set to zero, distinguishing this category of element from the single-octet category. The remainder of the first octet is dedicated to the Information Element Identifier. The second octet defines the length of the contents of the information element and the third and subsequent octets give the contents.

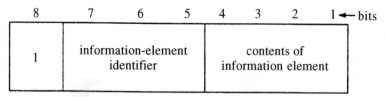

Fig. 12.3   Type 1 information element

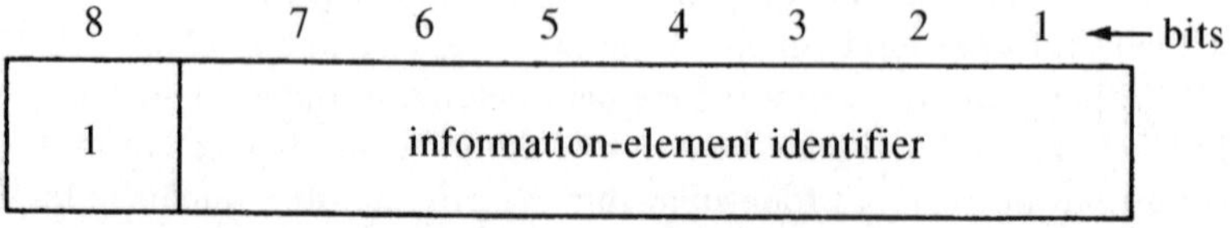

Fig. 12.4   Type 2 information element

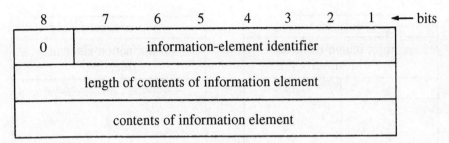

Fig. 12.5 *Variable Length Information Element*

Single Octet Information Elements can be placed at any point within the Information Element Field of a message. However, Variable Length Information Elements are placed in ascending order of the Information Element Identifier. This allows equipment at the user's premises, or at the local node, to detect the presence or absence of a particular piece of information without having to analyse the entire message.

12.2.5 *Codesets*

The format of the Variable Length Information Element allows 7 bits for the Information Element Identifier. Thus, up to 128 different information elements can be identified in this category. The number of bits allowed in Single Octet Information Elements depend upon the type adopted, with 3 bits being available in Type 1 and 7 bits in Type 2. Thus, at least 8 different information elements can be identified in this category. Combining the two categories, at least 136 Information Element Identifiers are available, although in practice this is reduced to 133 after allowing for reserved values. This group of 133 Information Element Identifiers is termed a 'codeset'.

The number of information elements that can be identified within a message can be increased substantially by re-using the Information Element Identifier Codes in different codesets. Thus, a given Information Element Identifier Code can define one information element in one codeset and another information element in another codeset. The principle is illustrated in Fig. 12.6.

Consider the Information Element Identifier in Codeset 0, termed Calling Party Number and coded 1101100. The same code, 1101100, can be used in another codeset (Codeset 5 in Fig. 12.6) to define a different information element. Codeset 5 is reserved for network operators to use for non-standard codings of information elements: in this example, the code 1101100 is used to define the Time Information Element, giving the time at which a message is sent. The same code can be used again in other codesets to define other Information Element Identifiers.

For this re-use approach to work, it is essential that any equipment sending or receiving a message is fully aware of the codeset being employed. This applies

codeset 0	
information-element identifier	
code	name
⋮	⋮
1101100	calling-party number
⋮	⋮

codeset 5	
information-element identifier	
code	name
⋮	⋮
1101100	time
⋮	⋮

Fig. 12.6   Principle of Codesets

to equipment at the user's premises and to equipment at the local node. The method adopted by DSS1 is to define Codeset 0 as the initial codeset used for analysing a message. If any equipment needs to use an information element within a different codeset (e.g. Codeset 5) then the frame of reference is shifted from Codeset 0 to Codeset 5. This is achieved by transmitting a Type 1 Single Octet Information Element termed 'Shift', which is illustrated in Fig. 12.7. Bit 8 of the Shift-Information Element is coded 1 to indicate a Single Octet Information Element. Bits 5 to 7 are the Information Element Identifier and are coded 001 to define the request to shift from the current codeset to a new codeset. The new codeset is defined in Bits 1-3.

Bit 4 is used to indicate whether the shift in codeset should be applied for the remainder of the message (Locking Shift Procedure) or temporarily (Non-Locking Shift Procedure).

In the Locking Shift Procedure, the new codeset in the Shift Information Element is applied for the remainder of the message (or until a further shift is requested). For example, Codeset 0 is used as the standard codeset in the analysis of a message. If a Locking Shift Information Element is received requesting a shift to Codeset 5, the remaining information elements of the

8	7	6	5	4	3	2	1	← bits
1	shift identifier			lock/ temporary bit	new codeset identification			

Fig. 12.7   Shift Information Element

message are interpreted according to the Codeset 5 (unless further shift information is received).

In the Non-Locking Shift Procedure, the new codeset in the Shift Information Element is applied only to the next information element. Thus, if a Non-Locking Shift Information Element is received requesting a shift to Codeset 5, only the next information element is interpreted according to Codeset 5. After this information element is interpreted, the codeset used to interpret subsequent information elements reverts to Codeset 0.

The Shift Information Element illustrated in Fig. 12.7 allocates 3 bits for the new codeset identification. Thus, up to 8 codesets can be accommodated. ITU-T has reserved the following values of codeset in addition to Codeset 0:

- Codeset 4: for use by other standards organisations, e.g. ISO;
- Codeset 5: for national use, thus allowing network operators to use codings that do not form part of the international specification;
- Codeset 6: for local networks (public and private);
- Codeset 7: for user-specific information, i.e. information defined by users.

12.3 Examples of Message Formats

12.3.1 General

The principles of the format technique for DSS1 are illustrated below by selecting some commonly used message types and information elements. The messages can be relevant to different portions of a call:

- 'Local significance' describes messages relevant in either (a) the originating access (i.e. the link between the originating user and local node) or (b) the terminating access (i.e. the link between the terminating user and local node).
- 'Access significance' means that messages are relevant in both the originating and terminating accesses (but not within the network).
- 'Dual significance' describes messages that are relevant in either (a) the originating access and the network or (b) the terminating access and the network.
- 'Global significance' describes messages that are relevant to all portions of a call, i.e. originating access, terminating access and within the network.

The information elements in the examples are all from Codeset 0. Each information element is either mandatory (i.e. it must be included in a particular message) or optional (i.e. it can be included or not, as appropriate). The length in octets of each information element is also given.

The message type examples chosen are appropriate for the control of circuit-switched connections (C), user-to-user signalling connections (U) and packet-switched connections (P). The examples indicate the applicability of information elements to the three cases. Other messages in the specification are applicable to only one type of call control.

12.3.2 Set-up Message

An example of the format of the Set-up Message is given in Table 12.2.

This message is sent from the calling user to the originating local node and from the terminating local node to the called user. The message initiates call establishment and is equivalent to the Initial Address Message in the ISUP of CCSS7 (Chapter 6). The Set-up Message has global significance and includes compatibility information elements that are used to ensure that the calling and called user terminals can communicate effectively.

Table 12.2   Set-up Message

Information Element	Type	Length (octets)	Control	Description
Protocol Discriminator	M	1	all	see Section 12.2.1
Call Reference	M	2+	all	see Section 12.2.2
Message Type	M	1	all	see Section 12.2.3
Bearer Capability	M	4–12	all	requests a specific bearer service to be provided by the network
Channel Identification	O	2+	all	identifies a channel within the interface controlled by DSS1
Display	O	2+	all	supplies information that can be displayed by the user
Calling Party Number	O	2+	all	gives address of the origin of the call
Called Party Number	O	2+	all	gives address of the destination of the call
High Layer Compatibility	O	2–5	C, U	utilised by the remote user for compatibility checking
Packet Size	O	4	P	indicates the requested packet size values to be used for the call

12.3.3 Connect Message

The Connect Message denotes that a call is accepted by the called user. It is sent by the called user to the network and by the network to the calling user. The message has global significance for circuit-switched and packet-switched connections and local significance for user-to-user signalling. The message is the equivalent of the answer message in the ISUP of CCSS7. An example of the format of the message is given in Table 12.3.

12.3.4 Disconnect Message

The Disconnect Message is sent by a user (either the calling or called user) to request the network to clear the call. It is also sent by the network to the other user to indicate that the call is cleared. The message has global significance for circuit-switched call control and local significance for packet-switched call control. The message is not applicable to user-to-user signalling. An example of the format of the message is given in Table 12.4.

12.4 Set-up Procedures for Circuit-Switched Calls

12.4.1 General

The procedures for the establishment of basic circuit-switched calls assume that a Layer-2 data-link connection (Chapter 11) is in operation between the calling user and the originating local node before a call is initiated. Layer 3 Messages are sent to the Data Link Layer using DL-Data Request Primitives.

The procedures vary according to whether en bloc or overlap operation is adopted by the calling user. They also vary according to whether the called user

Table 12.3   Connect Message

Information Element	Type	Length (octets)	Control	Description
Protocol Discriminator	M	1	All	see Section 12.2.1
Call Reference	M	2+	All	see Section 12.2.2
Message Type	M	1	All	see Section 12.2.3
Bearer Capability	M	4–12	C	indicates bearer service provided by the network
Channel Identification	O	2+	All	see Table 12.2
Signal	O	3	C	allows the network to convey tones/alerting information

Table 12.4 Disconnect Message

Information Element	Type	Length (octet)	Control	Description
Protocol Discriminator	M	1	All	see Section 12.2.1
Call Reference	M	2+	All	see Section 12.2.2
Message Type	M	1	All	see Section 12.2.3
Cause	M	7+	C, P	gives reason for the disconnect
Signal	O	3	C	see Table 12.3

has multiple terminals associated with a signalling connection or a single terminal. If the user has multiple terminals, the broadcast form of working of the Data Link Layer is used to initiate a call. If the destination local node is aware that a single terminal exists, and the terminal endpoint identifier is known, the point-to-point form of working is used.

Messages can contain two types of information element; Functional and/or Stimulus. Functional Information Elements require a degree of intelligent processing by the terminal in either their generation or analysis. Stimulus Information Elements, on the other hand, are either generated as a result of a single event or contain a basic instruction from the network to be executed by the terminal.

12.4.2 En bloc procedures with point-to-point working

In en bloc operation, a call is initiated by the calling user formulating a Set-up Message and sending it to the originating local node over the signalling channel (D channel). The procedures are illustrated in Fig. 12.8.

The Set-up Message includes all the information needed to route the call to the destination node. A Call Reference, chosen from a pool of numbers, is used to correlate all messages between the calling user and the originating local node for that call. The Set-up Message also includes a Bearer Capability Information Element (defining the type of connection required to fulfil the needs of the calling user) and information to check that the calling and called user terminals are compatible.

The user indicates, in the Channel Identification Information Element, either the identity of the traffic channel upon which the call should be made or an indication that any traffic channel will suffice. If a specific traffic channel is nominated it can be qualified as being essential or desirable.

Upon receipt of the Set-up Message from the calling user, the originating local node returns a Call Proceeding Message to acknowledge receipt of the

DSS1 Network Layer 213

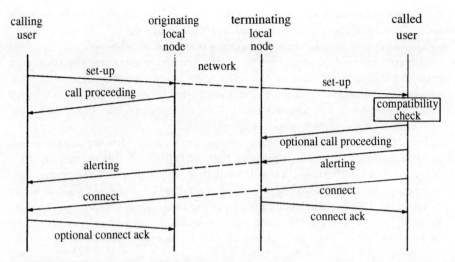

Fig. 12.8 Basic call establishment, circuit-switched, en bloc

Set-up Message and indicate that the call is being processed. The Call Proceeding Message confirms the identity of the channel to be used for the call. The originating local node also transmits the request to establish a call through the network to the terminating node by means of an ISUP Initial Address Message.

If the call leaves the ISDN environment (e.g. interworking with a non-ISDN network or a call to a non-ISDN user), this information is returned to the calling user within a general call control message or within a Progress Message. This informs the calling user about restrictions in features that are available.

Upon receipt of a request to set up a call, the terminating local node sends a Set-up Message to the called user over the D channel. In this example, the point-to-point data-link layer form of working is used. The Set-up Message includes a Call Reference generated by the terminating local node and a proposed channel that should be used for the call. The Set-up Message also includes the compatibility check information supplied by the calling user. Before any further action is taken, the compatibility information is checked by the called user to ensure that effective communication can take place between the calling and called users. If compatibility is not achieved, the call is rejected by the called user returning a Release Complete Message with an information element giving the reason for rejecting the call (e.g. incompatible terminals). If compatibility is achieved, the call set-up process continues.

At this stage, the called-user equipment can return a Call-Proceeding Message to the terminating local node, but this message is optional. The message informs the terminating local node that the call is being processed by the user equipment. However, the message results in no specific action at the local node other than to reset timers that are used by the local node to check for faulty conditions (e.g. non-response to a Set-up Message).

The next step in the set-up sequence depends on the type of terminal connected at the called user's premises. Some terminals automatically answer an incoming call without manual intervention (e.g. some data terminals) and these are termed 'automatic answering terminals'. Other terminals need manual intervention to answer a call (e.g. the lifting of the handset on a telephone) and these are termed 'non-automatic answering terminals'.

For a non-automatic answering terminal, an Alerting Message is returned to the terminating local node, indicating that the called user is being notified that there is an incoming call. This is equivalent to ringing the telephone in a telephony call, thus notifying the called user that an incoming call is being offered. The terminating local node passes the alerting information through the network to the originating node. The originating node sends an Alerting Message to the calling user to keep the calling user informed of call progress. When the call is answered by the called user (e.g. the telephone handset is lifted) a Connect Message is sent from the called user terminal to the terminating local node. This message is passed through the network to the originating local node and subsequently to the calling user. The terminating local node sends a Connect Acknowledge message to the called user to confirm that the call is now established. Optionally, the calling user terminal can send a Connect Acknowledge message to the originating node.

For an automatic answering terminal, the speed of response to an incoming call is generally much faster than for a non-automatic version and there is unlikely to be an extended alerting period. Thus, for an automatic answering terminal, the alerting message can be omitted from the set-up sequence. This means that the response to a Set-up Message from an automatic answering terminal is a Connect Message.

The Set-up Message to the called user includes the identity of the channel which it is proposed to use for the call. If it is not possible to select that channel, or the user wishes to utilise another channel, the user terminal can negotiate the use of another channel.

12.4.3 En bloc procedures in conjunction with broadcast working

The procedures adopted with broadcast working are similar to those described for point-to-point working. The differences, described below, are to cater for responses from multiple terminals.

In this case, the Set-up Message from the terminating local node to the called user is sent in broadcast form, i.e. it is sent to all terminals associated with the signalling channel (D channel). Each terminal checks the compatibility information supplied in the Set-up Message to determine whether or not it is compatible with the calling-user terminal. If a terminal determines that it is not compatible, then it can either ignore the Set-up Message and take no further action or it can return a Release Complete Message with a Cause Information Element indicating that it is not compatible with the calling-user terminal. If a terminal determines that it is compatible with the calling-user terminal, then it responds to the terminating local node with a Call Proceeding Message, an Alerting Message and/or a Connect Message, according to the circumstances of the call. The terminating local node keeps a record of each terminal that responds.

The first called-user terminal to respond to the terminating local node with a Connect Message is deemed to be the recipient of the call. A Connect Acknowledge Message is sent from the terminating local node to the recipient terminal to confirm that it has been selected to take the call. The recipient terminal can then connect to the appropriate channel. The terminating local node also sends a Release Message to every terminal that responded to the Set-up Message (except the recipient terminal) to indicate that the offer of the call has been withdrawn.

In the broadcast form of working, negotiation of the channel to be used between the terminating local node and the called user is not possible, unless specialised procedures are adopted. Thus, the call is made using the traffic channel selected by the terminating local node.

12.4.4 Overlap procedures

Overlap operation can be adopted in the originating access, the terminating access, within the network and in combinations of these three elements. To illustrate the principles involved, consider the case when the originating access and the network use overlap operation and the terminating access uses en bloc operation. The procedures in this case are illustrated in Fig. 12.9. The contents of the messages and the differences in procedures between point-to-point working and broadcast working are similar to those described above.

Overlap operation is initiated by the calling user sending a Set-up Message to the originating local node that does not include all the information to make a connection with the called user. The originating local node returns a Set-up Acknowledge Message to the calling user, indicating that further address information is required. Further address digits are provided to the originating local node in one or more Information Messages. When sufficient address digits have been received to route the call to the terminating node, the request to set-up a call is transmitted through the network in an ISUP Initial Address Message (IAM). In the meantime, further address digits can be received by the originating local node from the calling user in Information Messages: these are passed through the network to the terminating local node.

When the terminating local node has received sufficient address information to identify the called user, a Set-up Message is sent to initiate the call-establishment procedures on the terminating access link. The procedures on the terminating access link are the same as those described for en bloc operation.

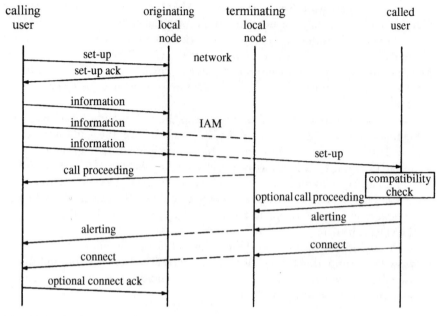

Fig. 12.9 Basic call establishment, circuit-switched, overlap

The terminating local node also informs the originating local node that sufficient address information has been received. In response to this information, the originating local node sends a Call Proceeding Message to the calling user. The Call Proceeding Message indicates that no further address information will be accepted. Procedures beyond this stage of the call are the same as for en bloc operation.

12.5 Call-Clearing Procedures for Circuit-Switched Calls

In these procedures, the term 'disconnected' means that the appropriate channel has been cleared down, but it is not yet ready to be used again for another call. The term 'released' means that the traffic channel has been cleared down and is available to be used again for another call. Similarly, when the term 'released' is applied to a Call Reference, the Call Reference is available to be used again for another call. The procedures for clearing a call are illustrated in Fig. 12.10.

The clearing procedures can be initiated by either the calling user or the called user: this example assumes that the calling user requests the call to be cleared. In this case, the calling user transmits a Disconnect Message over the signalling channel (D channel) to the originating local node. The calling user also disconnects the traffic channel (B channel). Upon receipt of the Disconnect

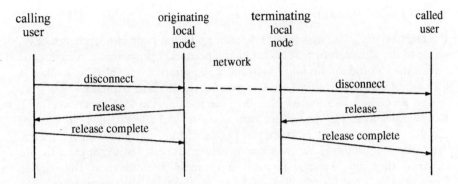

Fig. 12.10 Call clearing, circuit-switched

Message, the originating local node disconnects the traffic channel and, when this is completed, returns a Release Message to the calling user. The originating node also passes the request to clear the call to the terminating node through the network.

When the calling user receives the Release Message, the traffic channel is released, a Release Complete Message is sent to the originating node and then the Call Reference is released. Upon receipt of the Release Complete Message, the originating node releases the traffic channel and the Call Reference. Note that, whereas the traffic channel is first disconnected and then released, the Call Reference is held active until the last message in the clear sequence is sent. This is essential because the Call Reference is needed to correlate the messages with the call: the final act is to release the Call Reference.

When the terminating local node receives the request to clear the call from the originating node, the terminating local node disconnects the traffic channel to the called user and sends a Disconnect Message over the signalling channel. This message prompts the called user to disconnect the traffic channel and return a Release Message to the terminating local node. Upon receipt of the Release Message, the terminating local node releases the traffic channel, sends a Release Complete Message to the called user and releases the Call Reference. Receipt of the Release Complete Message causes the called user to release the traffic channel and Call Reference.

12.6 Other Features for Basic Circuit-Switched Calls

12.6.1 General

The specification of the basic procedures for DSS1 includes a range of features in addition to call establishment and call clearing. A summary of some of the important features is given below.

12.6.2 Restart procedure

If a fault occurs on an access link, the user and local node can lose track of the status of the channels in that access link. The restart procedure can be used to return the channels to an idle condition. The procedure can also be invoked if, for example, a user terminal fails to respond to the clearing sequence.

The procedure is initiated by either the user or the local node sending a Restart Message. The procedure can relate to a single channel, to a single interface or to all interfaces. A Restart Indicator Information Element is used to determine the channel or interface involved. If the restart applies to one channel, then the message includes a Channel Identification Information Element. Upon receipt of a Restart Message, the recipient clears the relevant channel(s) and Call Reference(s) and returns a Restart Acknowledge Message. Receipt of the Restart Acknowledge Message causes the initiator of the procedure to release the appropriate channel(s) and Call Reference(s).

12.6.3 Call re-arrangement procedure

This procedure allows a user to suspend a call, make changes to the terminal being used and then resume the call. The changes can include physically replacing one terminal with another, physically moving from one terminal to another and disconnecting/reconnecting a terminal.

The procedure is initiated by the user sending a Suspend Message to the local node. The message includes a Call Identity that replaces the Call Reference, thus allowing the local node to release the Call Reference. The local node confirms suspension by returning a Suspend Acknowledge Message. When the user wishes to resume the call, a Resume Message is sent to the local node, including a new Call Reference and the Call Identity. The local node correlates the Call Identity with that included in the Suspend Message and returns a Resume Acknowledge Message to allow the call to continue.

12.6.4 Error conditions

A range of procedures for handling error conditions is specified. This is essential to ensure that the user and local node have procedures to cover any error, or corrupted message or information element, that might occur. The procedures for handling error conditions are left to the specification.

12.6.5 User notification

This procedure allows the network to notify a user, or one user to notify another user, of any appropriate call-related event. The procedure between users works by sending a Notify Message to the network, which sends a Notify Message containing the same information to the other user involved in the call.

12.7 Procedures for Packet-Data Calls

12.7.1 General

Packet data is a form of information transfer in which data is divided into blocks called packets. Each packet is routed independently through the network, the blocks being recombined when they have been received at the intended destination. Although details of packet-data calls are beyond the scope of this book, it is worth giving a brief outline of the role of DSS1 in supporting packet communications in an ISDN. Very similar procedures and formats apply to Frame Mode (Frame Relay or Frame Switching) connections, which can be considered as a simplified version of a packet-data service.

There are two ways of providing packet communications from a DSS1 perspective. The first method is to provide a circuit-switched access to a packet-switched public data network (PSPDN)[2]. A PSPDN is a network dedicated to the provision of packet communications. PSPDNs use protocols optimised for packet data. The second method is to provide packet-switched access, using equipment that gives an inherent packet-data capability (a Packet Handler) within an ISDN. The user can decide whether to use a traffic channel (B channel) or the signalling channel (D channel) to gain access to the Packet Handler.

In each case, a packet-switched virtual call is set-up using procedures for the packet network.

12.7.2 Circuit-switched access

In circuit-switched access, the user gains access to a PSPDN by a circuit-switched channel. In this case, a circuit-switched traffic channel is established between the user and a PSPDN. The point at which access is gained to the PSPDN is termed the Access Unit. The configuration is illustrated in Fig. 12.11.

The procedures adopted on the access link and within the ISDN are the same as for a normal circuit-switched call. Thus, as far as the access signalling channel (D channel) is concerned, the procedures explained in Section 12.4 are applied to establish a traffic channel between the user and the Access Unit. The

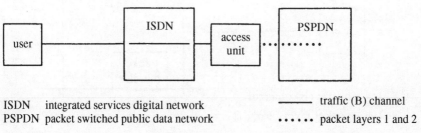

ISDN   integrated services digital network   ——— traffic (B) channel
PSPDN   packet switched public data network   •••••• packet layers 1 and 2

Fig. 12.11 Circuit-switched access to packet network

Set-up Message includes the called number of the appropriate Access Unit. The Bearer Capability Information Element includes an indication that the call is being established in circuit-switched mode.

Once the traffic channel has been established, the protocols adopted for the PSPDN are used over the traffic channel between the user and the PSPDN. The procedures described in Section 12.5 are used to clear the traffic channel upon completion of the call.

12.7.3 *Packet-switched access – B channel*

In packet-switched access, the user gains access to a Packet Handler, for example in a local node. If the traffic channel (B channel) is used to gain access, as shown in Fig. 12.12, then the signalling procedures adopted on the D channel are those necessary to establish a B channel between the user and the packet handler.

The procedures are the same as those described in Section 12.4, except that:

(*a*) en bloc operation is used and the procedures for overlap operation do not apply;
(*b*) an Alerting Message is not sent from the local node to the calling user to indicate that the called user is being alerted;
(*c*) the Bearer Capability Information Element in the Set-up Message includes appropriate values to indicate packet-data mode.

Once established, the traffic channel is used to convey packet protocols between the user and the packet handler. The procedures described in Section 12.5 are used to clear the B channel upon completion of the call.

12.7.4 *Packet-switched access – D channel*

If access is gained using the signalling channel (D channel), as shown in Fig. 12.13, the D channel provides a Data Link Layer connection between the user and the Packet Handler in the node. The Data Link Layer provides an Acknowledged Information Transfer Service (I-frames), as explained in

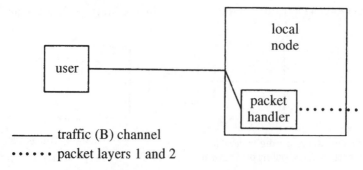

Fig. 12.12 *Packet access using B channel*

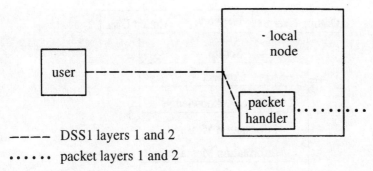

–––– DSS1 layers 1 and 2
······ packet layers 1 and 2

Fig. 12.13 Packet access using D channel

Chapter 11. Thus, packet protocols can be conveyed directly between the user and the Packet Handler over the D channel. This approach is based on the D channel providing the services of Layers 1 and 2 of the OSI model (see Chapter 3) and the packet protocol providing the Layer 3 service.

Similarly, the Layer 3 clearing procedures are those adopted for the packet protocol, whereas the Layer 2 clearing procedures are the standard DSS1 procedures described in Chapter 11.

12.8 User Signalling Bearer Service

12.8.1 General

The User Signalling Bearer Service allows users to exchange user-to-user signalling information without setting up a circuit-switched connection. A temporary signalling connection is established and cleared in a manner similar to the procedures for circuit-switched connections. The user-to-user information is exchanged between the calling and called users without the information being analysed by the network.

12.8.2 Procedures

The procedures for the User Signalling Bearer Service are illustrated in Fig. 12.14.

To initiate the service, the calling user sends a Set-up Message to the called user. The Bearer Capability and Channel Identification Information Elements are set to values that request a temporary signalling connection to be established. The called user accepts the temporary signalling request by sending a Connect Message to the calling user. The calling user returns a Connect Acknowledge Message to the called user.

When the calling user receives the Connect Message, user-to-user information can be sent to the called user. Similarly, receipt of the Connect Acknowledge

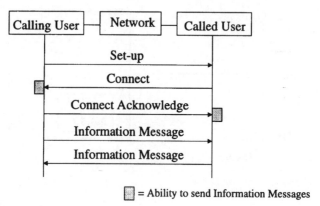

Fig. 12.14   User Signalling Bearer Service

Message by the called user allows user-to-user information to be sent to the calling user.

Signalling information is transferred between users in Information Messages. Each Information Message contains the Call Reference, Protocol Discriminator and the user-to-user information elements. The connection is cleared by using standard release procedures.

12.8.3 Congestion control

The exchange of user-to-user information can place the network at risk of overload. Hence, a Congestion Control Procedure is implemented in which the transmission of Information Messages is curtailed if overload is being threatened. The instruction 'Receive Not Ready' prohibits the sending of Information Messages. The instruction 'Receive Ready' is used to permit the continuation of user-to-user information.

12.9 Circuit Mode Multi-Rate Procedures

The Circuit Mode Multi-Rate procedures are used to support the provision of a number of 64kbit/s connections. The procedures are similar to those for normal circuit-switched connections. The number of channels required for the connection is identified in the Set-up Message and is reflected in the information transfer rate identified in the Bearer Capability Information Element. The channels can be assigned on a contiguous basis or non-contiguous basis. In the contiguous form, the capacity is provided in a block of adjacent channels. In the non-contiguous form, the channels can be interleaved with other traffic.

12.10 Supplementary Services

12.10.1 General

Generic procedures are defined to invoke and operate DSS1 supplementary services[3]. Three generic protocols are defined to control supplementary services: Keypad, Feature Key Management and Functional.

Keypad and Feature Key Management Protocols adopt a 'stimulus' approach, in which the user terminal is not expected to exhibit knowledge of the supplementary service being invoked. Terminals using the stimulus approach react to specific instructions and exhibit a low level of processing capability. The functional approach is characterised by requiring a terminal to exhibit a degree of intelligent processing in terms of the supplementary service being invoked.

12.10.2 Keypad Protocol

The Keypad Protocol is appropriate for basic-rate and primary-rate access. It involves the calling user requesting supplementary services by using a Keypad to generate alphanumeric codes. The codes are analysed by the network (e.g. in the originating local node) to determine the intention of the calling user. The allocation of codes to supplementary services is not internationally standardised and the codes are therefore network dependent.

The request to invoke a supplementary service is included within a Set-up or Information Message using a Keypad Facility Information Element. If the invocation is requested during call establishment, the Keypad Facility Information Element is included in the Set-up Message. If the invocation is requested during any other phase of the call, the information element is included in an Information Message. The network can respond to the calling user in a Call Proceeding Message or an Information Message.

To help the calling and called users understand the progress being made with the supplementary service, information can be supplied to be displayed on the user terminals. This information is provided by the network in Display Information Elements included in a range of messages. An example of the procedures is given in Fig. 12.15.

12.10.3 Feature Key Management Protocol

The Feature Key Management Protocol is applicable to the basic-rate access. It relies upon establishing a Service Profile for each participating user. A service profile consists of a file of information, which characterises the services offered by the network to that user. The service profile is agreed between the user and the network operator upon subscription to the service. The user requests a supplementary service by sending a feature identifier to the network. The feature identifier is mapped to the corresponding supplementary service by consulting the service profile of the calling user.

224 *Telecommunications Signalling*

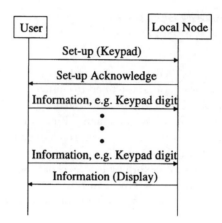

Fig. 12.15 Keypad Protocol

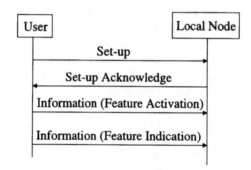

Fig. 12.16 Feature Key Management Protocol

The feature identifier is included in a Feature Activation Information Element, contained within either a Set-up Message or an Information Message. The network responds to the user by including a Feature Indication Information Element in a range of call control messages. Information that can be displayed on the user's terminal can be sent in a Display Information Element in a range of call-control messages. An example of the procedures is given in Fig. 12.16.

12.10.4 Functional Protocol

The Functional Protocol requires that the user terminal supporting a supplementary service has a knowledge of that supplementary service. The protocol applies to basic rate and primary rate access. There are two categories of Functional Protocol; Separate Message Category and Common Information Element Category. If it is necessary to synchronise the availability of equipment at the user's premises and at the local node, then it is necessary to define specific messages to control supplementary services. These messages belong to the

Separate Message Category. If it is not necessary to synchronise such user and local node resources, then a Common Information Element approach can be adopted.

Messages specified in the Separate Message Category are defined to enact specific functions in the control of supplementary services. The two major functions are Hold and Retrieve. If a user sends a Hold Message to the local node during a call, the local node reserves the traffic channel (B channel) and the corresponding call reference. A Hold Acknowledge Message is returned to the user to confirm that the call has been held. The call is returned to an active state by the user sending a Retrieve Message to the local node. This is confirmed by the return of a Retrieve Acknowledge Message.

In the Common Information Element Category, a Facility Information Element is used in a generic manner to provide invocation and control of supplementary services. This approach allows new services to be introduced more easily, it supports supplementary services with a large number of variants without a proliferation of messages and it allows multiple supplementary service invocations in one message. The Facility Information Element contains a number of components, including the Invoke and Return-Result Components described in Chapter 7 for Transaction Capabilities. Indeed, the formatting technique adopted in the Facility Information Element is the same as that described in Chapter 7.

The Facility Information Element can be applied to supplementary services either involving an existing call or not related to an existing call. For the case when a call exists, or is being established, the Facility Information Element is included in a Facility Message or in a Set-up Message. If a call does not exist, a Register Message is used to establish an appropriate signalling connection: Facility Messages can be used once a signalling connection is established. An example of the procedures is given in Fig. 12.17.

12.10.5 Message format

The messages used to facilitate supplementary services are similar in format for the three protocol types and a typical example is the Facility Message, illustrated

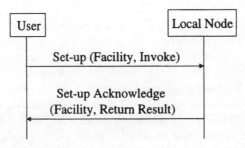

Fig. 12.17 Functional Protocol

Table 12.5   Facility Message format

Information element	Type	Length (octets)	Description
Protocol Discriminator	M	1	see Section 12.2.1
Call Reference	M	2+	see Section 12.2.2
Message Type	M	1	see Section 12.2.3
Facility	M	8+	indicates the invocation and operation of supplementary service
Display	O	2+	supplies information to be displayed by the user

in Table 12.5. The Facility Message is used to request or acknowledge a supplementary service.

12.11 Chapter Summary

Layer 3 of DSS1 is responsible for setting-up, maintaining and clearing-down circuit-switched calls in an ISDN. It is also responsible for providing access to packet-data facilities and for providing user-to-user signalling capability.

A Layer 3 message consists of a Protocol Discriminator, a Call Reference, a Message type and Information Elements. The Protocol Discriminator is used to discriminate between circuit-switched call control messages and other types of message. The Call Reference is a number that is used to identify the call to which the message pertains. The Call Reference is chosen by the user or node initiating the call. The Message Type defines the function of the message being sent.

The information elements provide a range of information pertinent to the message, e.g. the Bearer Capability Information Element defines a range of features required for a particular call. Information elements can be a single octet in length or of variable length. Each information element is defined by an Information Element Identifier.

Procedures for setting up circuit-switched calls can use a broadcast or a point-to-point form of working. In broadcast working, a call is established by sending a Set-Up Message to all terminals associated with the signalling link. All terminals conduct a compatibility check to ensure that communication with the calling terminal can take place. Those terminals that can successfully communicate with the calling terminal respond with an Alerting or Connect Message. The first terminal to respond with a Connect Message is selected as the receiving terminal. In point-to-point working, a node can identify a particular terminal and messages are sent to that terminal.

The clearing of circuit-switched calls is achieved by a three message sequence using Disconnect, Release and Release Complete Messages. Clearing can be initiated by either the calling or the called user.

Access to packet-data facilities can be circuit-switched or packet-switched. In circuit-switched access, a traffic channel is established to an Access Unit that provides an interface to a packet-switched public data network. In this case, normal set-up and clear-down procedures apply. In packet-switched access, a Packet Handler within an ISDN node is used to provide packet data facilities. In this case, access is gained to the Packet Handler either by establishing a traffic channel or by using the signalling channel.

The User Signalling Bearer Service allows users to exchange user-to-user signalling information without setting up a circuit-switched connection. A temporary signalling connection is established and cleared in a manner similar to the procedures for circuit-switched connections. Signalling information is transferred between users in Information Messages, which are not analysed by the network. Each Information Message contains a Call Reference, Protocol Discriminator and the user-to-user Information Elements.

Three generic procedures are described for DSS1 supplementary services. The Keypad Protocol is a procedure in which alphanumeric codes are dialled by the user, the local node analysing the information and acting accordingly. The Feature Key Management Protocol relies on a service profile being generated for each participating user, a particular supplementary service being enacted by a user dialling a code referenced in that user's service profile. The Functional Protocol relies on a user's terminal to exhibit a degree of intelligent processing in generating and analysing messages. In the Functional Protocol, the terminal is expected to understand a supplementary service and participate in its operation.

12.12 References

1 ITU-T Recommendation Q.931: 'ISDN User-Network Interface Layer 3 Specification for Basic Call Control' (ITU, Geneva)
2 ITU-T Recommendations X.25 and X.75: 'Interface between Data Terminal Equipment and Data Circuit Terminating Equipment' and 'Packet Switched Signalling System Between Public Networks' (ITU, Geneva)
3 ITU-T Recommendation Q.932: 'Generic Procedures for the Control of ISDN Supplementary Services' (ITU, Geneva)

Chapter 13
Private Networks

13.1 Introduction

13.1.1 General

Communications are a critical component in the efficiency of business organisations. The ability to connect geographically separate sites is a key requirement. For example, consider a banking organisation that wishes to connect its branches with its head office. Its branches can vary in size tremendously and can be located throughout the world. The ability to make voice calls and transfer data amongst all locations in a secure manner is critical to the success of the business. Other requirements include the ability to:

- make and receive calls with members of the public;
- adopt flexible numbering plans;
- use equipment from more than one supplier;
- adopt a flexible network architecture;
- control and customise services.

Requirements of this nature can be met in two ways. One approach is to provide a set of network elements and resources that are available to a customer on a dedicated basis. This is termed a 'Private network'. In this case, the customer can own or rent the network elements. Using private automatic branch exchanges (PABXs) at the various sites, a wide range of features can be employed to improve communication between users at the various locations. In the past, the range of supplementary services available on private networks has been greater than those on public networks. Private networks also permitted additional control and customisation of services.

However, dedicated private networks can be expensive to operate, because many customers have periods when a large proportion of capacity is not being used. Efficient capacity allocation between sites is also difficult to achieve and the operations and maintenance aspects of the network can become a great burden.

A second approach to meet the requirements is to use a Virtual Private Network (VPN). In a VPN, the characteristics of a private network are emulated (hence the term virtual) in a public network. Capacity is guaranteed to be provided on demand. Operations and maintenance is incorporated within the public network systems. Although VPNs are not able to provide the full range of capabilities of dedicated networks at this stage, further evolution of public networks is expected to lessen the need for dedicated networks in the future.

13.1.2 Signalling

(a) Inter-PABX

For many years, the additional features provided by dedicated private networks outweighed their disadvantages. A number of countries produced signalling systems that specialised in connecting PABXs to provide services. An example of this type of signalling system is summarised in Appendix 6.

Further standardisation activities have now taken place for inter-PABX signalling and the Private Signalling System No.1 (PSS1)[1] has evolved. PSS1 is also known as 'QSIG', reflecting the terminology of the ISDN interface at which the system is adopted. PSS1 has been defined by a collaboration of the European Association for Standardising Information & Communications Systems (ECMA), the International Standards Organisation (ISO) and the European Standards Institute (ETSI). These organisations have involved manufacturers and operators from around the world in a joint approach to inter-PABX signalling. The principles of PSS1 are outlined in Sections 13.2 to 13.5.

(b) Virtual Private Networks

The comprehensive solution to private network emulation lies in the specification of the necessary capabilities in appropriate modern CCS systems, e.g. DSS1 and CCSS7. Although much work remains to be done, the ITU has specified a Global Virtual Network Service[2] (GVNS), which provides private network functions (e.g. private network addressing) to users at geographically dispersed locations with the minimum use of dedicated network resource.

In today's liberalised environment, communication takes place over a wide range of national and international networks. GVNS is specified for use within such a multiple network environment and can therefore be applied globally as a basis for providing a worldwide VPN. Sections 13.6 to 13.9 give a brief description of GVNS and its signalling implications.

13.2 PSS1 (QSIG) – General

13.2.1 Introduction

PSS1 is designed to operate in dedicated private networks comprising a number of private telecommunication network exchanges (PTNXs). A PTNX is a private automatic branch exchange (PABX) designed for a digital environment.

The term 'private integrated services network exchange' (PINX) is also used within an ISDN environment. A typical dedicated private network is illustrated in Fig. 13.1. In this case, each site has a PTNX and the PTNXs are connected to each other, either directly or via a Transit PTNX.

PSS1 is based upon the access signalling system DSS1, explained in Chapter 12, and is designed to:

- be compatible with public network capabilities;
- work with PTNXs manufactured by differing vendors, thus permitting competitive tendering for PTNX provision;
- allow a wide range of network architectures;
- provide feature transparency between two sites, even if intermediate PTNXs are of lower capability.

PSS1 is defined for narrowband operation. Corresponding protocols are planned for broadband operation.

13.2.2 Architecture

PSS1 is defined for use between PTNXs. The interface between PTNXs is termed the Q Reference Point in ITU ISDN standards. This explains the use of the alternative name (QSIG). Fig. 13.2 illustrates the Protocol Model for PSS1. Layers 1 and 2 are identical to those for DSS1 (explained in Chapter 3). Layer 3 defines protocols for:

- establishing and releasing basic calls;
- carrying supplementary services information.

Basic calls[3] are established by the Protocol Control Functions.

Information related to supplementary services is handled by the Generic Function (GF) Transport Control, with additional functions provided in the Protocol Control. The Generic Functional (GF) Procedures[4] are used to transfer information on supplementary services from one PTNX to another. This enables supplementary services to be handled in an efficient manner using PTNXs manufactured by a variety of suppliers.

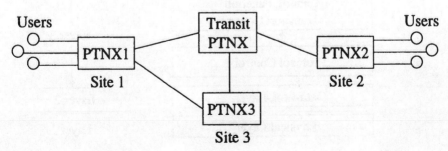

Fig. 13.1 Typical private network

232 *Telecommunications Signalling*

Supplementary services themselves are not specified as part of Layer 3: these use the Layer 7 Application Service Element (ASE) structure explained in Chapter 3. To implement supplementary services between two PTNXs, an association is required between the respective supplementary service control functions. In PSS1, this can be provided either by using a Network Layer Connection (based on Protocol Control and Generic Transport) or by means of the Dialogue Service Element (DSE). The DSE can establish associations that are independent of the Network Layer.

The supplementary service control functions make use of the services provided by Remote Operations Service Element (ROSE) and DSE, which in turn make use of the services offered by the GF and Protocol Control.

The Layer 7 boundary in Fig. 13.2 depends upon the actions performed by the co-ordination and supplementary services control functions, which are service-specific.

13.2.3 Services

A range of services can be implemented by Layer 3 of PSS1. These include setting up and clearing basic calls. For supplementary services, they involve

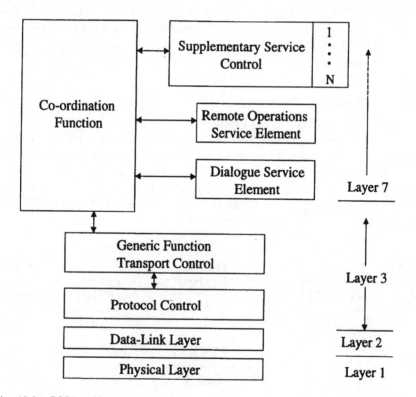

Fig. 13.2 PSS1 architecture

transferring ROSE and DSE Application Protocol Data Units (APDUs) between Layer 7 peers. The APDUs can be transferred:

- during call set-up. This is termed Call Related and uses basic call messages;
- when a call has been established. This is termed Active Call and uses a Facility Message;
- by establishing a separate signalling connection. This is termed Call Independent.

Examples of these services are given in the following sections.

13.3 PSS1 Basic Call Procedures

The procedures for establishing a simple call are illustrated in Fig. 13.3.

The procedures for basic calls are symmetrical, i.e. the procedures on the user side of the Q interface are the same as those on the network side. A call is initiated by PTNX 1 identifying that a call is required from a user. PTNX 1 selects an appropriate channel for the call and indicates whether or not an alternative channel is acceptable. PTNX 1 then sends a Set-up Message to PTNX 2 (the Transit PTNX). The Set-up Message contains a call reference and the relevant address digits (en bloc and overlap sending can be used, as explained in Chapter 6). PTNX 2 acknowledges that the call is being processed by returning a Call Proceeding Message to PTNX 1.

After routing analysis, PTNX 2 sends a Set-up Message to PTNX 3 and this is acknowledged by the return of a Call Proceeding Message. When PTNX 3 detects that the called user is being alerted (e.g. the sending of ring current to the called user), PTNX 3 returns an Alerting Message to PTNX 2, which is subse-

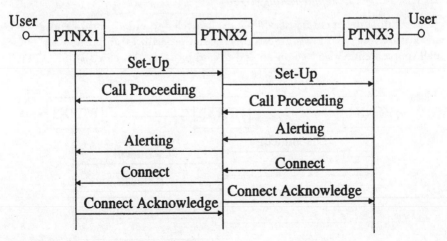

Fig. 13.3 Simple basic call set-up

quently sent to PTNX 1. Upon answer of the call by the called user, a Connect Message is sent from PTNX 3 to PTNX 2. This message is passed on to PTNX 1. Each Connect Message is acknowledged by returning a Connect Acknowledge Message.

Basic call release procedures are illustrated in Fig. 13.4.

In this case, the calling user indicates that the call has finished and PTNX 1 sends a Disconnect Message to PTNX 2. PTNX 2 disconnects the channel (meaning that the channel is no longer available for use by the call) and returns a Release Message. When PTNX 1 receives the Release Message, it releases the call (i.e. makes the channel and call reference available for use by another call) and issues a Release Complete Message. Upon receipt of the Release Complete Message, PTNX 2 releases the channel and call reference. The same procedures apply between PTNXs 2 and 3.

13.4 PSS1 Supplementary Services

13.4.1 Call-related supplementary services

Call-related supplementary services can refer to (a) the originating and terminating PTNXs (i.e. an end-to-end service invocation) or (b) adjacent PTNXs (i.e. relating to a link). Fig. 13.5 gives an example of the procedures supporting an end-to-end call related supplementary service.

In this case, ROSE APDUs need to be transferred between PTNXs 1 and 3 at the same time as a call is being established. The APDUs (Invoke and Return Result) are included within the PSS1 call set-up messages (Set-Up and Alerting). The transfer inherently uses connection-oriented procedures.

13.4.2 Active call supplementary services

Active call supplementary services can also refer to either end-to-end or link invocation. Fig. 13.6 gives an example of the procedures supporting an active call supplementary service on an end-to-end basis. In this case, ROSE APDUs

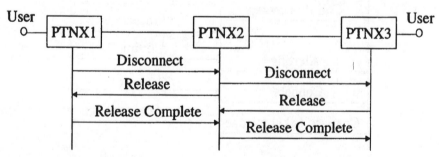

Fig. 13.4   Basic call release

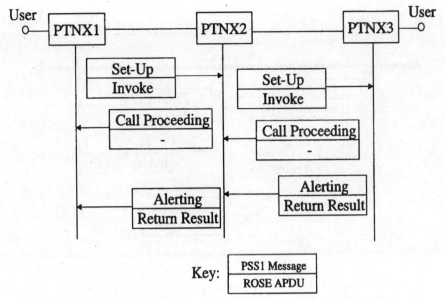

Fig. 13.5   Call-related procedures

need to be transferred between PTNXs 1 and 3 during the active phase of a call. The APDUs (Invoke and Return Result) are included within PSS1 Facility Messages.

13.4.3 *Call-independent supplementary services*

Call-independent supplementary services can rely on either connection-oriented or connectionless transfer of APDUs. Fig. 13.7 gives an example of the procedures supporting a call independent supplementary service on a connectionless basis. In this case, ROSE APDUs need to be transferred between PTNXs 1 and 3 in the absence of a call by a user. The APDUs (Invoke and Return Result) are included within PSS1 Facility Messages.

13.5 PSS1 Message Format

The message format for PSS1 is very similar to that for DSS1, explained in Chapter 12. This includes the use of information elements, such as the mandatory elements called Protocol Discriminator, Call Reference and Message Type. The use of codesets is also the same.

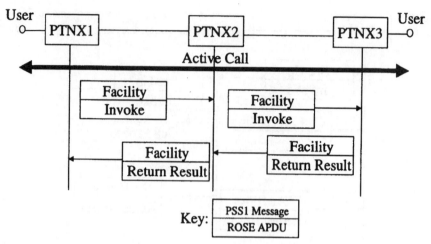

Fig. 13.6 Active call procedures

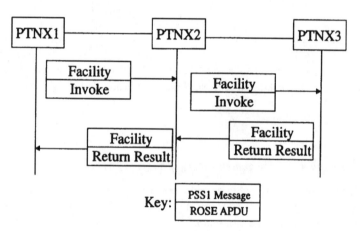

Fig. 13.7 Call-independent procedures

13.6 GVNS – General

13.6.1 Virtual network

GVNS provides the functions typically associated with private networks. The service is provided by a series of interconnected physical networks using, for example, the CCSS7 ISUP. The ISUP establishes connections across the physical networks. These connections form a virtual network that, from a customer perspective, permits end users to communicate as if they are connected to a single PABX. The physical and virtual networks are illustrated in Fig. 13.8.

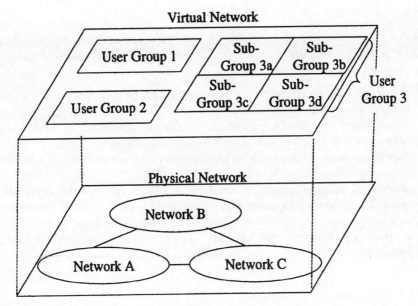

Fig. 13.8   *Virtual and physical networks*

A customer can allocate users to groups and sub-groups. Users within a group or sub-group are allocated certain features and calling restrictions. In Fig. 13.8, for example, User Group 1 could be permitted all facilities available, whereas User Group 2 might be prevented from using a feature like Executive Intrusion (where a caller can break into an existing call). Similarly, all Sub-Group 3 Users could have a common set of features available, with different sub-groups having varying restrictions. For example, Sub-Group 3b might be prevented from making international calls.

When subscribing to the service, the customer specifies the locations that are to form part of the virtual network. These locations are referred to as 'on-net'. GVNS calls can then be made on-net, i.e. within the virtual network, or off-net, i.e. to locations external to the virtual network.

13.6.2 Access types

The customer can choose how users gain access to GVNS from a range of access types. These include:

- direct access from on-net locations, in which Calling Line Identity Procedures or physical network connection identities are used to confirm authority to use GVNS;
- indirect access from on-net, in which a user inserts a Service Access Code to gain access to GVNS;
- indirect access from off-net, in which a user inserts a Service Access Code and Authorisation Code to gain access to GVNS.

13.6.3 GVNS features

GVNS allows a customer to configure network characteristics by providing customer profile data sets within the physical networks. GVNS offers features that include:

- call screening, e.g. to implement customer-defined call types that can be made from on-net;
- customer-defined numbering plans, to permit customers to use public network numbering schemes or privately-defined schemes;
- standard announcements, e.g. to provide information to a user on a feature being implemented;
- customised announcements, in which the customer specifies the announcement and the conditions under which it is played;
- authorisation codes to determine a level of approved functionality;
- overflow, in which incoming calls are routed to other locations if the called location is busy.

13.6.4 Operation and maintenance

One of the advantages of the GVNS is that the burden of operation and maintenance can be devolved from the customer to a network operator. Examples of operations and maintenance facilities are:

- provisioning, which includes providing the service to the customer;
- configuration management, which permits a customer to change dialling plans, activate authorisation codes, change routing plans, etc;
- performance monitoring, e.g. providing data on traffic levels, capacity usage, etc.

13.7 GVNS Functional Model

13.7.1 GVNS functions

When specifying services and supplementary services, it is essential to be comprehensive and unambiguous, but it is also desirable to avoid unnecessary implementation aspects. This allows flexibility in implementation. In some cases, the performance of a function needs to be carried out at a particular network element, e.g. a local exchange, and this is defined as part of the specification.

In other cases, a function can be performed by one of a number of network elements, e.g. a local exchange, Transit Exchange or Gateway International Exchange. In these cases, the specification needs to avoid unnecessary constraints. This is achieved by using a functional model that describes Functional

Entities (FEs) and the functions that they perform, but does not specify the type of network element that applies to each FE.

A functional model[5] is used to specify GVNS. The model is particularly important for GVNS because certain functions need to be performed before others for correct operation. Consider an outgoing call in which a set of functions (Function Set A) needs to be performed before another set of functions (Function Set B). If Function Set B is performed in a Transit Exchange, then Function Set A must be performed either in the Originating Local Exchange or in the Transit Exchange. It is not possible for Function Set A to be performed in an International Gateway Exchange. The functional model describes this dependency while avoiding constraining functions to particular network elements.

13.7.2 Model

The GVNS functional model consists of five FEs incorporated within three categories. The model is illustrated in Fig. 13.9 and shows the relationship between FEs in solid lines. The Figure also shows the call control functions associated with a basic call and these relationships are shown as dotted lines.

FEs 1 and 3 belong to the Service Switching and Resource Control (SSRC) Category. These FEs establish and release GVNS calls upon the request of other FEs and interface with the call control functions within a node.

FEs 2 and 4 belong to the Service Logic and Data Control (SLDC) category. These FEs store and process the service logic and data required to handle a GVNS call, e.g. call screening and call routing for GVNS. These FEs

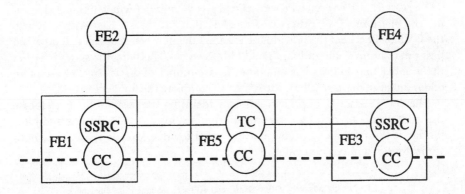

Fig. 13.9   *GVNS functional model*
        ----Basic Call Relationship
        —GVNS Relationship
        CC = Call Control
        SSRC = Service Switching & Resource Control
        TC = Transit Control

also direct the SSRC FEs in call establishment. If appropriate, FEs 2 and 4 can communicate with each other to provide additional capabilities.

FE 5 belongs to the Transit Control (TC) category and provides the functions associated with transit capabilities.

13.7.3 GVNS call types

Two types of GVNS call are currently specified:

- Type A, in which customer-specific information is stored in FE2, and
- Type B, in which customer-specific information is stored in both FE2 and FE 4, but these FEs are not directly connected.

A third type of call (Type C) is being considered for specification. Type C is similar to Type B, but FEs 2 and 4 are directly connected.

13.8 GVNS Formats

The ISUP contains a range of information that can be used by the GVNS. In addition, specific GVNS parameters are included in the forward and backward directions.

13.8.1 Forward GVNS parameter

When a GVNS call is made, the Initial Address Message (IAM) of the ISUP includes a Forward GVNS Parameter, illustrated in Fig. 13.10.

The Originating Service Provider Field gives a number which uniquely identifies the operator (GVNS Service Provider) that provides a user with access to the GVNS. The Length Indicator (LI) shows the number of octets to follow and has a maximum value of four. The Odd/Even (O/E) Indicator states whether the number that follows has an odd or even number of digits. In the case of an odd number of digits, a filler code is used to complete the final octet.

The GVNS User Group Field uniquely identifies the customer, e.g. to locate the address of the database record for the customer. The field has a similar format to that for the Originating Service Provider, except that the Length Indicator has a maximum value of eight.

LI = Length Indicator, O/E = Odd/Even, NPI = Numbering Plan Indicator

Terminating Network Routing Number				GVNS User Group			Originating Service Provider		
Digits	O/E	NPI	LI	Digits	O/E	LI	Digits	O/E	LI

Fig. 13.10 Forward GVNS parameter

The Terminating Network Routing Number is used by the termination service provider to route the call. The number can either be an ISDN number or a network-specific number. The field has a similar format to that for the Originating Service Provider, except that:

- the Length Indicator has a maximum value of nine;
- a Numbering Plan Indicator is included to identify the type of number, e.g. an ISDN Number.

13.8.2 Backward GVNS parameter

In the backward direction, the Answer and Connect messages contain a Backward GVNS Parameter, illustrated in Fig. 13.11. The Terminating Access Indicator identifies the type of access that the terminating service provider used to complete the call. It shows whether the terminating access is provided by switched means or by dedicated transmission. The Extension Bit provides for future expansion into further octets.

13.9 GVNS Procedures

Three procedures apply to GVNS that are additional to basic call set-up, namely The Access Procedure, the Routing Procedure and the Information Transfer Procedure.

The Access Procedure can be performed by the Originating Local Exchange, a Transit Exchange or the Outgoing International Gateway Exchange. Upon receipt of a request for GVNS service, the exchange checks for GVNS access validity, identifies the GVNS User Group and screens the call for feature authorisation. If GVNS access is not permitted, the call is released.

The Routing Procedure can also be performed at one of the three exchange types, but the Access Procedure must be implemented first.

The Routing Procedure uses GVNS information that can be stored at the exchange performing the procedure or at a remote database. In the terms of the GVNS model (Fig. 13.9), FE1 is located within the exchange and FE2 is located either within the exchange or at a remote site. In both cases, information provided by the calling user generates GVNS Routing Information. The information exchanged depends on a number of factors, including the type of

Extension	Spare	Terminating Access Indicator	
8	3-7	1-2	Bits

Fig. 13.11   Backward GVNS parameter

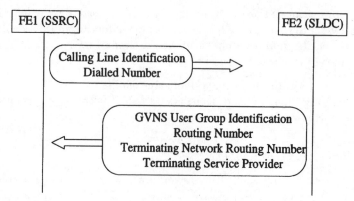

Fig. 13.12 *Typical Routing Information generation*

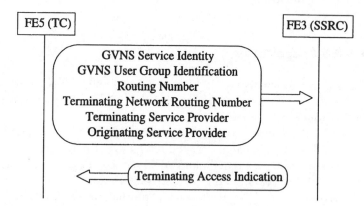

Fig. 13.13 *Typical information transfer*

access being used, but a typical exchange of information is illustrated in Fig. 13.12.

The Routing Number is an ISDN number that forms the basis of routing the call to the Terminating Service Provider. The Routing Number is included in the Called Party Number Parameter of the Initial Address Message formulated by the ISUP. The Terminating Network Routing Number can be an ISDN number or a network-specific number. It is used within the terminating network to complete the call.

To permit operation of GVNS, the originating and terminating networks need elements of information. Typical information transferred is illustrated in Fig. 13.13.

13.10 Chapter Summary

13.10.1 Meeting the requirements

Communications are a critical component in the competitiveness of business organisations. Such organisations have demanding requirements and these can be met in two ways.

One approach is to provide a dedicated private network. Using private automatic branch exchanges (PABXs) at the various sites, a wide range of features can be employed between users at various locations. Over the years, a number of operators have produced signalling systems that specialised in connecting PABXs, e.g. the Digital Private Network Signalling System (DPNSS), produced by British Telecommunications in conjunction with several manufacturers.

The Private Signalling System No.1 (PSS1), also known as 'QSIG', has now been specified for inter-PABX signalling by a collaboration of the European Association for Standardising Information and Communications Systems (ECMA), the International Standards Organisation (ISO) and the European Standards Institute (ETSI).

A second approach to meet the requirements of business organisations is to use a virtual private network (VPN). In a VPN, the characteristics of a private network are emulated in a public network. Capacity is guaranteed to be provided on demand. Operations and maintenance are incorporated within the public network systems.

The VPN solution lies in the specification of the necessary capabilities in appropriate modern CCS systems, e.g. DSS1 and CCSS7. The ITU has specified a Global Virtual Network Service (GVNS) on this basis. GVNS is specified for use within multiple networks and can therefore be applied globally as a basis for providing a worldwide VPN.

13.10.2 PSS1 (QSIG)

PSS1 is designed to operate in dedicated private networks comprising a number of Private Telecommunication Network Exchanges (PTNXs). It is defined in terms of a layered architecture. Layers 1 and 2 are identical to those for DSS1. Layer 3 defines protocols for establishing and releasing basic calls and carrying supplementary services information.

Basic calls are established by the Protocol Control Functions. Basic call setup is achieved using Set-up, Alerting and Connect Messages. Call Proceeding and Connect Acknowledge Messages are used as acknowledgements. A three-message release sequence (Disconnect, Release and Release Complete) is used to clear calls.

Information related to supplementary services is handled by the Generic Function (GF) Transport Control, with additional functions provided in the Protocol Control.

Supplementary services themselves are not specified as part of Layer 3: these use the OSI Layer 7 Application Service Element (ASE) structure. The Remote Operations Service Element (ROSE) and Dialogue Service Element (DSE) make use of the services offered by PSS1 to enable supplementary service information to be transferred between peers. The information is in the form of Application Protocol Data Units (APDUs) and these can be transferred as:

- Call Related, using basic call messages (Set-up and Alerting);
- Active Call, using a Facility Message in conjunction with an existing call;
- Call Independent, using a separate signalling connection.

The message format for PSS1 is very similar that for DSS1, including use of information elements and codesets

13.10.3 GVNS

GVNS is provided by a series of interconnected physical networks using the CCSS7 ISUP. The connections across the physical networks form a virtual network that, from a customer perspective, permits end users to communicate as if they are connected to a single PABX.

Users can be allocated to groups with common features and restrictions. Locations that form part of the virtual network are referred to as 'on-net'. Users can gain access to GVNS directly from on-net, indirectly from on-net and indirectly from off-net. Features provided by GVNS include call screening, customer-defined numbering plans, customised announcements and overflow arrangements.

Operation and maintenance capabilities are provided by the network operator and include provisioning, configuration management and performance monitoring.

The GVNS Functional Model consists of five Functional Entities (FEs). Two FEs perform Service Switching and Resource Control Functions. One FE performs Transit Functions. Two FEs perform Service Logic and Data Control Functions.

The CCSS7 ISUP contains a range of information that can be used by the GVNS. In addition, specific GVNS parameters are included in the Initial Address, Answer and Connect Messages.

Three procedures apply to GVNS that are additional to basic call set-up and they can be performed by the Originating Local Exchange, a Transit Exchange or an Outgoing International Gateway Exchange. The Access Procedure checks for GVNS access validity, identifies the GVNS User Group and screens the call for feature authorisation. In the Routing Procedure, an FE translates information from the user into a Routing Number and a Terminating Network Routing Number. The Routing Number is included in the Called Party Number Parameter of the Initial Address Message formulated by the ISUP. The

Information Transfer Procedure is used to inform the Originating and Terminating Networks of the status of the GVNS connections.

13.11 References

1 'QSIG Handbook' (ETSI)
2 ITU Recommendation Q.735.6, Stage 3: 'Global Virtual Network Service (GVNS)' (ITU, Geneva)
3 ETS 300 172: 'Circuit Mode Basic Services' (ETSI)
4 ETS 300 239: 'Generic Functional Protocol' (ETSI)
5 ITU Recommendation Q.85, Stage 2: 'Global Virtual Network Service' (ITU, Geneva)

Chapter 14
Broadband Signalling Platform

14.1 Introduction

Customer requirements are becoming more exacting as quality expectations rise and the demand for new services escalates. With the increasing range of services, bandwidth requirements are also increasing. Although 64 kbit/s circuits are adequate for telephony, the emergence of multi-media services is spawning massive increases in bandwidth requirements. The convergence of the computing, IP, broadcast/media and telecommunications industries is consistent with this demand. The term 'broadband' is used to describe high-bandwidth capabilities. Although the term is not strictly defined, it is generally applied to bandwidths of greater than 2 Mbit/s. The term 'Narrowband' is used to describe lower-bandwidth capabilities, typically 2 Mbit/s or less.

Networks are responding to these rapidly changing requirements by establishing secure and flexible high-bandwidth transport capabilities. These capabilities can be used as a solid platform upon which the increasing range of services can be carried. New switching systems are based on evolving standards for asynchronous transfer mode (ATM) technologies.[1] ATM will provide the backbone for high-bandwidth communications.

This Chapter describes the ITU-T signalling platform for broadband networks. Section 14.2 summarises the signalling relations that are available. Section 14.3 describes the architecture for access and inter-nodal signalling. The rest of the chapter focuses on the lower layers of the architecture that provide a platform for broadband signalling, namely the ATM layer and ATM Adaptation layer (AAL). The higher layers (B-ISUP and DSS2) are described in Chapters 15 and 16, respectively.

Other standards for broadband signalling are also being proposed. The ITU-T standards are used here to describe the principles of evolving towards broadband operation.

14.2 Broadband Signalling Relations

Signalling relations for broadband networks are based on similar principles to those described for narrowband in previous chapters. Fig. 14.1 summarises the broadband signalling relations.

248 *Telecommunications Signalling*

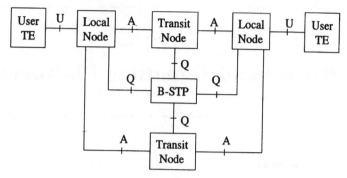

Fig. 14.1 Broadband signalling relations
Source: (ITU-T Recommendation Q.2010)

Users have a wide range of terminal equipment (TE), from PCs to high-capacity processing devices. Signalling between these devices and the local node is via a User-Network Interface (UNI), denoted in Fig. 14.1 by 'U'. Broadband signalling between nodes can adopt the associated and quasi-associated modes of operation, as described in Chapter 2.

Associated signalling between local nodes and transit nodes is via a Network Node Interface (NNI), shown as Interface A in Fig. 14.1. Quasi-associated signalling is via an NNI shown as Interface Q. The B-STP is a broadband version of the Signal Transfer Point (STP), also explained in Chapter 2.

14.3 Architecture

The signalling for broadband communications is structured into layers, as explained in Chapter 3. This approach permits each layer to evolve while maintaining consistent interfaces between layers. Several configurations are possible. The use of ATM technologies improves routing and quality aspects. If such technologies are adopted, the layers are configured as shown in Fig. 14.2. Peer-to-peer communications refer to a layer in one entity (e.g. a node) exchanging information with a corresponding layer in another entity.

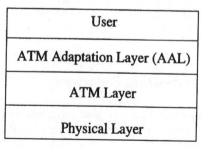

Fig. 14.2 Architecture for broadband signalling

The AAL user is the higher-layer signalling for broadband. The AAL user can be the Digital Subscriber Signalling System No.2 (DSS2) for access signalling or the Broadband ISDN User Part (B-ISUP) for inter-nodal signalling. These are equivalent to DSS1 and the ISDN User Part for narrowband signalling.

The ATM Adaptation Layer (AAL) bridges the functions of the ATM layer and the higher layers. The functions performed by the AAL depend upon the type of user and application. The functions are grouped into services that are offered to the higher layers.

The ATM layer provides a service-independent transport function. Any requirements that are specific to a particular application are covered by higher layer functions (e.g. the AAL). The ATM layer transports information supplied by higher layers in packets of data called cells. This information is transported transparently by the ATM layer, without any processing of the information field.

The Physical Layer defines the physical characteristics of the transmission links and corresponds to the Physical Layer in the OSI Model.

The architectures for the UNI and NNI are based on the same model, but they differ to take account of the varying nature of access and inter-nodal signalling.

14.3.1 UNI architecture

The architecture for broadband signalling at the UNI is illustrated in Fig. 14.3.

The Digital Subscriber Signalling System No. 2 (DSS2) is the broadband access signalling system. DSS2 provides equivalent functions for broadband signalling that DSS1 performs for narrowband. Layer 3 of DSS2 is the highest protocol in the UNI architecture. DSS2 is described further in Chapter 16.

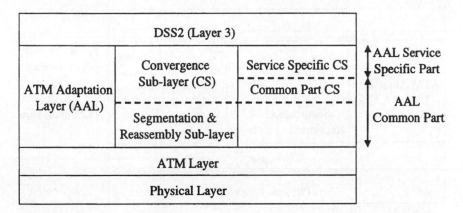

Fig. 14.3   UNI signalling architecture

The AAL is divided into a Convergence Sub-Layer (CS) and a Segmentation and Reassembly (SAR) Sub-Layer. The CS is further divided into a Service Specific Part and a Common Part to take account of the AAL dependence on higher layer requirements. The AAL Service Specific Part performs functions that are specific to the needs of the higher layers. The AAL Common Part provides common functions that can be used by a range of Service Specific Parts.

The SAR sub-layer segments long datastreams into packets of data that can be transmitted to another user. The destination reassembles the data into its original form. SAR is described further in Section 14.6.1. The UNI supports the point-to-point and broadcast forms of operation described in Chapter 3. Management protocols are required to establish the signalling capabilities for the broadcast form of operation. The management protocols are known as 'Meta-signalling'.[2] The Meta-Signalling Protocol has procedures for the assignment, removal and checking of signalling channels.

14.3.2 NNI architecture

The architecture for broadband signalling at the NNI is illustrated in Fig. 14.4. In this case, the highest layer of the architecture is the broadband version of the ISDN User Part (B-ISUP). This is supported by Level 3 of the Message Transfer Part (MTP). The remaining layers correspond to the UNI interface.

14.4 Platform Format Convention

Each layer of the architecture has formatting rules. These vary greatly at the higher layers (B-ISUP and DSS2). However, for AAL and ATM, the peer-to-peer communications adopt the general format illustrated in Fig. 14.5.

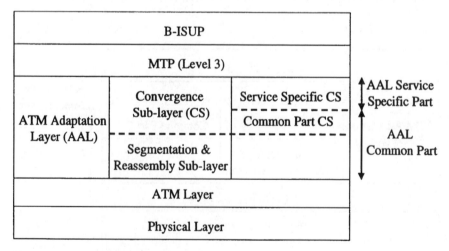

Fig. 14.4 NNI signalling architecture

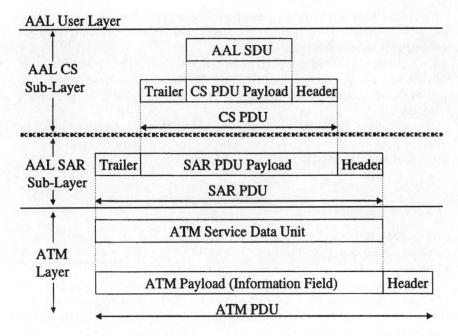

Fig. 14.5   ATM and AAL general format
Source: (ITU-T Recommendation I.363)

Each layer and sub-layer forms a Protocol Data Unit (PDU), which consists of a Payload, a Header and a Trailer. The Payload is the information field that contains the data to be transferred between peer entities. The Header is used to route the PDU through the network. The Trailer can have various functions, depending on the application.

The information provided by an adjacent layer is included in a packet termed a Service Data Unit (SDU). The AAL SDU consists of the information that needs to be transferred from one AAL user to another and forms the payload for the CS PDU. Similarly, the ATM SDU consists of the information to be transferred from the AAL in one entity to the AAL of another entity and forms the payload of the ATM PDU.

14.5 Asynchronous Transfer Mode (ATM) Layer

ATM[3] is a packet-oriented transfer mechanism that uses time division multiplexing techniques, as explained in Appendix 1. Information to be transferred from one ATM user to another is split into PDUs (termed cells) of a fixed size. These cells are transmitted over transmission links and the datastream is recombined upon reception at the destination. ATM provides Virtual Circuits, i.e. the transmission links are shared on a real-time basis and are utilised by a particular user only when there is information to transfer. The links are used for other calls when there is no information to transfer.

Each cell consists of a Header and an Information Field. The Header identifies cells belonging to the same Virtual Channel. The Information Field contains the data to be transferred. Cell sequence integrity is maintained over the Virtual Channel.

The signalling for ATM is carried on an ATM layer connection that is separate from the user information. This is consistent with the principles of Common Channel Signalling outlined in Chapter 2.

14.5.1 Cell structure

Each ATM cell comprises a header of five octets and an information field (payload) of 48 octets, as illustrated in Fig. 14.6. The format of the header is different for the UNI and NNI. These are illustrated in Fig. 14.7.

The combination of Virtual Path Identifier (VPI) and Virtual Channel Identifier (VCI) forms the Routing Field and is used to route cells through the network. There are 24 bits available at the UNI and 28 bits available at the NNI for the Routing Field.

The Generic Flow Control (GFC) Field consists of four bits and is used to assist in the control of information flow and ATM connection queues. The Payload Type (PT) Field consists of three bits and indicates the form of data being carried. Examples are:

- user data cell;
- operations and maintenance cell;
- resource management cell.

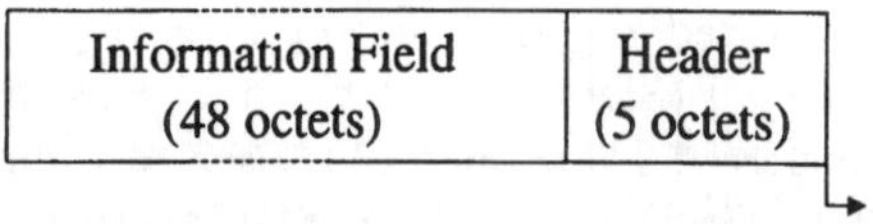

Information Field (48 octets)	Header (5 octets)

Fig. 14.6   ATM cell structure

Octet	User-Network Interface			Network Node Interface		
1	GFC	VPI		VPI		
2	VPI	VCI		VPI	VCI	
3	VCI			VCI		
4	VCI	PT	CLP	VCI	PT	CLP
5	Header Error Control			Header Error Control		

Fig. 14.7   Header format

GFC = Generic Flow Control; VCI = Virtual Channel Identifier; CLP = Cell Loss Priority; VPI = Virtual Path Identifier; PT = Payload Type

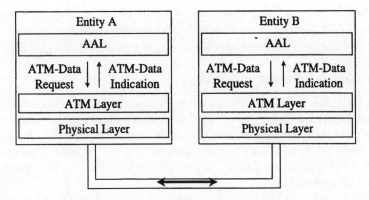

Fig. 14.8 Use of primitives

The Cell Loss Priority (CLP) Field is a bit that indicates the priority of the cell. In abnormal network conditions, lower priority cells are discarded before higher priority cells. The Header Error Control (HEC) Field is a means of checking the validity of the header. The receiving node or user can perform several modes of error detection and correction under fault conditions.

14.5.2 Primitives

Communication between the ATM Layer and the AAL is by ATM-Data Request and ATM-Data Indication Primitives. The ATM-Data Request Primitive is used by the AAL in one entity to request the transfer of information to a corresponding AAL in another entity. The ATM- Data Indication Primitive is used by the ATM Layer to indicate to the AAL the arrival of an ATM SDU. These are illustrated in Fig. 14.8.

Each primitive includes a Data parameter that carries the information to be transferred between AALs. Other parameters are used to establish the appropriate values for the CLP and PT fields generated in the ATM layer. Submitted Loss Priority and Receive Loss Priority Parameters indicate the priority requested and allocated, respectively. Similar primitives are defined for communication between the ATM Layer and the Physical Layer.

14.6 ATM Adaptation Layer (AAL)

The ATM Adaptation Layer (AAL)[4] bridges the functions of the ATM Layer and the users of the AAL. The users of the AAL have a range of requirements and the AAL provides a corresponding range of services to meet their needs. Two generic functions are provided to facilitate services; namely Segmentation and Reassembly, and Blocking/Deblocking. These are described briefly to establish the principles involved. Five types of AAL Service were originally identified, but Types 3 and 4 were subsequently combined. Type 5 is similar in nature to Type 3/4 and these are described in the same section.

14.6.1 Segmentation and reassembly

When the information to be transmitted between peer entities is too long, the Segmentation and Reassembly Function is used to break down (segment) the information into manageable portions, as illustrated in Fig. 14.9. In this case, each portion is formed into a PDU and the original payload (Payload A) is segmented into Payloads 1 and 2. Each PDU needs a Header for routing to the destination peer entity.

Upon reception at the destination entity, Payloads 1 and 2 are recombined (assembled) to reconstitute the data into its original form (Payload A).

14.6.2 Blocking and deblocking

Blocking is used to describe the aggregation of small amounts of information to form a single unit of data. The concept is illustrated in Fig. 14.10. In this case, Payloads 1-3 can be transferred to the destination within a single PDU with Payload A. Upon reception at the destination entity, Payload A is deblocked to reconstitute the data into its original form (Payloads 1-3).

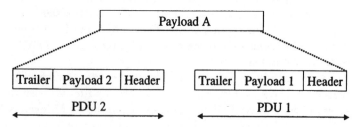

Fig. 14.9   Segmentation

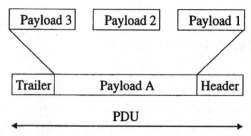

Fig. 14.10   Blocking

14.7 AAL Type 1 Services

14.7.1 Services

AAL Type 1 Services transfer Service Data Units (SDUs) with a constant bit rate and deliver them to the destination at the same bit rate. They also permit the transfer of timing and structure information between the source and destination entities.

14.7.2 Architecture

The functions of AAL Type 1 are divided into two sub-layers, namely the Segmentation and Reassembly (SAR) Sub-Layer and the Convergence Sub-Layer (CS), as illustrated in Figs. 14.3 and 14.4. The sub-division of the CS (into Specific and Common Parts) is not used for Type 1 Services.

14.7.3 Primitives

The primitives between the AAL and its users are the AAL Unitdata Request and AAL Unitdata Indication, as illustrated in Fig. 14.11.

The AAL Unitdata Request Primitive is used to request the transfer of an AAL SDU from one entity to another. A parameter includes the data to be transferred. The AAL Unitdata Indication Primitive is used to notify an AAL user of the receipt of a Request Primitive. In the Type 1 service, the length of the SDU is constant and the time interval between two consecutive primitives is also constant. This reflects the constant source bit rate information transfer offered by this service. Three parameters are associated with the primitives:

- Data Parameter, which is used to transfer the appropriate information between the AAL and its user;

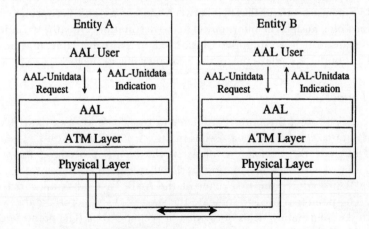

Fig. 14.11   AAL Type 1 primitives

- Structure Parameter, which is used to transfer datastreams that are organised into multiples of octets, e.g. for 64kbit/s circuit-switched data;
- Status Parameter, which is used to indicate the detection of errors.

14.7.4 Functions

The CS analyses the requirements of the AAL user and organises the AAL user information accordingly. The CS also provides a range of additional functions, including:

- handling of cell delay variation by means of a buffer;
- processing the SAR sequence count mechanism to handle lost and corrupted cells;
- handling timing information for asynchronous transport (termed Source Clock Frequency Recovery).

The SAR Sub-Layer accepts a 47 octet block of data from the Convergence Sub-Layer and adds a one-octet SAR Header. The Header comprises a 4 bit Sequence Number (SN) Field and a 4 bit Sequence Number Protection (SNP) Field. The SN provides a sequential identifier for each SDU. The SNP provides an error detection and correction capability for the SAR Header, protecting the sequence numbering of SDUs.

The Sequence Count Procedure specifies the manner in which the SN and SNP are analysed. The results of the analysis indicate the arrival of a normal Payload or highlight a loss/mis-insertion of Payload. Corrective procedures depend on the service offered to the AAL user.

The services offered to the AAL user require differing timing arrangements. If there is a need to synchronise a local (service) clock with a standard (network) clock, the Synchronous Residual Time Stamp (SRTS) Procedure can be used. The SRTS measures the difference between the service and network clocks at the source and transfers the information to the destination. Only the variation in timing is transmitted. An Adaptive Clock Method is also available, in which the fill level of a buffer is used to control the frequency of a local clock.

14.7.5 Format

Type 1 Services use a Structure Parameter to transfer AAL user data. The 47 octets of data used by the CS has two formats, termed P and Non-P. Their use depends on how the Payload is filled. In the Non-P procedure, the whole Payload is filled with user information, as shown in Fig. 14.12.

In the P procedure, the first octet of the SAR Payload is allocated as a pointer. The pointer is used to indicate the start of the structured data, which can be in the information fields of the first or second PDUs. The pointer field is illustrated in Fig. 14.13.

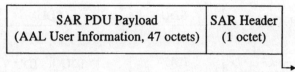

Fig. 14.12   Non-P format

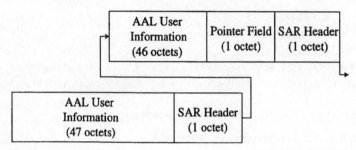

Fig. 14.13   P format

14.8   AAL Type 2

AAL Type 2 provides services with a variable source bit rate. It also permits the transfer of timing information between the source and destination entities. The functions performed by AAL Type 2 are similar to those for Type 1, but tailored to handle a variable source bit rate. Many aspects of AAL Type 2 are being derived and are not yet specified.

14.9   AAL Types 3/4 and 5

14.9.1   Services

Service Types 3/4 and 5 can transfer both Fixed Length and Variable Length SDUs. Two modes of operation can be adopted; Message Mode and Streaming Mode. The modes relate to how the information from the AAL user is passed to the AAL.

For Message Mode Services, the information from the user (the AAL SDU) is passed across the AAL interface in a single data unit, termed an Interface Data Unit (IDU). This is illustrated in Fig. 14.14. For Streaming Mode Services, the SDU is passed across the AAL interface in one or more IDUs.

Both modes of operation make use of the segmentation & reassembly and blocking/deblocking techniques to vary the nature of the service offered. They also include Assured and Non-Assured transfer of information. In Assured Transfer, every SDU is delivered in sequence and with guaranteed data content.

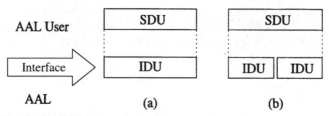

Fig. 14.14 Service modes
         a Messaging; b Streaming

Retransmission and flow control are provided. In Non-Assured Operation, SDUs may be lost or corrupted without retransmission and flow control is optional.

Service Types 3/4 and 5 differ in the following respects:

- details of connections between peer entities;
- data transfer parameters;
- format structure.

14.9.2 Architecture

In AAL Types 3/4 and 5, the CS is divided into a Common Part (CPCS) and a Service Specific Part (SSCS), as illustrated in Figs. 14.3 and 14.4. The CPCS and SAR form the AAL Common Part outlined in Section 14.3.

14.9.3 Primitives

The primitives between the AAL user and the AAL are service dependent and have not yet been specified.

Primitives between sub-layers are not accorded the same status as those between full layers. To avoid confusion, the primitives between the sub-layers are termed Invoke and Signal, instead of the conventional Request and Indication, as illustrated in Fig. 14.15.

14.9.4 Functions

AAL Types 3/4 and 5 provide capabilities to establish connections on both a point-to-point basis and a point-to-multi-point basis. For point-to-point connections, a user can exchange SDUs directly with another user, as illustrated in Fig. 14.16.

For point-to-multi-point connections, a user can exchange SDUs with more than one other user, as illustrated in Fig. 14.17. These multiple AAL connections can be made using a single ATM connection.

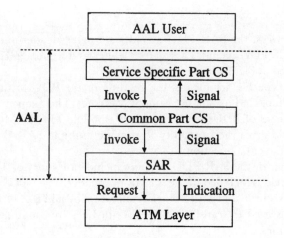

Fig. 14.15   Type 3/4 and 5 primitives

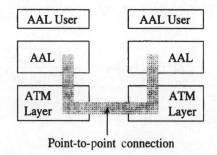

Fig. 14.16   Point-to-point connection
Source: (ITU-T Recommendation I.363)

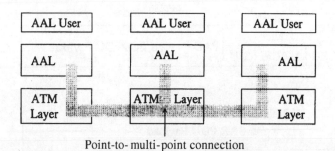

Fig. 14.17   Point-to-multipoint connection
Source: (ITU-T Recommendation I.363)

14.9.5 Format

For Type 3/4 Services, the SAR PDU includes a 2 octet Header and a 2 octet Trailer. The CPCS PDU includes a 4 octet Header and a 4 octet Trailer. The format is illustrated in Fig. 14.18.

The Segment Type Field indicates the relation of the PDU to the message (Start, Continuation, End or Single Segment Message). The Sequence Number Field allows a stream of PDUs to be numbered, modulo 16. The Multiplexing Identification Field gives the capability to provide multiple SAR Connections on a single ATM Layer Connection.

For Type 5 Services, the SAR PDU does not include a Header or Trailer, but a Payload Type Field is included in the ATM Cell Header. The CPCS PDU does not have a Header, but includes an 8 octet Trailer, as shown in Fig. 14.19.

The UU field is used to transfer CPCS User-to-User Information transparently between entities.

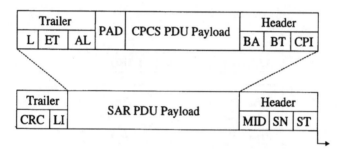

Fig. 14.18 Format for type 3/4 services

    CPI = Common Part Indicator   L = Length
    BT = Beginning Tag   ST = Segment Type
    BA = Buffer Allocation Size   SN = Sequence Number
    PAD = Padding   MID = Multiplexing Identification
    AL = Alignment   LI = Length Indicator
    ET = End Tag   CRC = Cyclic Redundancy Check

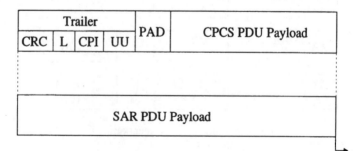

Fig. 14.19 Format for type 5 services

14.10 Chapter Summary

Although 64kbit/s circuits are adequate for telephony, the emergence of multimedia services is spawning massive increases in bandwidth requirements. Networks are responding by establishing secure and flexible high-bandwidth transport capabilities, e.g. based on asynchronous transfer mode (ATM) technologies. Several forms of signalling are possible for broadband networks. This chapter focuses on the ITU-T recommendations and ATM technologies to explain the principles.

The ITU-T signalling for broadband communications is structured into layers. The ATM Adaptation Layer (AAL), the ATM Layer and the Physical Layer establish a broadband platform. For the User-Network Interface (UNI), the User is DSS2 Layer 3. For the Network Node Interface (NNI), the User is the B-ISUP and MTP Level 3.

Each layer forms a Protocol Data Unit (PDU), which consists of a Payload, a Header and a Trailer. The Payload contains the data to be transferred between peer entities. The Header is used to route the PDU through the network. The Trailer can have various functions, depending on the application. The information provided by an adjacent layer is included in a Service Data Unit (SDU).

The ATM Layer provides a service-independent transport function. ATM is a packet-oriented transfer mechanism that carries PDUs (cells) transparently over Virtual Circuits. Each cell has a Header of 5 octets for routing and a Payload of 48 octets. ATM-Data Request and Indication Primitives are used between the ATM Layer and the AAL. The signalling for ATM is carried on a separate ATM Layer Connection from the user information, consistent with the principles of common channel signalling.

The AAL is divided into a Convergence Sub-Layer (CS) and a Segmentation and Reassembly (SAR) Sub-Layer. The CS is further divided into a Service Specific Part and a Common Part to take account of the AAL dependence on higher layer requirements.

The functions performed by the AAL depend upon the type of user and application. The functions are grouped into services that are offered to the higher layers. Type 1 Services transfer SDUs with a constant bit rate and also permit the transfer of timing and structure information. The AAL-Unitdata Request and Indication Primitives are transferred between the AAL and its User. AAL Type 2 provides Services with a variable bit rate. The functions performed by Type 2 Services are still being specified.

AAL Service Types 3, 4 and 5 can transfer both fixed length and variable length SDUs. The Primitives used between Sub-Layers are Invoke and Signal, reflecting the semi-formal nature of the Sub-Layers. In Message Mode Services, the AAL SDU is passed across the AAL Interface in a single Interface Data Unit (IDU). In Streaming Mode Services, the SDU is passed across the AAL Interface in one or more IDUs. Types 3, 4 and 5 are similar in nature, but differ in the details of the connections used, the data transfer parameters and format structure.

In summary, the Broadband Platform based on the AAL, ATM Layer and Physical Layer provides a foundation for the higher-layer signalling functions of DSS2 and B-ISUP. The combination of these functions establishes the future basis for networks in their evolution to meet the ever increasing demands of its customers.

14.11 References

1. ITU-T Recommendation Q.2010: 'Broadband ISDN Overview' (ITU, Geneva)
2. ITU-T Recommendation Q.2120: 'B-ISDN Meta-Signalling Protocol' (ITU, Geneva)
3. ITU-T Recommendation I.361: 'B-ISDN ATM Layer Specification' (ITU, Geneva)
4. ITU-T Recommendation I.363: 'B-ISDN ATM Adaptation Layer Specification' (ITU, Geneva)

Chapter 15
Broadband ISDN User Part for CCSS7

15.1 Introduction

The ITU-T signalling standards for broadband services are structured according to the layers outlined in Chapter 14. This chapter focuses on the higher layers for broadband inter-nodal signalling, namely the Broadband ISDN User Part (B-ISUP)[1]. The specification follows the trend outlined in Chapter 3 for a greater degree of modularity and a more open approach.

Other standards for broadband signalling are also being proposed. The ITU-T standards are used here to describe the principles of evolving towards broadband operation.

15.2 Architecture

The ISUP for narrowband (N-ISUP) takes a step towards the goal of modularity by separating call control from bearer control, using appropriate parameters to provide discrimination. The B-ISUP takes the modular approach to specification much further and separates key functions more comprehensively.

The specification model illustrated in Fig. 15.1 is used to achieve the requisite degree of modularity for basic calls. The model is based on the OSI Application Layer Model (described in Chapter 3). The model separates the actions that a node performs (Nodal Functions) from the communication protocols that occur between peer entities. The Nodal Functions contain procedures for Call Control, Maintenance and Compatibility. The communications protocols are sub-divided into the Single Association Control Function (SACF) and four Application Service Elements (ASEs), namely Call Control ASE, Bearer Control ASE, Maintenance Control ASE and Unrecognised Information ASE.

The B-ISUP Application Entity (AE) provides all the communication capabilities required by the nodal functions. Each ASE offers a service through the Single Association Control Function (SACF). The SACF offers a group of services to the Nodal Functions and Network Interface by defining the rules applicable to the ASEs.

The Call Control and Bearer Control ASEs consist of two sets of functions. One set of functions relates to the incoming side of the node and supports a

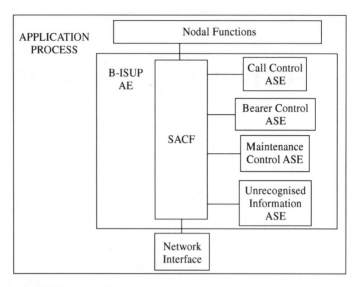

Fig. 15.1   B-ISUP basic call model

signalling association with a preceding node. The other set relates to the outgoing side of the node and supports the signalling association with a subsequent node. Each time a signalling association is required, an instance of the B-ISUP AE is created. Each instance within a node is identified by a Signalling Identifier (SID). The SID is used to distribute messages to the appropriate AE.

The Network Interface provides access to and from the MTP Level 3.

15.3 Application of the Architecture

Fig. 15.2 illustrates the application of the architecture to an intermediate node and shows the B-ISUP AEs for incoming and outgoing virtual circuits.

15.3.1 Dynamic modelling

Dynamic modelling refers to the set-up and clear-down of B-ISUP AEs. At an originating node, a nodal function initiates a signalling association in order, for example, to establish a call. The nodal function decides that B-ISUP is required and it creates an instance of the B-ISUP AE. When the signalling association is no longer required, the B-ISUP instance is deleted. At an intermediate or destination node, receipt of a message initiates an analysis of the SID. If the SID does not correspond to an existing instance, then an appropriate B-ISUP AE instance is created.

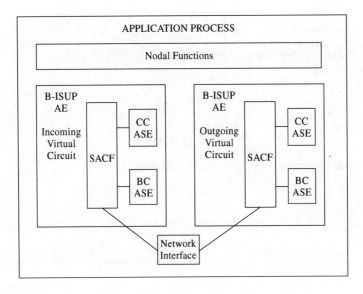

Fig. 15.2   Architecture applied to intermediate node

15.3.2 Static modelling

Static modelling refers to the use of existing B-ISUP instances, i.e. when a signalling association already exists. At an originating node, a nodal function requiring to communicate with a peer entity uses the services of the appropriate AE instances. Fig. 15.3 gives an example of the information flows when a nodal function needs a signalling association with call control and bearer control significance.

The SACF accepts a request from the Nodal Function and determines that both the Call Control and Bearer Control ASEs need to be involved. Two separate requests are passed to the ASEs and the responses are combined to

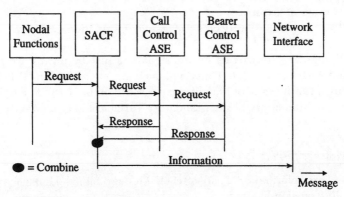

Fig. 15.3   Originating node information flows

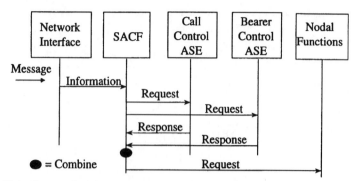

Fig. 15.4   Intermediate/destination node information flows

permit the formation of appropriate information to be passed to the Network Interface.

At an intermediate or destination node, receipt of a message initiates an analysis of the SID. If the SID corresponds to an existing instance of B-ISUP AE, the message is routed to that instance. Fig. 15.4 gives an example of the information flows when a message with call control and bearer control significance is received. In this case, the SACF passes the appropriate information to the ASEs, combines the responses and forms a primitive to pass to the nodal functions. Further information on the primitives is given in Section 15.5.

15.4 B-ISUP Formats

15.4.1 Principles

The format principles for B-ISUP[2] are similar to those for the narrowband version (N-ISUP), but they differ to take account of:

- the nature of the connection being established (virtual connection rather than physical circuit);
- a wider range of features;
- more focus on compatibility of messages and information.

B-ISUP messages are carried on the signalling link by means of the MTP Level 3 and AAL Service Data Units explained in Chapter 14. The B-ISUP uses the concept of variable and optional fields described in the N-ISUP, but the categorisation into Mandatory Fixed, Mandatory Variable and Optional Parts is not adopted.

15.4.2 Message format

The Signalling Information Field of each Message Signal Unit encompasses the parts illustrated in Fig. 15.5.

Routing Label
Message Type Code
Message Length
Message Compatibility Information
Message Content

Fig. 15.5   B-ISUP message format

The Routing Label is described in Chapter 4. The Message Type Code has the same function as for the N-ISUP, each type code uniquely defining the function and format of each message. The Message Compatibility Information defines the behaviour of the signalling system if a message is considered unreasonable, either by being out of context or by being unrecognised. The Message Length Field defines the number of octets in the Message Compatibility Information and Message Content Fields.

15.4.3   Message content

The Message Content of each message encompasses appropriate parameters. A parameter can be of fixed or variable length and an indicator is included to define its length. Every parameter also includes compatibility information. The general format is illustrated in Fig. 15.6.

15.4.4   Name codes

To ensure maximum compatibility between B-ISUP and N-ISUP, four guidelines have been adopted for the allocation of message and parameter name codes. The guidelines are:

- messages and parameters that are used in both N-ISUP and B-ISUP have the same name codes;

Parameter Name
Length Indicator
Parameter Compatibility Information
Parameter Content

Fig. 15.6   Message content

268 *Telecommunications Signalling*

- message and parameter codes that are used in N-ISUP but not in the B-ISUP, are marked as reserved in B-ISUP;
- message and parameter codes that are used in B-ISUP but not in N-ISUP, are marked as reserved in N-ISUP;
- a code is allocated in B-ISUP to indicate the extension of the name code.

15.4.5 Examples of message formats

(a) Initial address message (IAM)

The Initial Address Message (IAM) is the first message to be sent during call set-up and it initiates seizure of an outgoing virtual channel. It contains the address digits (e.g. digits dialled by the customer to route the call) and other information relating to the routing and handling of a call. Examples of parameters are shown in Table 15.1.

(b) IAM Acknowledgement (IAA)

The IAA is sent in the backward direction in response to an IAM message. The IAA indicates that the IAM has been accepted and the requested

Table 15.1 Format of Initial Address Message

Parameter	Length (octets)	Description
AAL Parameters	Up to 22	information indicating the requested/proposed ATM adaptation layer attribute values
ATM Cell Rate	8–21	gives the number of cells per second that are required for the call
Broadband Bearer Capability	7–11	information in the forward direction indicating the type of broadband connection oriented bearer service required
Broadband High Layer Information	Up to 17	information for compatibility checking by the remote user
Called Party Number	7–15	as N-ISUP
Connection Element Identifier	8–9	identifies the ATM virtual connection and includes the VPCI and the VCI

Table 15.2 Format of IAM Acknowledgement

Parameter	Length (octets)	Description
Connection Element Identifier	8–9	see IAM
Destination Signalling Identifier	8–9	identifies the call control or maintenance association at the receiving end
Origination Signalling Identifier	8–9	assigned by a node to identify the signalling association at that node

bandwidth on the incoming connection is available. Examples of parameters are shown in Table 15.2.

(c) Address Complete Message (ACM)

The Address Complete Message (ACM) is sent by the destination node to indicate successful receipt of sufficient digits to route the call to the called customer. Examples of parameters are shown in Table 15.3.

(d) Answer (ANS) Message

The Answer (ANS) Message is sent by the destination node to indicate that the called customer has answered the call. Examples of parameters are shown in Table 15.4.

(e) Release (REL) Message.

The Release (REL) Message can be sent by the originating or destination nodes to clear-down the connection. Either the calling or the called customer can initiate the release of the connection. Examples of parameters are shown in Table 15.5.

Table 15.3 Format of Address Complete Message

Parameter	Length (octets)	Description
Destination Signalling Identifier	8–9	see IAA
Called Party's Indicators	5–6	information on the called party's status and category
Cause Indicators	6 upwards	information indicating where and why a call failed or was cleared

Table 15.4   Format of Answer Message

Parameter	Length (octets)	Description
AAL Parameters	Up to 22	see IAM
Additional Connected Number	6–15	an address pertaining to a supplementary service
Call History Information	6–7	information indicating the accumulated propagation delay of a connection

Table 15.5   Format of Release Message

Parameter	Length (octets)	Description
Automatic Congestion Level	5–6	information indicating that a particular level of congestion exists at the sending node
Cause indicators	6 upwards	see ACM
Destination Signalling Identifier	8–9	see IAM

15.4.6 Message Segmentation

Long messages can exceed the maximum message length of 272 octets that the MTP is capable of carrying. In this case, it is possible for an originating node to split the original message and send it in two parts. See the information on N-ISUP in Chapter 6 for more details.

15.5 B-ISUP Procedures

The procedures for the B-ISUP are based on describing the message flows between nodes and the primitive flows between elements of the application process. A summary of each is given to explain the principles involved.

15.5.1 Call Control – Nodal Functions

The Nodal Functions for Call Control are similar to those for the N-ISUP. This Section focuses on the differences that apply to the B-ISUP to make it applicable in a broadband environment.

Table 15.6   Call Control − example primitives

Primitive name	Corresponding message
Set-up	Initial Address
Address Complete	Address Complete
Incoming Resources Accepted	IAM Acknowledgement
Release	Release, Release Complete
Answer	Answer

(a) Primitive Interface

The Nodal Functions concerning Call Control use the services provided by the SACF. Examples of the primitives over this interface and the corresponding B-ISUP Messages are given in Table 15.6.

(b) Bandwidth and identifier assignment

When a route between two nodes is brought into service, the Virtual Path Connections (VPCs) between the two nodes are split into two ranges for call set-up purposes. One node becomes the Assigning Node for the first range of connections and the other node becomes the Assigning Node for the second range of connections. Each VPC is allocated a Virtual Path Connection Identifier (VPCI).

The Assigning Node for a particular VPC has control of its set-up: if the other node needs to use that VPC, it must request the permission of the Assigning Node. In this way, dual seizure of a connection is avoided entirely.

(c) Call/connection set-up

The procedures for call set-up are similar to those for narrowband and are illustrated in Fig. 15.7. Focus is placed on the differences that are applicable to B-ISUP. Both en bloc and overlap operation are specified: this text concentrates on the en bloc mode.

Upon receipt of a request to set up a call from a calling party, the originating node selects a route based on the called party number, bearer capability requirements and ATM cell rate. The originating node establishes an instance of the B-ISUP AE and issues a Set-Up Primitive to the instance. The primitive includes a Connection Element Identifier Parameter to identify the VPC and Virtual Channel. An IAM is sent to the intermediate node.

The intermediate node assigns the appropriate bandwidth and VPCI and returns an IAM Acknowledgement (IAA) Message to the originating node. This stimulates the originating node to 'switch through' (i.e. permit cell transfer through) the connection in the backward direction.

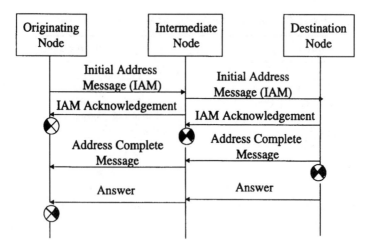

Fig. 15.7 Call/connection set-up

The intermediate node also analyses the routing information, creates an instance of the B-ISUP AE and issues a Set-up Primitive to the instance. This stimulates the sending of an IAM to the destination node.

The destination node receives the IAM and returns an IAA Message, which causes the intermediate node to switch through the connection. Analysis of the routing information indicates that sufficient information is available to connect to the called party and an Address Complete Message (ACM) is returned through the network.

The sending of an Awaiting Answer Indication (e.g. ring tone) from the destination node depends on the type of connection. For connections involving speech, etc. appropriate tones/messages are applied. Receipt of a connection indication from the called party causes the destination node to switch through.

The Answer Message is similar to that for narrowband. The latest point at which the originating node can switch through is upon receipt of the Answer Message.

(d) Call/connection release

The release procedures are similar to those for narrowband. Upon receipt of a Release Indication, a node makes the VPCI and bandwidth available for new traffic. After return of the Release Complete Message, the signalling association is terminated.

(e) Other procedures

Other procedures are similar to those for the narrowband ISUP, including:

- Segmentation, to handle messages that are too long to be carried by underlying layers (e.g. the MTP);

- Suspend/Resume, but these are only applicable when B-ISUP is interworking with narrowband ISUP;
- Progress, indicating that an event has occurred that should be drawn to the attention of the calling party;
- Propagation Delay, in which information is accumulated during set-up and returned to give a history of the connection set-up characteristics.

15.5.2 Compatibility – Nodal Functions

The N-ISUP has procedures for handling unreasonable information, e.g. when an updated version of the User Part sends a new message to an earlier version of the User Part. Recognition of such compatibility issues has grown significantly and the B-ISUP places more focus on appropriate procedures.

The procedures for B-ISUP are similar to those for the N-ISUP. However, the procedures are made more rigorous by establishing a general requirement that every message and parameter includes a Compatibility Information Field. Thus, the rules for receiving unreasonable information are defined for all messages and parameters, and the interworking of versions of User Parts are defined comprehensively.

15.5.3 Maintenance Control – Nodal Functions

These Nodal Functions relate to the management of calls and connections and include:

- resetting resources/connections;
- blocking and unblocking virtual paths;
- virtual path consistency check.

The Reset and Blocking Procedures are very similar to those described for the N-ISUP. The Virtual Path Consistency Check Procedure is used to confirm that user information can flow over a virtual path between two nodes. The procedure involves verifying the correct allocation of Virtual Path Connection Identifiers and the performance of a loop-back test to ensure that an appropriate connection is available.

15.5.4 Application Service Elements (ASEs)

(a) Primitive Flow

Each Application Service Element (ASE) offers a service, or a set of services, to the SACF. The SACF offers corresponding services to the Nodal Functions. These services are described by outlining the primitives that flow over the associated interfaces.

For messages outgoing from the node, the Nodal Functions decide on the next action and issue a pertinent primitive to the SACF. The SACF sends a corresponding primitive (or primitives) to the ASE(s), as illustrated in Fig. 15.8 for Call Control and Bearer Control. The parameters associated with the primitives to the ASEs are formed from the information received by the SACF from the Nodal Functions. The outputs from the ASEs are received by the SACF in a Transfer Request Primitive. These primitives are used to populate a parameter (User Data Field) in a Transfer Request Primitive sent to the Network Interface.

For incoming messages, the sequence for the outgoing messages is reversed. In this case, the Network Interface receives a message and sends a Transfer Indication Primitive to the SACF. The SACF distributes the information based on the message type and parameter values. After receiving replies from the ASEs, the SACF issues an appropriate primitive to the Nodal Functions.

(b) Primitive mapping

The services offered by the ASEs result from the analysis of received primitives. Table 15.7 gives examples of the primitives that are exchanged between the SACF and the Call Control/Bearer Control ASEs for successful transactions.

(c) Message mapping

The primitives from the ASEs are mapped onto corresponding B-ISUP messages that are exchanged between nodes. Examples of the mapping that takes place are given in Table 15.8.

(d) Example procedure

To illustrate the use of the primitives, consider a request by the Nodal Functions to set up and clear down a call to an adjacent node. Example procedures

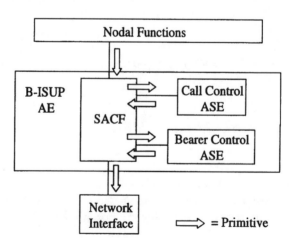

Fig. 15.8   Outgoing messages

Table 15.7   Primitive mapping

Primitive from nodal functions	Primitive to call control ASE	Primitive to bearer control ASE
Set Up Request	Call Set Up Request	Link Set Up Request
Address Complete Request	Call Address Complete	Link Information Request*
Answer Request	Call Answer Request	Link Information Request*
Release Request	Call Release Request	Link Release Request*

* = depends on parameter values

Table 15.8   Message mapping

Call control ASE primitive	Bearer control ASE primitive	B-ISUP message
Call Set Up Request	Link Set Up Request	Initial Address
Call Address Complete Request	Link Information Request*	Address Complete
Call Release Request	Link Release Request	Release
Call Answer Request	Link Information Request*	Answer

* = depends on parameter values

are illustrated in Fig. 15.9. In this case, the call also requires a connection to be established and both the Call Control (CC) and Bearer Control (BC) ASEs are involved. A Set Up Request Primitive received by the SACF causes corresponding primitives to be sent to the two ASEs.

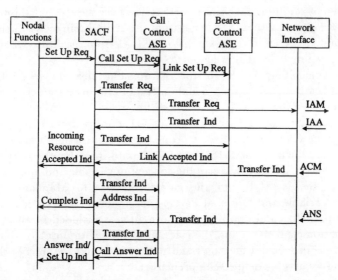

Fig. 15.9   Set-up procedure

The parameters received by the BC ASE in the Link Set Up Request Primitive are sent to the SACF in a Transfer Request Primitive for passing to the subsequent node in an Initial Address Message (IAM). If an IAM Acknowledgement (IAA) Message is received, this is passed to the BC ASE within a Transfer Indication Primitive. Receipt of such an acknowledgement indicates the successful set-up of a bearer connection and the contents of the IAM Acknowledgement Message are passed to the SACF within a Link Accepted Indication Primitive.

Similar procedures apply for receipt of the Address Complete (ACM) and Answer (ANS) Messages, but in these cases the CC ASE is responsible for handling the information (unless specific information is included, in which case then BC ASE is also involved).

15.6 Chapter Summary

The Broadband ISDN User Part (B-ISUP) makes use of the MTP Level 3 and the ATM and ATM Adaptation Layers to control the set-up, operation and release of broadband connections.

The specification model is modular to improve flexibility and the open approach to the signalling system. The model comprises Nodal Functions, a Single Association Control Function (SACF) and four Application Service Elements (ASEs). The ASEs are Call Control, Bearer Control, Maintenance Control and Unrecognised Information.

In the model, each ASE offers a service, or a set of services, to the SACF. The SACF offers corresponding services to the Nodal Functions. These services are described by outlining the primitives that flow over the associated interfaces. The services offered by the ASEs result from the analysis of received primitives and examples are given in the text. The primitives from the ASEs are mapped onto corresponding B-ISUP messages.

Dynamic modelling refers to the set-up and clear-down of B-ISUP Application Entities (AEs), e.g. a nodal function initiating a signalling association in order to establish a call. Static modelling refers to the use of existing B-ISUP instances, e.g. a nodal function needing to communicate with a peer entity using information flows with call control and bearer control significance.

The format principles for B-ISUP are similar to those for the narrowband version (N-ISUP). The B-ISUP uses the concept of variable and optional fields described in the N-ISUP, but the categorisation into Mandatory Fixed, Mandatory Variable and Optional Parts is not adopted. Focus is placed upon system compatibility and appropriate information is included in all messages and information elements. Thus, the rules for dealing with unreasonable information are defined for all messages and parameters, and the interworking of versions of user parts are defined comprehensively.

B-ISUP adopts a Message Segmentation Procedure to overcome the 272 octets limitation that the MTP imposes.

The procedures for call set-up are similar to those for narrowband. Each node selects a route based on the called party number, bearer capability requirements and ATM cell rate. A Connection Element Identifier Parameter identifies the Virtual Path Connection (VPC) and Virtual Channel (VC). The sequence uses an Initial Address Message (IAM), an Address Complete Message (ACM) and an Answer Message in a similar manner to N-ISUP, but an IAM Acknowledgement (IAA) Message is also used to permit cell flow.

The release procedures are similar to those for narrowband. Upon receipt of a Release Indication, a node makes the VPC Identifier and bandwidth available for new traffic. After return of the Release Complete Message, the signalling association is terminated.

Maintenance Nodal Functions relate to the management of calls and connections and include:

- resetting resources/connections;
- blocking and unblocking virtual paths;
- virtual path consistency check.

15.7 References

1 ITU-T Recommendation Q.2764: 'B-ISDN User Part Procedures' (ITU, Geneva)
2 ITU-T Recommendation Q.2763: 'B-ISDN User Part Formats and Codes Specification' (ITU, Geneva)

Chapter 16
Broadband Access Signalling, DSS2

16.1 Introduction

This chapter describes the Digital Subscriber Signalling System No.2 (DSS2), which defines the ITU-T procedures and formats for establishing and clearing broadband connections in the access network.[1] DSS2 is the broadband equivalent of DSS1, which is described in Chapter 12. DSS2, in conjunction with the CCSS7 Broadband ISDN User Part (B-ISUP), provides the opportunity for users to invoke broadband services.

DSS2 applies to the control of broadband ISDN (B-ISDN) point-to-point calls on virtual channels. The point-to-multipoint mode of operation that is an option for DSS1 is not used in DSS2 at this stage. The procedures are specified at the interface between B-ISDN terminal equipment and the network. The signalling system is allocated its own virtual channel. The procedures and formats for DSS2 are similar to those for DSS1. This Chapter focuses in the differences applicable to DSS2.

16.2 Capabilities

This Section describes the capabilities provided by DSS2 at present. Further capabilities will be added as part of the evolution strategy for the system.

16.2.1 Connections

The main purpose of the signalling system is to support the setting-up, operation and clearing-down of connections. The connections are of a switched nature, are established in real time and are not automatically re-established in the event of network failure. They operate on a point-to-point basis and represent a collection of ATM Virtual Channel Links that connect two end points (e.g. two user terminals). Such virtual channel connections are explained in Chapters 2 and 14.

The connections are bidirectional, but the bandwidth allocated to each direction can be different. This permits a large bandwidth to be provided in one direction (e.g. to the calling user) and a lower bandwidth to be provided in the other direction (e.g. to the called user). End-to-end compatibility parameters are specified to ensure that users can communicate effectively. Checks are made, for example, on the AAL Service Type (Types 1, 3/4 or 5) and the high-layer protocols that can be adopted.

In the future, it is possible that DSS2 will support more than one connection in relation to a call. However, at this stage, DSS2 supports one connection per call.

16.2.2 ATM services

DSS2 supports three forms of connection-oriented ATM Transport Service:

- constant bit rate with end-to-end timing requirements. The user requests the network to provide a desired bandwidth and the appropriate quality of service parameters when establishing the connection;
- variable bit rate without end-to-end timing requirements;
- AAL, traffic type and timing requirements are defined by the user. The user chooses only the desired bandwidth and quality of service when establishing the connection and negotiates the remaining parameters with the other user.

16.2.3 Error recovery

The error recovery capabilities of DSS2 include:

- detailed error handling procedures, including a means for one entity to inform another entity when it has encountered errors, such as message format/content and procedural errors;
- procedures for recovery from AAL Reset;
- capability to force calls, VCCs, and interfaces to an idle state;
- cause and diagnostic information for fault resolution;
- mechanisms to recover from loss of individual messages.

16.2.4 Narrowband interworking

The primary role of DSS2 is to support broadband services. However, narrowband services will continue to be provided and DSS2 needs to be compatible with their existence. This is specified in two ways:

(i) DSS2 supports interworking with narrowband ISDNs. Hence DSS2 can be used to establish connections that span both narrowband and broadband networks.

(ii) DSS2 directly supports narrowband services, i.e. signalling procedures are defined to establish and clear connections for narrowband services.

The following are underlying assumptions for broadband ISDNs providing narrowband services:

- The originating user does not know the type of network that will terminate the call (i.e. broadband or narrowband).
- Signalling interworking should be made as simple as possible.
- Information items that have end-to-end significance should be identified using the narrowband indicators.
- Information items that have global significance about the requested service should be identified using both narrowband and broadband indicators.

Further information on interworking is given in Section 16.7.6.

16.3 Primitives

The use of primitives in system definition is explained in Chapter 3. The primitives used between DSS2 and the ATM Adaptation Layer (AAL) are given in Table 16.1.

16.4 Message format

16.4.1 General format

The message format for DSS2 is very similar to that for DSS1, except that the Message Type Field is longer and a Message Length Field is included. The message format for DSS2 is illustrated in Fig. 16.1.

To avoid duplication of text, Table 16.2 gives a comparison of the message format for DSS1 and DSS2. The Table indicates the increased size of the fields to take account of an anticipated higher degree of connection activity and to increase flexibility.

Table 16.1   DSS2/AAL primitives

	Request	Indication	Response	Confirmation
Establish	Y	Y	N	Y
Release	Y	Y	N	Y
Data	Y	Y	N	N
Unit-Data	Y	Y	N	N

Y=Yes (supported), N=No (not supported)

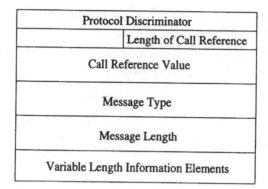

Fig. 16.1   Message format

Table 16.2   Comparison of formats for DSS1 and DSS2 messages

Field	DSS1	DSS2	Comment
Protocol Discriminator	Length 1 octet, Coded 00010000	Length 1 octet, Coded 10010000	identifies type of protocol and distinguishes DSS1 and DSS2
Call Reference	Default length 3 octets	Length 4 octets	longer for DSS2 to cater for more connection activities. Both use flag to indicate the source of the reference
Message Type	Length 1 octet	Length 2 octets	see Table 16.3
Message Length	Not provided	Length 2 octets	indicates length of message contents
Information Elements	Single octet and variable length	Variable length only	single octet not applicable to DSS2

16.4.2 Message type

The Message Type Field for DSS2 consists of two octets and is illustrated in Fig. 16.2.

The Flag Field indicates the form of error handling procedures to be adopted. Value 0 indicates that normal error handling procedures should be adopted, whereas Value 1 initiates specific instructions. The Action Indicator gives information on the handling of unrecognised messages (clear the call, ignore the message, etc.). The Extension Bit permits the length of the Message Type Field to be expanded.

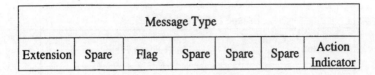

Fig. 16.2   Format for message Type Field

Table 16.3   Comparison of Message types for DSS1 and DSS2

Message Type	DSS1	DSS2/N-ISDN	DSS2/B-ISDN
Set-up	Y	Y	Y
Alerting	Y	Y	Y
Call Proceeding	Y	Y	Y
Connect	Y	Y	Y
Progress	Y	Y	N
Suspend	Y	N	N
Resume	Y	N	N
User Information	Y	Y	Y
Disconnect	Y	N	N
Release	Y	Y	Y
Release complete	Y	Y	Y
Notify	Y	Y	Y

Y=Yes (supported), N=No (not supported)

The Message Types for DSS1 and DSS2 are very similar, but some messages are deleted from the DSS2 specification. Table 16.3 compares the key messages. DSS2 has a set of messages for broadband call/connection control, indicated by B-ISDN in the Table. DSS2 also has a set of messages to facilitate interworking with narrowband services, indicated by N-ISDN in the Table.

It can be seen that most Message Types are common to DSS1 and DSS2. However, DSS2 does not support the Disconnect Message or the Suspend/Resume Messages. The Progress Message is not supported for B-ISDN.

16.4.3 Message Length Field

The Message Length Field identifies the length of the contents of the message, i.e. excluding the octets used for Protocol Discriminator, Call Reference, Message Type and the Message Length Indication itself. The Message Length Field consists of 2 octets.

16.4.4 Variable Length Information Elements

The Variable Length Information Elements have the format illustrated in Fig. 16.3. The Single Octet Information Elements used in DSS1 are not applicable to

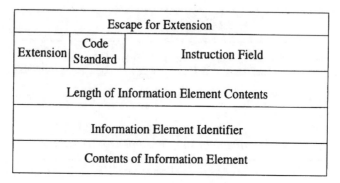

Fig. 16.3 *Format for Variable Length Information Elements*

DSS2. The Escape for Extension Field is an octet reserved to expand the number of Information Elements in the future.

The Instruction Field is an octet with a similar role to the Flag and Action Indicator in the Message Type Field. The Instruction Field indicates the actions that should be taken if unrecognised information elements are received.

The Length of Information Element Contents Field and the Information Element Identifier Field are both two octets in length. The information elements can appear in any order within the Message Contents Field.

Table 16.4 gives a comparison of example information elements for DSS1 and DSS2. The information elements in DSS2 are either designed for operation in a broadband environment using AAL and ATM Layers, or to facilitate the operation of narrowband services.

Table 16.4 DSS1/DSS2 information elements

Information element	DSS1	DSS2
Narrowband Bearer Capability	Y	Y
Broadband Bearer Capability	N	Y
Channel Identification	Y	N
Connection Identifier	N	Y
Signal	Y	N
Display	Y	N
Calling Party Number	Y	Y
Called Party Number	Y	Y
ATM Traffic Descriptor	N	Y
AAL Parameters	N	Y
High Layer Compatibility	Y	Y
Cause	Y	Y

Y=Yes (supported), N=No (not supported)

16.5 Message Examples

This Section gives examples of the Set-up and Release Messages to illustrate the use of information elements. The messages have the same role as in DSS1.

16.5.1 Set-up Message

Table 16.5 Example of Set-up Message

Information element	Type	Length (octets)	Description
Protocol Discriminator	M	1	see Section 16.4
Call Reference	M	4	see Section 16.4
Message Type	M	2	see Section 16.4
Message Length	M	2	see Section 16.4
ATM Traffic Descriptor	M	12–20	indicates peak ATM cell rate
Broadband Bearer Capability	M	6–7	requests a Broadband Connection-Oriented Bearer Service
Connection Identifier	O	9	identifies the local ATM connection resources
Called/Calling Party Number	O	4+	—
AAL Parameters	O	4–21	specifies a range of requested parameter values for AAL procedures
Broadband High Layer Information	O	4–13	used for compatibility checking at interworking point or called user
End-to-End Transit delay	O	4–10	indicates maximum transit delay acceptable/expected

16.5.2 Release Message

Table 16.6 Example of Release Message

Information element	Type	Length (octets)	Description
Protocol Discriminator	M	1	see Section 16.4
Call Reference	M	4	see Section 16.4
Message Type	M	2	see Section 16.4
Message Length	M	2	see Section 16.4
Cause	M	6–34	gives reason for release
Notification Indicator	O	4+	gives information pertaining to a call

16.6 Codesets

The extension of codesets uses the same technique as described for DSS1 in Chapter 12. A broadband Locking and Non-Locking Procedure is specified.

16.7 Call/Connection Procedures

16.7.1 Message sequence – establishment

The message sequence to establish a call/connection between the user terminal and the network is very similar to that for DSS1. The sequence uses:

- a Set-up Message to initiate the establishment of the call;
- a Call Proceeding Message to indicate that the call is being processed;
- an Alerting Message to indicate that the called user has been notified of an incoming call;
- a Connect Message to indicate acceptance (answer) of the call by the called user.

The Set-up Message includes a Call Reference to identify messages pertaining to the call. The message also includes ATM Traffic Descriptor, Broadband Bearer Capability and Quality of Service Information Elements defining the requirements for the call. DSS2 supports both the en bloc and overlap provision of called party address digits.

Upon receiving a Set-up Message at the destination node, the point-to-point method of alerting the called user is implemented. The broadcast form of alerting is not supported at this stage. Further details of the message sequence and procedures are given in Chapter 12 for DSS1.

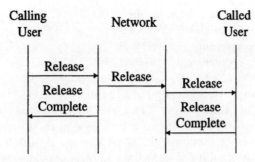

Fig. 16.4   *Release sequence*

16.7.2 *Message sequence – release*

The message sequence for releasing the call/connection is similar to that for DSS1, but DSS2 adopts a two-message sequence. Only the Release and Release Complete Messages are used and the Disconnect Message is not supported. The sequence for a calling user initiating release using DSS2 is illustrated in Fig. 16.4.

The calling user sends a Release Message and disconnects the virtual channel to the network. In the disconnected state, the call/connection is not available for use, but the Call Reference still exists.

The network disconnects the virtual channel and returns a Release Complete Message to the Calling User. The network then releases the virtual channel and its call reference. In the released state, the Call Reference and virtual channel can be re-used for another call.

Upon receipt of the Release Complete Message, the calling user releases the virtual channel and Call Reference. The Release Complete Message has local significance and does not relate to the state of the called user.

16.7.3 *Connection identification*

There are two cases for connection identification, Associated Signalling and Non-Associated Signalling. However, it is desirable to recall the terminology in Chapter 14 to assist in explaining the connection identification. A Virtual Channel (VC) provides bandwidth for the transfer of data between two entities (e.g. a user and the network). A Virtual Channel Identifier (VCI) relates the data with the VC. A Virtual Path (VP) is a grouping of one or more VCs. A Virtual Path Identifier (VPI) relates the VCs with the VP. A Virtual Path Connection (VPC) is a concatenation of VP links and it is identified by a Virtual Path Connection Identifier (VPCI).

DSS2 entities, e.g. a user terminal and a network termination, exchange signalling information over a VC within a VPC. In Associated Signalling, a DSS2 entity exclusively controls the VCs within that VPC. There can exist one or more VPCs between the two DSS2 entities. In Non-Associated Signalling, a DSS2 entity again controls the VCs in the VPC containing the signalling VC. However, the entity can also control VCs in other VPCs. Thus, more flexibility and capability are required.

The standard requirement specified for DSS2 is the Non-Associated Signalling case. Associated Signalling is available as an option. For Non-Associated Signalling, the allocation of a channel depends on the user's request. The user can either indicate the VPCI and VCI required, or can leave the selection to the network. The choice is indicated in the Connection Identifier Information Element within the Call Set-up Message. As a result of checking the request and the availability of channels, the network informs the user of the allocated channel by means of the Connection Identifier Information Element within the first message returned to the user, e.g. the Call Proceeding Message.

For Associated Signalling, the procedures are similar but the choice of channel available is more restricted.

16.7.4 Compatibility checking

Upon receiving a Set-up Message, the called user performs a compatibility check based on the compatibility information in the message. There are two categories of compatibility information.

In Category 1, compatibility information is provided for both the network and the called user to determine the attributes of the ATM connection. This is based on the following information elements:

- Broadband Bearer Capability;
- End-to-End Transit Delay;
- ATM Traffic Descriptor;
- Quality of Service.

In Category 2, compatibility information is provided for the called user, based on the following information elements:

- ATM Adaptation Layer;
- Broadband Low-Layer Information;
- Broadband High-Layer Information.

If the compatibility check succeeds, the call proceeds as normal. If compatibility fails, the called user returns a Release Complete Message with appropriate information as to the cause of the failure.

16.7.5 Support of N-ISDN Services

The term 'N-ISDN Services' refers to the basic 64 kbit/s circuit-switched ISDN services specified as part of DSS1. The aim is to allow the provision of N-ISDN services in a broadband environment.

To achieve N-ISDN Services capability, the narrowband Bearer Capability, High-Layer Compatibility and Low-Layer Compatibility Information Elements are defined in DSS2. To work in the broadband environment, the information elements are converted to the DSS2 format technique, e.g. adding the DSS2 Instruction Field and increasing the Length Field to two octets.

When N-ISDN Services are requested, the information elements of DSS2 are used in the same manner as for broadband services. However, some information elements from N-ISDN need to be considered e.g. for compatibility checking. Table 16.7 illustrates the information elements that need attention for the provision of N-ISDN Services.

The following concepts are adopted for the bearer service:

- Both the Broadband and Narrowband Bearer Capability Information Elements are included in the Set-up Message, because a calling user does not always know if the called user implements DSS1 or DSS2.
- The ATM cell rate within the Traffic Descriptor Information Element is set at a suitable value to transport the circuit-switched bit rate.
- The Quality of Service parameter is set as unqualified.

For low-layer compatibility, either the Broadband or the Narrowband Information Elements are used (not both), depending on the nature of the service being requested. For high-layer compatibility, the Narrowband Information Element is included in the Set-up Message and transported transparently through the broadband network. No specific broadband Information Element is required in this case.

16.7.6 Interworking of N-ISDN and broadband

When a call crosses both a narrowband network and a broadband network, an Interworking Function (IWF), e.g. a terminal adapter, is required. The IWF

Table 16.7 Information elements to support N-ISDN

	Bearer attributes	Low-layer attributes	High-layer attributes
N-ISDN	Narrowband Bearer Capability	Narrowband Low-Layer Compatibility	Narrowband High-Layer Compatibility
B-ISDN	Broadband Bearer Capability ATM Traffic Descriptor Quality of Service End-to-End Transit Delay	AAL Parameters	—

modifies the format of the information elements according to the new network type. Hence, when a call is routed from a narrowband environment to a broadband environment, the IWF inserts an Instruction Field and a second octet into the Length Field of Information Elements. Conversely, the Instruction Field and the second octet of the Length Field are deleted when a call is routed from a broadband environment to a narrowband environment.

The IWF also makes appropriate changes to the information elements to ensure that they are compatible with the new network. When a call is routed from a narrowband environment to a broadband environment, the IWF inserts the following information elements:

- Broadband Bearer Capability;
- ATM Traffic Descriptor;
- Quality of Service Parameter;
- AAL Parameters.

When a call is routed from a broadband environment to a narrowband environment, the IWF deletes the information elements above (and the End-to-End Transit Delay Information Element).

Interworking requires DSS2 to support the Progress Indicator specified for DSS1. This allows the incident of interworking to be reported to the users.

16.8 Chapter Summary

The Digital Subscriber Signalling System No.2 (DSS2) defines the procedures and formats for establishing and clearing broadband connections in the access network. The procedures and formats for DSS2 are similar to those for DSS1 and this chapter focuses on the differences applicable in the broadband environment. DSS2 provides the capability to:

- support the setting-up, operation and clearing-down of broadband point-to-point connections. The bandwidth provided in each direction can be different;
- support a range of ATM Transport Services;
- handle errors and implement error recovery;
- support narrowband (N-ISDN) services;
- interwork with N-ISDN networks.

The message format for DSS2 is similar to that for DSS1 and includes a Protocol Discriminator, Call Reference and Message Type. The Call Reference is four octets in length. The Message Type includes an Action Indicator and Flag to give instructions upon receipt of unrecognised messages. A 2 octet Message Length Field is included and Information Elements are of Variable Length.

The message sequence for establishing connections is similar to that for DSS1 and includes Set-up, Call Proceeding, Alerting and Connect Messages. The information elements include information to define the type of connection required and its quality of service. Such information elements include Broadband Bearer Capability, ATM Traffic Descriptor, End-to-End Transit Delay and Quality of Service.

The message sequence for releasing connections is based on the Release and Release Complete Messages. The Disconnect Message for DSS1 is not supported.

There are two cases for connection identification. DSS2 entities exchange signalling information over a Virtual Channel (VC) within a Virtual Path Connection (VPC). In Associated Signalling, a DSS2 entity exclusively controls the VCs within that VPC. In Non-Associated Signalling, a DSS2 entity again controls the VCs in the VPC containing the Signalling VC. However, the entity can also control VCs in other VPCs. The standard connection identification is the Non-Associated form.

Compatibility checking is performed at the destination end of the connection. Category 1 checks are performed by both the user and the network. Category 2 checks are performed by the user only.

DSS2 supports the provision of N-ISDN services by including Narrowband Information Elements for Bearer Capability, High-Layer Compatibility and Low-Layer Compatibility. The information elements are converted to the DSS2 format technique, e.g. adding the DSS2 Instruction Field. The ATM traffic Descriptor and Quality of Service Information Elements are coded according to the narrowband service.

DSS2 supports interworking with N-ISDNs by means of an Interworking Function (IWF). The IWF converts the format of information elements to the appropriate network. The IWF also inserts/deletes a range of information elements to ensure correct routing in the new network.

DSS2 will continue to evolve to permit broadband services to be as flexible as possible in the quest to meet the increasing communications demands of customers.

16.9 Reference

1 ITU-T Recommendation: Q.2931: 'B-ISDN User-Network Interface Layer 3 Specification' (ITU, Geneva).

Chapter 17
Interworking of CCS Systems

17.1 Introduction

Previous chapters have described the architecture, formats and procedures of the main CCS systems, namely CCSS7, DSS1 and DSS2. There are many principles that are common to the CCS systems. However, Chapter 2 explains that, because DSS1 and DSS2 are optimised for use in the access network and CCSS7 is optimised for inter-nodal use, the two types of system are different when considered at a detailed level. This chapter establishes the principles of interworking CCS systems by describing the interworking[1] of CCSS7 and DSS1. Examples of interworking for basic and more complex calls are given.

17.2 Interworking Principles

17.2.1 Constituents

Signalling facilitates the transfer of information (a) between users, (b) between users and nodes in the network and (c) between nodes in the network. In this context, users and nodes can be regarded as 'entities' wishing to communicate.

Chapter 3 describes the architecture of modern CCS systems and, in particular, the tiered structure adopted for specification purposes. The structure allows CCS systems to be defined in terms of:

- the primitives transferred between adjacent tiers within an entity, reflecting the services offered by one tier to another tier;
- the procedures used between corresponding tiers in different entities to transfer information;
- the formats of messages exchanged between corresponding tiers in different entities.

These constituents are illustrated in Fig. 17.1.

Provided there is an understanding of the relationship between the tiers in different signalling systems, the structure can be used as a basis for specifying interworking. The interworking of DSS1 and CCSS7 takes place at the highest tier of each system. For circuit-related signalling, this is Layer 3 for DSS1 and

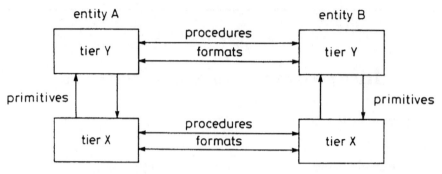

Fig. 17.1 Constituents of interworking

Level 4 for CCSS7. This Chapter concentrates on this aspect of interworking to illustrate the principles and therefore focuses on Layer 3 of DSS1 and the ISDN User Part (ISUP) of CCSS7.

This interworking typically takes place in a local exchange operating in an ISDN. Although the constituents of interworking concentrate upon operation within an ISDN, they also have to take into account the need to connect to analogue networks (e.g. an analogue public switched network) to complete some calls.

The means of defining the interworking of DSS1 and CCSS7 is to address each of the constituents mentioned above and document the relationship between the two signalling systems.

17.2.2 Primitive constituent

For CCSS7, the ISUP has an interface with the call control of a node and also with the MTP. Call control acts as a user of the ISUP and the MTP provides a service to the ISUP. Similarly for DSS1, Layer 3 has an interface with call control (user) and DSS1 Layer 2 (service provider).

In the case of basic circuit-switched calls, the primitive constituent of the interworking definition is covered by considering the relationship between each signalling system and the call control of the node (local exchange in this case). Fig. 17.2 gives an example of a model for defining the primitive constituent. The model consists of three groups of functions: call control, DSS1 Layer 3 and CCSS7 ISUP. In this case, DSS1 is treated as the incoming signalling system (i.e. the corresponding user has initiated the call).

The call control acts as an intermediary between the two signalling systems. Each signalling system communicates with the call control by means of primitives. When a signalling system transfers a primitive to call control, analysis takes place and this results in a primitive being returned to the same signalling system or to the other signalling system. There are four types of primitive, summarised in Table 17.1.

To avoid confusion, it is worth noting that the primitive constituent model is intended to provide a structured approach to the interworking process. The

Interworking of CCS Systems 295

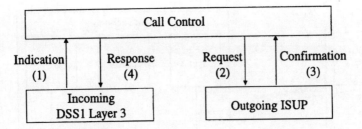

Fig. 17.2 Primitive constituent model

*Note: The numbers in brackets indicate the sending sequence (Source: ITU-T Recommendation Q. 699)

Table 17.1 Primitives

Primitive	Issued by	Purpose
Request	Call control	invokes a signalling procedure and transfers information to a peer entity
Indication	Signalling System	invokes a call control procedure indicates that a procedure has been invoked by a peer call control entity
Response	Call control	shows completion of a procedure previously invoked by an Indication
Confirmation	Signalling System	shows completion of a procedure previously invoked by a Request from the same call control

model does not imply that interworking is performed in this way in practical implementations.

17.2.3 Procedure constituent

Procedures can be documented by means of time sequence diagrams or the 'Specification and Description Language' (SDL). Time sequence diagrams show the flow of messages in a logical sequence and are used in previous Chapters of this book. The diagrams are supported by descriptive text.

The SDL[2] is a system that documents the status of an interface using symbols to denote incoming and outgoing messages and call processing decisions. The SDL is a very effective means of specifying procedures. It is a very comprehensive approach, particularly under failure conditions. However, this book continues to use the time sequence diagram as the clearest means of explaining message flows. The general format of a time sequence diagram used for interworking is given in Fig. 17.3.

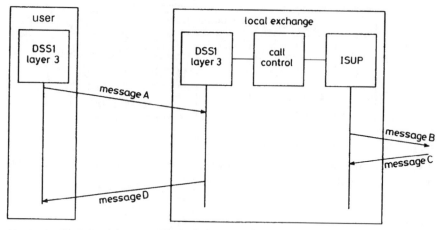

Fig. 17.3 *General time sequence diagram*

Message A is sent from the user's DSS1 Layer 3 to the corresponding function at the local exchange. This results in a primitive being sent to the local exchange call control. The call control processes the information provided by the primitive and decides on an appropriate action to take.

In this case, the appropriate action is to send a message to another exchange. Hence, a primitive is sent to the local exchange ISUP requesting a message to be formulated and routed accordingly. The ISUP thus sends Message B to the next exchange involved in the call.

Similar activities apply to Messages C and D. Although primitives are generated within the local exchange, the procedure constituent generally sees only an exchange of messages in a time sequence.

17.2.4 *Format constituent*

The type of message and the content of each message in DSS1 Layer 3 is compared with the equivalent information for the ISUP in a 'mapping table'. Thus, the information elements of DSS1 are mapped into the parameters of the ISUP. In some cases, there is a direct correlation between an information element and a parameter, whereas in other cases only a subset of an information element is mapped into a parameter. Information elements of DSS1 that are of local significance (i.e. relevant only to the access link) are not considered.

17.3 Example of Interworking for a Basic Call

17.3.1 *General*

Consider the example of User A (connected to Local Exchange A) initiating a call to User B (connected to Local Exchange B). Both users are connected to respective local exchanges by digital transmission links and the terminals at the users' premises both operate DSS1. The network connecting the two users is based on ISUP and uses en bloc procedures. The establishment and release of the call can be described in terms of the three constituents.

17.3.2 Basic call procedure constituent

The procedure constituent for basic call interworking is illustrated in Fig. 17.4.

The entities involved in the call are Users A and B, the respective local exchanges and a trunk (transit) exchange. In this example, both users operate the en bloc form of working and User B has a non-automatic answering terminal (this means an Alerting Message applies).

(a) Call set-up

User A initiates the call by instructing the terminal to establish a call to User B and providing the appropriate address of User B. If the call is a telephony call, this is equivalent to User A picking up a telephone handset and dialling (or keying) the telephone number of User B. These activities result in a Set-up

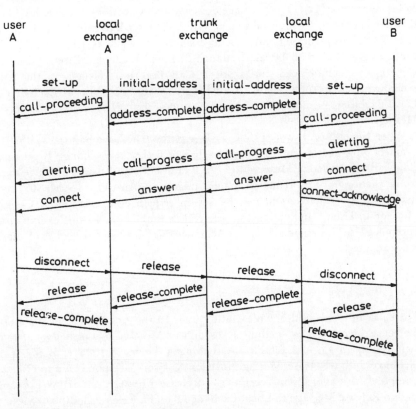

Fig. 17.4 Basic call, procedure constituent

Message being sent to Local Exchange A. The Set-up Message includes the address of User B and the type of connection required.

Local Exchange A analyses the Set-up Message and recognises that the call needs to be routed via the trunk exchange. Thus, the ISUP of Local Exchange A formulates an Initial Address Message (IAM) and sends it to the trunk exchange. Local Exchange A also returns a Call Proceeding Message to User A to indicate that the call is being processed.

Upon receipt of the IAM, the trunk exchange analyses the address of User B and determines that the call needs to be routed to Local Exchange B. Thus, the trunk exchange formulates an appropriate IAM and sends it to Local Exchange B. Local Exchange B analyses the information contained within the IAM and determines the identity of User B. A Set-up Message is sent to User B and an Address Complete Message is returned to the trunk exchange to indicate that enough information has been received to identify User B. In this example, Local Exchange B recognises that User B does not have a multiterminal configuration, which would require the broadcast form of working. Thus, the point-to-point form of working is adopted.

Upon receipt of the Set-up Message, the User B terminal adopts the option of returning a Call Proceeding Message to Local Exchange B to indicate that the call is being processed. This message does not involve actions at Local Exchange B, other than to re-set internal timers. When the User B terminal begins to alert User B that an incoming call is being made (e.g. begins to ring), an Alerting Message is returned to Local Exchange B. Local Exchange B recognises that alerting has commenced and sends a Call Progress Message to the trunk exchange: this, in turn, sends a Call Progress Message to Local Exchange A. Local Exchange A informs User A of the status of the call by sending an Alerting Message.

When the call is answered by User B (e.g. the telephone handset is lifted), a Connect Message is sent to Local Exchange B. Local Exchange B returns a Connect Acknowledge Message to User B and sends an Answer Message to the trunk exchange. The trunk exchange passes an Answer Message to Local Exchange A, which completes the set-up sequence by sending a Connect Message to User A indicating that the call is established. In this example, the option for User A to send a Connect Acknowledge Message to Local Exchange A is not adopted.

(b) Call clearing

The call can be cleared-down by either user. In this example, User A initiates the release sequence by sending a Disconnect Message to Local Exchange A. This results in Local Exchange A sending a Release Message to the trunk exchange and a Release Message to User A. User A responds with a Release Complete Message. Upon receipt of a Release Message, the trunk exchange sends a Release Message to Local Exchange B and a Release Complete Message to Local Exchange A. Upon receipt of a Release Message at Local Exchange B, a Disconnect Message is sent to User B and a Release Complete Message to the trunk exchange. Finally, upon receipt of a Release Message from User B, Local Exchange B sends a Release Complete Message to User B.

17.3.3 Basic call primitive constituent

For explanation purposes, consider the portion of the call involving User A and Local Exchange A, as illustrated in Fig. 17.5.

Exchange A has an 'incoming' signalling system (defined as receiving a set-up or Initial Address Message), an 'outgoing' signalling system (defined as sending a Set-up or Initial Address Message) and a call control. In this example, User A is initiating a call and has an outgoing DSS1. In the Figure, dotted lines indicate the transfer of primitives and solid lines indicate the transfer of messages.

(a) Call set-up

User A initiates the call by sending a Set-up Request Primitive to the outgoing DSS1 of User A. The outgoing DSS1 formulates a Set-up Message, including the address of User B and the type of connection required. The Set-up Message is sent to the incoming DSS1 of Local Exchange A, which causes a Set-up Indication Primitive to be sent to the call control of Local Exchange A.

The call control analyses the information provided in the primitive and takes three courses of action. First, it returns a Proceeding Request Primitive to the incoming DSS1, thus causing a Call Proceeding Message to be returned to User A. Secondly, the call control recognises that a call needs to be established via a trunk exchange and it requests the outgoing ISUP to formulate an IAM by sending a Set-up Request Primitive. The outgoing ISUP acts upon the request by formulating an IAM and sending it to the appropriate trunk exchange.

The third action of call control is to instruct the switch block to connect-through the traffic circuit involved in the call in the backward direction (so that User A can hear any tones provided by the network, if provided).

When the outgoing ISUP receives an Address Complete Message from the trunk exchange, a Proceeding Indication Primitive is sent to the call control. Receipt of this primitive does not result in the call control issuing a corresponding primitive; it allows the call control to release certain specialised information that is held in short-term memory for the call.

The next message to be received by the outgoing ISUP is Call Progress, indicating that User B is being alerted. This results in an Alerting Indication Primitive being passed to the call control. The call control recognises that User B is being alerted and that User A should be informed of the status of the call. An Alerting Request Primitive is sent to the incoming DSS1, resulting in an Alerting Message being sent to User A.

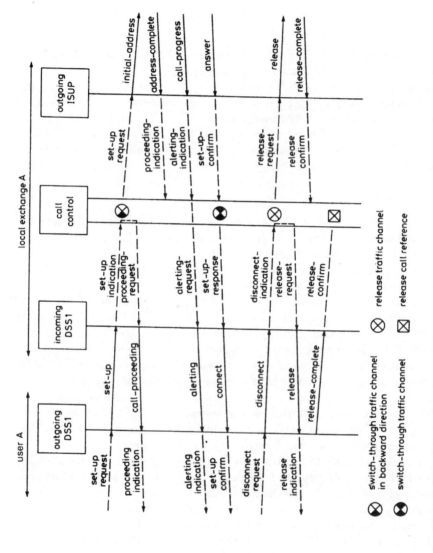

Fig. 17.5 Basic call, primitive constituent

When User B answers the call, an Answer Message is returned through the network to the outgoing ISUP at Local Exchange A. This results in a Set-up Confirm Primitive being passed to the call control. The call control recognises that User B has answered the call and instructs the switch-block to connect through the traffic circuit in the forward direction; this completes the traffic circuit connection. The call control also sends a Set-up Response Primitive to the incoming DSS1, causing the transmission of a Connect Message to User A.

The Connect Message indicates that the set-up sequence is complete.

(b) Call clearing

In this example, the release of the call is initiated by User A. This results in a Disconnect Request Primitive being sent to the outgoing DSS1 at User A. This, in turn, results in a Disconnect Message being sent to the incoming DSS1 at Local Exchange A. The receipt of the Disconnect Message causes a Disconnect Indication Primitive to be sent to the call control.

Upon receipt of the Disconnect Indication Primitive, the call control performs three activities. First, the call control recognises that the call needs to be cleared down in the network. Thus, the call control issues a Release Request Primitive to the outgoing ISUP, causing the sending of a Release Message to the trunk exchange. Secondly, the call control instructs the switch-block to release the speech circuit. Thirdly, the call control recognises that the access link needs to be cleared down. It issues a Release Request Primitive to the incoming DSS1, thus causing a Release Message to be sent to User A. When the release sequence is completed, the call control receives a Release Confirm Primitive from the incoming DSS1 and outgoing ISUP. Upon receipt of the Release Confirm Primitive from the incoming DSS1, the call control recognises that there is no further use for the access call reference and the final act is to release it and return it to the common pool, ready for allocation to another call.

17.3.4 Basic call format constituent

In this constituent, the DSS1 messages used in the procedure constituent are mapped onto the ISUP messages. The constituent thus applies to the two points of interworking, i.e. the two local exchanges. The mapping is presented in terms of the messages sent from User A to the network and from the network to User B. The mappings given below are illustrative and are not intended to be exhaustive, e.g. only some of the optional information elements and parameters are described.

(a) Set-up and IAM

The Set-up Message is mapped onto the Initial Address Message, as shown in Table 17.2.

Table 17.2 Mapping of set-up and initial-address messages

User A Set-up message	Network Initial-address message	User B Set-up message
Information element	Parameter	Information element
Bearer capability	User-service information Transmission-medium requirements	Bearer capability
No mapping	Forward-call indicator	No mapping
Progress indicator	Access transport	Progress indicator
Calling-party number	Calling-party number	Calling-party number
Called-party number	Called-party number	Called-party number

The Bearer Capability Information Element defines a large range of features relating to the type of connection and service required by User A. The element is mapped unchanged into an optional IAM parameter called User-Service Information. This parameter is transported through the network to Local Exchange B, thus allowing it to be mapped back into the Bearer Capability Information Element in the Set-up Message to User B.

However, a sub-set of the information in the Bearer Capability Information Element is required to enable the correct type of connection to be chosen when routing the call through the network. The sub-set chosen is incorporated into the Transmission Medium Requirements Parameter. For example, if it is necessary that a 64 kbit/s traffic circuit is provided between Users A and B, then the Transmission Medium Requirements Parameter in the IAM ensures that such a circuit is selected as the call is routed through the network. The mapping described allows the network to transport transparently (i.e. without analysis) the User Service Information Parameter while using a sub-set (in the form of the Transmission Medium Requirements Parameter) to route the call.

The Forward Call Indicator Parameter is used in the IAM to indicate the signalling capability of the call being established, e.g. whether or not end-to-end signalling is possible. This is an example of a parameter that is relevant to the network, but for which there is not a mapping to a DSS1 information element.

The Progress Indicator Information Element is used to keep the users informed of events that occur during the call. This information is not directly relevant to the network. It is therefore carried in a parameter called Access Transport, which is transferred between the local exchanges without being analysed.

The calling and called party numbers are examples of information elements that are mapped directly into IAM parameters. There is no mapping for the

Table 17.3 Mapping of connect and answer messages

User A Connect	Network Answer	User B Connect
Information element	Parameter	Information element
Progress indicator	Access transport	Progress indicator

Table 17.4 Mapping of disconnect and release messages

User A Disconnect	Network Release	User B Disconnect
Information element	Parameter	Information element
Cause	Cause	Cause

Call Proceeding Message because it is of local significance, i.e. it applies only to the access links.

(b) Connect and Answer

The mapping for the Connect and Answer Messages is given in Table 17.3.

(c) Disconnect and Release

The mapping of the Disconnect and Release Messages is given in Table 17.4.

The Cause Information Element indicates the reason for sending the Disconnect or Release Message. In the example of Users A and B, the cause is coded as Normal Call Clearing.

17.4 Example of Interworking for an Unsuccessful Basic Call

It is useful to consider an example of an unsuccessful call but, for brevity, only the procedure constituent of interworking is described. Consider the case described in Section 17.3.2 of User A attempting to establish a call to User B. However, in this example, User B is already busy with another call and cannot accept the call attempt by User A. The procedures are illustrated in Fig. 17.6.

The call is initiated by User A sending a Set-up Message to Local Exchange A. This causes the sending of an IAM through the network. Local Exchange B

304   Telecommunications Signalling

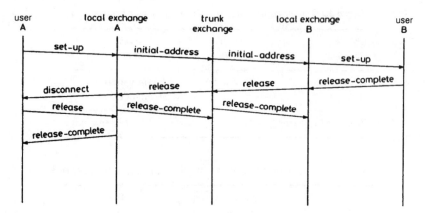

Fig. 17.6   Unsuccessful call, procedure constituent

sends a Set-up Message to User B. However, User B is busy. Hence, instead of responding with a Call Proceeding, Alerting or Connect Message, User B sends a Release Complete Message to Local Exchange B. The Release Complete Message contains an information element termed User Busy to indicate the reason for refusing the call.

The Release Complete Message is mapped into a Release Message to initiate the normal clear-down procedure in the network. The Release Message includes a cause parameter coded as User Busy. Upon receipt of a Release Message at Local Exchange A, the normal release procedures are enacted. The Disconnect Message includes the Cause Information Element to inform User A of the reason for the call being unsuccessful.

17.5 Example of Database Access Call

Exchanges within a network store a great deal of information to allow calls to be routed on demand. However, some routing information is specialised in nature and it can be more economic to store such information in a database rather than duplicate it in a large number of exchanges. In this case, exchanges that are nominated to gain access to specialised routing information need to recognise that a particular call requires the interrogation of a database.

17.5.1 Network structure

A typical network structure, in which a database contains specialised routing information, is illustrated in Fig. 17.7.

The normal routing of a call involves the originating local exchange, the trunk (transit) exchange and the terminating local exchange. However, in this case, the trunk exchange recognises that specialised routing information is required to complete the call and that it is stored in a database. The trunk

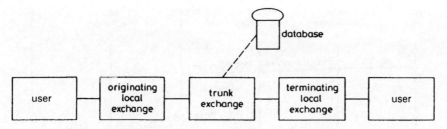

Fig. 17.7   Typical network structure

exchange thus gains access to the database, to determine the routing information. Using the specialised routing information, the trunk exchange completes the call by routing it to the appropriate terminating local exchange.

17.5.2 Typical service

An example of the database approach is a call that is charged to the called user rather than the calling user. This service is often called 'Freephone' or 'Toll-free' because each call is free to the calling user. An effective way of implementing such a service is to dedicate a National Number Group (NNG) code to the service. A NNG code is the number that is normally associated with a geographic area when dialling a trunk (or long-distance) call. A typical value of NNG for toll-free service is 800.

Thus, a calling user typically dials a code of +800XXXXXX, where the initial + is the prefix associated with a trunk or long-distance call (typically 0 or 1), the 800 code identifies the toll-free service and the remainder of the number identifies the called user. In the structure shown in Fig. 17.7, the originating local exchange only needs to recognise the prefix or the 800 NNG and pass the call to the trunk exchange. The trunk exchange recognises the 800 NNG and initiates an interrogation of the database to determine the routing of the call.

17.5.3 Procedure constituent

The procedure constituent is illustrated in Fig. 17.8. User A dials (or keys) the number +800XXXXXX. The user terminal formulates a DSS1 Set-up Message with the dialled number as the Called Party Number Information Element. This message is sent to Local Exchange A. Local Exchange A analyses the Set-up Message and, identifying either the prefix or the 800 NNG code, recognises that the call needs to be passed to the trunk exchange. Thus, the Local Exchange A ISUP formulates an Initial Address Message (IAM), including a parameter containing the Called Party Number, and sends it to the trunk exchange. Local Exchange A also establishes a traffic circuit to User A and to the trunk exchange.

Upon receipt of the IAM, the trunk exchange recognises that the call is a Freephone call and that specialised routing information is required from the database before the call can be established. The trunk exchange therefore estab-

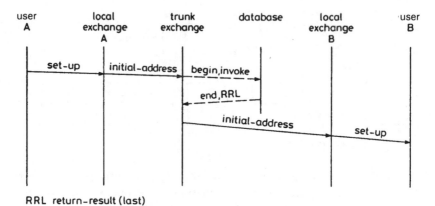

Fig. 17.8 *Simple database call, procedure constituent*

lishes a communication with the database. In this communication, there is no need for a traffic circuit: only an exchange of signalling information is required. The non-circuit-related form of CCSS7 (Transaction Capabilities) is therefore used to conduct the communication with the database.

The trunk exchange formulates a Begin Message, to establish the communication, including an Invoke Component. Upon receipt of the Begin Message, the database analyses the Invoke Component to determine the nature of the request. It recognises that the operation code requests routing information for the Called Party Number and it consults its records to provide the required information. When the routing information has been retrieved, it is included in an End Message with a Return Result (Last) Component. The End Message is used to signify that the communication is completed.

In this example, the routing information is supplied in the form of a new Called Party Number, with the 800 NNG code being translated into a traditional geographically based NNG and the XXXXXX being translated into a new user number.

When the trunk exchange receives the End Message, it extracts the new Called Party Number and recognises that the communication with the database is now completed for this call. The trunk exchange analyses the location of User B and formulates an ISUP IAM, including the new Called Party Number. The IAM is sent to Local Exchange B and a traffic circuit is also established to Local Exchange B. As far as Local Exchange B is concerned, the incoming call is a normal circuit-related call and it is treated accordingly to complete the set-up sequence.

17.5.4 Format constituent

For brevity, the format constituent is restricted here to giving an example of a typical Begin Message in Figure 17.9.

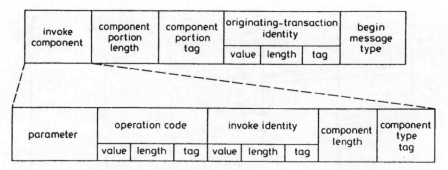

Fig. 17.9 *Typical Begin Message format*

The Begin Message includes a Transaction Identity that is used to correlate the messages used with the call. The Invoke Component contains an Invoke Identity that is used to correlate the response to the Invoke Component. The Invoke Component also includes an operation code that requests the database to examine the Called Party Number and provide routing information. The Called Party Number (i.e. the value received by the trunk exchange in the incoming IAM) is included in the parameter field following the operation code.

17.6 Example of a Complex Database Access Call

In this example, consider the same network structure as that described in Section 17.5. However, in this case, the sequence is made more complex by the database needing additional information about User A before it is able to supply the routing information to the trunk exchange.

For example, User B has two locations, one dealing with enquiries from users in one area and the other dealing with enquiries from another area. In this example, the set-up sequence is the same as that described in Section 17.5, up to and including the trunk exchange establishing a communication with the database. The additional sequences are shown in Fig. 17.10.

A Begin Message with an Invoke Component (Invoke 1) is sent to the database. In this case, the database needs to know to which location to route the call. To ascertain this additional information, the database issues a Continue Message that includes an Invoke Component (Invoke 2). The Continue Message is used to indicate that the message refers to the communication under consideration. The Invoke Component (Invoke 2) includes an operation code that requests the trunk exchange to supply the Calling Party Number.

Upon receipt of the Continue Message, the trunk exchange analyses the request for the Calling Party Number and supplies it in a Continue Message that includes a Return Result (Last) Component (reply to Invoke 2).

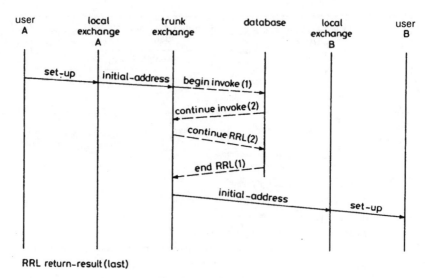

Fig. 17.10 *Complex database call, procedure constituent*

Receipt of the Calling Party Number allows the database to select the appropriate location of the User B and supply the corresponding Called Party Number in an End Message with a Return Result (Last) Component (reply to Invoke 1). The call set-up process continues as described in Section 17.5.

17.7 Chapter Summary

Modern CCS systems are specified in terms of:

- the primitives transferred between adjacent tiers within an entity, reflecting the services offered by one tier to another tier;
- the procedures used between corresponding tiers in different entities;
- the formats of messages exchanged between corresponding tiers in different entities.

These constituents are used to define the interworking of CCS systems. The principles are explained by considering the interworking of DSS1 and CCSS7 ISUP. The interworking takes place at Layer 3 for DSS1 and Level 4 for ISUP.

For the primitive constituent, a model has been devised comprising call control, DSS1 and ISUP. The call control acts as a processor, taking inputs from the signalling systems in the form of primitives, deciding upon action to be taken and issuing responses to the signalling systems. All communication between call control and the signalling systems in the model is conducted using the Indication, Response, Request and Confirm primitives. The model provides a disciplined approach to interworking but it does not define practical implementations.

The procedure constituent uses time sequence diagrams similar to those used in earlier Chapters. Message flows to and from entities illustrate the logical sequence of events that can take place for various types of call.

The format constituent is defined by mapping tables. Each DSS1 message is mapped onto an ISUP message, unless a particular message is of local significance only. Similarly, information elements of DSS1 are mapped onto parameters of ISUP.

Examples of the interworking of DSS1 and ISUP are given for a basic call, an unsuccessful call and calls needing the support of remote-network databases.

17.8 References

1 ITU-T Recommendations Q.699: 'Interworking Between DSS1 and the CCSS7 ISUP' (ITU, Geneva)
2 ITU-T Recommendation Z.100: 'Functional Specification and Description Language Criteria for Formal Description Techniques' (ITU, Geneva)

Chapter 18
Internet Protocols

18.1 Introduction

18.1.1 General

As described in Chapter 1, the telecommunications, broadcast/media, computing and Internet Protocol (IP) industries have a mutual goal to generate, store and manipulate information. As the Information Age evolves, the amount of information that is transferred across networks continues to increase substantially and there will be much more focus on data communications. The IP technologies are a key part of this evolution. Hence, whereas the focus of this book is on telecommunications signalling, this chapter gives a summary of IP signalling systems.

IP technologies were introduced to support open interfaces and facilitate the inter-operability of computers. IP was originally intended to provide a transparent transport capability over packet-switched data networks, particularly in wide area networks (WANs). A WAN is a data network designed for use over a large geographical area, thus imposing constraints and requirements over the way communications are provided. IP has now expanded to become a provider of communications to a wide range of users. IP is defined to operate primarily in a multiple-network environment.

The structure, organisation and routing approaches used in telecommunications and IP networks are different. However, the signalling systems in the two forms of network have many principles in common.

18.1.2 Terminology

Before entering the IP world, it is worth reflecting on some confusing terminology. The reader is advised to take notice of the context in which terms are used. However, the terminology issues do not negate the common principles of signalling systems in the two forms of network.

(a) Within IP

The term IP (Internet Protocol) is used in two ways. In the first way of usage, IP is a general description of Internet-type technologies. In this case, IP describes a technology that applies to a range of protocols, users, switches, etc.

In the second way of usage, IP refers to a specific protocol that provides network services.

Similarly, the term Transmission Control Protocol (TCP) can be used in different ways. The phrase TCP/IP is used to describe a technology type or a combination of protocols.

(b) Between telecommunications and IP

There is also disparity in terminology between telecommunications and IP. For example, the term 'Protocol' in telecommunications refers to the peer-to-peer signalling communication between entities (Chapter 3). In the IP world, the term can be used in the same way (i.e. as a signalling communication). However, an IP can also include a description of functions within the entity itself.

18.2 IP Signalling Architecture

18.2.1 Structure

Chapter 3 describes the tiered structure adopted by CCS systems in telecommunications networks. The IP signalling systems are also defined within a layered structure, which is illustrated in Fig. 18.1.

IP uses a four-tier structure, the tiers being called 'layers'. The Application Layer provides the interface to the users. It performs functions that are application-specific and provides data in a suitable format to be applied by the users.

A 'Host' is a node or computer that originates or terminates IP traffic. This is in contrast to nodes that forward IP traffic between hosts, which are known as Routers. A host, for example, allows users to communicate with other computers on a network. The IP Host-to-Host Layer provides a communication function between host computers. It is the role of this layer to facilitate the transfer of Application Layer data either by setting up end-to-end connections or by connectionless message transfer.

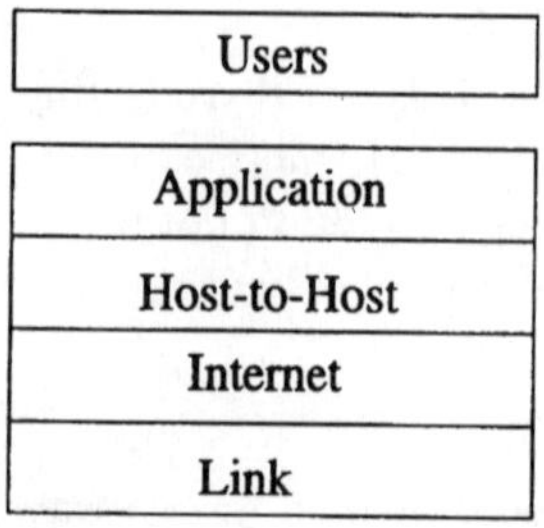

Users
Application
Host-to-Host
Internet
Link

Fig. 18.1   IP signalling architecture

The Internet Layer is responsible for routing data across networks from one end-user to another.

The Link Layer hides the details of the network implementation from the higher layers by defining the physical and link-related characteristics of the network. The attributes of the physical wiring, data rates and methods of access are defined.

18.2.2 Relationship with the OSI Model

Readers familiar with the OSI Model explained in Chapter 3 will recognise strong similarities with the 4-Layer IP architecture. Further work is needed to ensure that a comprehensive alignment is achieved, but Fig. 18.2 gives a guide to the relationship.

It can be seen that there is a moderate alignment between the models and the arrangement should not cause confusion, provided the context of the layers is clear.

The IP Application Layer combines the functions of the OSI Application, Presentation and Session Layers. This is not an issue in the short term, because existing telecommunications signalling systems aligning with the OSI Model (e.g. Transaction Capabilities, explained in Chapter 7) treat the Session and Presentation Layers as transparent. However, in the future, these layers will need further consideration if connection-oriented signalling becomes more ubiquitous.

18.2.3 Protocols

IP technologies incorporate a wide range of protocols. For clarity, focus is placed on some commonly used protocols. This allows the principles of IP tech-

IP Model	OSI Model
Application	Application
Application	Presentation
Application	Session
Host-to-Host	Transport
Internet	Network
Link	Data-Link
Link	Physical

Fig. 18.2 Relationships of OSI and IP models

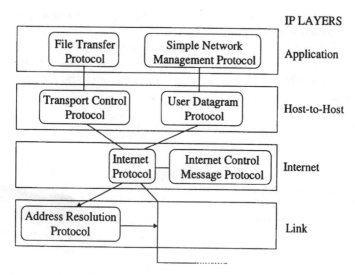

Fig. 18.3   Examples of IP protocols

nologies to be established without entering detailed specifications. Examples of such IP technology protocols and their fit within the IP Model are shown in Fig. 18.3.

A brief description of the examples is given in Table 18.1. More details of the examples are given in following sections.

Table 18.1   Example protocol descriptions

Protocol	Description
Address Resolution Protocol (ARP)	translates an IP address into a physical address
Internet Protocol (IP)	used to route IP packets
Internet Control Message Protocol (ICMP)	detects and reports routing errors
Transmission Control Protocol (TCP)	transfers information between users on a connection-oriented basis
User Datagram Protocol (UDP)	transfers information between users on a connectionless basis
File Transfer Protocol (FTP)	provides the ability to transfer files between users
Simple Network Management Protocol (SNMP)	facilitates the collection and alteration of management data

18.3 Address Resolution Protocol (ARP)

The Address Resolution Protocol (ARP)[1] fits within the Link Layer of the IP Model, as illustrated in Fig. 18.3.

A unit of data transferred between the Internet and Link Layers is termed an 'IP Packet'. The Address Resolution Protocol (ARP) translates an IP address for an outgoing IP Packet into the physical address of the destination point of the network. A typical physical network is an Ethernet, which is a means of connecting hosts to a cable at a standard bit rate.

The translations are held in a table listing all computers or users on a network. The translation table is filled with data as required. If a translation is not available for an IP Packet with a particular IP address, the IP Packet is stored or discarded and a procedure is invoked to determine the appropriate translation. The procedure is illustrated in Fig. 18.4.

The originating ARP broadcasts an ARP Request Packet to all entities connected to the network. The broadcast packet includes the IP and physical addresses of the originating ARP and the IP address for which a translation is being sought.

Upon receipt of the broadcast packet, the ARP functions in each entity compare the IP Address given in the packet with their own IP address. In this example, ARP 2 has the IP address that is requested in the broadcast message. ARP 2 therefore returns an ARP Response Packet to the originating ARP.

The ARP Response Packet consists of the IP and Physical Addresses of ARP 2, and the IP and Physical Addresses of the originating ARP. As the Physical Address of ARP 2 is included, the originating ARP can now perform the translation from the IP Address to the Physical Address. The address translation is stored in the originating ARP's routing table. The IP packet that invoked the procedure can be routed now (if previously stored) or upon retransmission (if previously discarded).

18.4 Internet Protocol (IP)

18.4.1 IP service

The Internet Protocol (IP)[2] fits within the Internet Layer of the IP Model, as illustrated in Fig. 18.3. IP makes use of the Link Layer to provide suitable media,

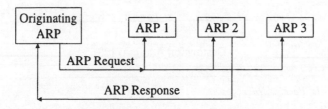

Fig. 18.4   Example of an ARP procedure

etc. IP offers a routing service to the higher layers, routing data between nodes in a connectionless manner. IP is kept simple and it does not guarantee a quality of service to its users. For example, the users of IP have to determine if IP packets are corrupted or mis-sequenced and take appropriate actions.

18.4.2 IP formats

The format of an IP Packet consists of a Header followed by data. The header format is based on 32-bit fields and is illustrated in Fig. 18.5.

The Version Number indicates the form of IP Protocol that is being used, e.g. Version 6. Thus, hosts and routers can discriminate between old and newer versions of the IP Protocol.

The Traffic Class is used by hosts and routers to determine the priority of the IP Packet. The value of the Traffic Class is selected by a higher-layer function at the originating host, but the value can be changed by routers as the packet is transferred through the network.

A 'flow' is a sequence of packets that an originating host wishes to be treated in a special manner by routers. A flow is applicable between a particular originating host and a particular destination host. A flow is identified by the Source Address and the Flow Label. The Flow Label is generated randomly by the originating host. The Flow Label value of zero is reserved to identify packets that do not belong to a flow.

The Payload Length gives the size of the packet (excluding the size of the header) in octets.

The IP Header can be followed by a range of other headers that identify options. Such headers are termed 'Extension Headers'. The Next Header Field indicates the type of Extension Header that follows the IP Header. If no options are adopted, the next header would be that of the higher-layer user (e.g. the TCP or the UDP). Further details are given in the following section.

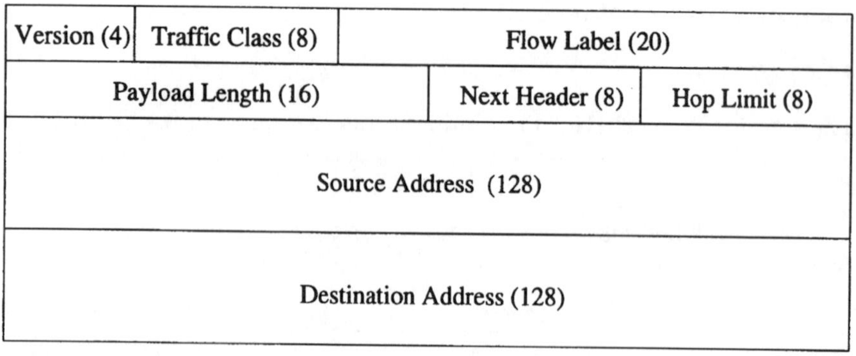

Version (4)	Traffic Class (8)	Flow Label (20)	
Payload Length (16)		Next Header (8)	Hop Limit (8)
Source Address (128)			
Destination Address (128)			

Fig. 18.5 IP Packet Header format
( ) = number of bits

The Hop Limit is used to prevent a packet encountering too many routers, e.g. avoiding loop routing. The value of the field is decremented by one by each router. If the value is decremented to zero, the packet is discarded.

The Source and Destination Addresses are the identities of the originating and destination hosts. (Note: the Destination Address is not necessarily the host that is ultimately required.)

18.4.3 Extension Headers

Optional information can be incorporated within Extension Headers. These Extension Headers are placed between the IP Header and the higher-layer Header (e.g. the TCP Header). In general, Extension Headers are not analysed at intermediate nodes: only hosts need to determine their contents. An exception is the Hop-by Hop Header, which does require analysis by intermediate nodes. If it is present, the Hop-by-Hop Header is always placed next to the IP Header, so that it is easily available for analysis.

Table 18.2 gives a summary of the Extension Headers and their purpose.

The Fragment Header allows large packets to be fragmented into smaller packets, thus permitting their transfer over low capacity media. The packets are reassembled at the destination node or a gateway node. This is very similar to

Table 18.2   Summary of Extension Headers

Extension Header	Description
Hop-by-Hop	carries information that must be analysed by each node in a packet's path
Destination Options	carries information that is of relevance only to the destination node
Routing	used by the source host to list one or more intermediate nodes through which the packet should be routed on its way to the destination host
Fragment	used by the source host to send a packet that is larger than the maximum frame length that can be sent on a physical medium. A packet is fragmented into smaller Fragment Packets
Authentication	provides a means of verifying the identity and integrity of the source of packets
Encapsulating Security Payload	provides a mix of security services, including confidentiality and data origin verification. It can be applied alone or in combination with the Authentication Header

the Segmentation and Reassembly function used in, for example, the SCCP (explained in Chapter 5). The IP Header, and any Extension Headers used to route the call, should not be fragmented. These fields form the first part of each of the fragmented packets.

18.4.4 Internet Control Message Protocol (ICMP)

IP offers a connectionless service without a guaranteed quality of performance. The Internet Control Message Protocol (ICMP)[3] is designed to report difficulties in transferring IP Packets to their destination. ICMP is not intended to improve the quality of IP, but rather to report on problems in the communications environment. ICMP is an integral part of IP and its implementation is mandatory.

ICMP is designed to operate between gateway nodes (interconnecting networks) and originating hosts. ICMP is invoked when, for example:

- an IP Packet cannot be delivered;
- an IP Packet is too big for an outgoing link;
- a problem is identified with parameters within the message that prevents processing.

Every ICMP message is preceded by an IP header. Extension Headers can also be present, e.g. the Authentication Header can be used to improve security. The general format of an ICMP message is illustrated in Fig. 18.6.

The Type Field indicates the type of the message and its value determines the format of the remaining part of the message. The Code Field expands the granularity of the Message Type.

The Checksum Field is used to detect data corruption in the ICMP Message and parts of the IP Header.

The first 32-bit field of the Message Body is either unused or it is specific to the message. For example, this field can be used to provide a pointer or information relating to the capacity of an outgoing link. The remainder of the Message Body is used to return all or part of the invoking packet.

Table 18.3 gives a summary of the Error Messages defined by ICMP. Other messages (Informational Messages) relate to echo control.

Type (8)	Code (8)	Checksum (16)
Message Body		

Fig. 18.6 Format of ICMP message

Table 18.3 Summary of ICMP error messages

Message name	Description
Destination Unreachable	generated when a packet cannot be delivered (except for congestion). Reasons are (a) no route, (b) destination prohibited, (c) address unreachable, (d) port unreachable
Packet Too Big	sent by a router when a packet exceeds the Maximum Transmission Unit of an outgoing link
Time Exceeded	Sent by a router if a packet has a Hop Limit of zero or is decremented to zero. Indicates a routing loop or an original hop limit value that was too small
Parameter Problem	generated when a problem with a field in an IP Header (or extension) prevents processing. A pointer identifies the octet where the error was detected

18.5 Transmission Control Protocol (TCP)

Whereas the Internet Protocol (IP) is kept very simple, with little error detection capability, the Transmission Control Protocol (TCP)[4] is designed to provide a higher degree of reliability and availability. TCP is required to provide services for a wide variety of users. TCP is a connection-oriented protocol and is designed to provide communications between hosts belonging to interconnected networks.

18.5.1 Architecture

TCP fits within the Host-to-Host Layer of the IP Model, as illustrated in Fig. 18.3. It makes use of the Internet Layer to provide network functions (typically the IP). It supplies services to the Application Layer and users.

The interfaces to higher layers are described in detail. The interfaces to lower layers (e.g. to IP) are only described in terms of being able to transfer data. This assumes that the interfaces to lower layers are specified by those lower layers. This allows a flexible approach to implementation, but it must be ensured that the lower layers do include the definition of the interface: the efficacy of the layered architecture is significantly reduced if the interfaces are not clearly defined.

18.5.2 Functions

(a) Segments

Data is transferred between the TCP and the lower layer (IP in this example) in 'TCP Segments' (also called 'Transport Messages'). Each segment has a header that contains fields covering addresses, sequence numbers, etc. The format of the header is explained in Section 18.5.3.

(b) Data transfer

In general, information supplied by a user is transferred by the TCP according to its own algorithms, e.g. a TCP might wait until a certain amount of data is available before sending it to the destination TCP. An additional capability, known as a 'push' procedure, is built into TCP. The push procedure allows a user to request a prompt delivery of outstanding information to the destination. TCP also has a flag to indicate that urgent data is available to the receiving TCP, with a pointer to indicate the location of the urgent data.

(c) Error detection and correction

TCP uses check bits and a positive acknowledgement non-compelled system for error detection and correction. The system has some common principles with CCSS7 (e.g. as described in Chapter 4 for the MTP), but is different in its implementation.

Check bits are generated by the TCP sending a segment (the sender). The sender applies an algorithm to the information to be transmitted. The check bits are analysed by the TCP receiving the segment (the receiver) by reversing the algorithm procedure. If the check bits do not generate the expected value when the receiver applies the algorithm, it is assumed that the segment is corrupted and it is discarded.

Error correction is based on allocating a sequence number to each octet transmitted. A timer period is started upon transmission of a segment from the sender and a copy of the segment is stored in a re-transmission buffer. If an acknowledgement is received by the sender within the timer period, then the segment is assumed to have been delivered successfully and the segment is deleted from the retransmission buffer. If an acknowledgement is not received within the timer period, it is assumed that the segment has been lost or discarded and the segment is re-transmitted.

(d) Flow control

The receiver can regulate the number of segments transmitted by the sender. This can be used to avoid the receiver becoming overloaded. This is achieved by including data flow information in each acknowledgement. Each acknowledgement includes a field that indicates the range of sequence numbers that can be satisfactorily received in the next segment. The sender is not permitted to transmit more than this amount of sequence numbers without obtaining permission in the next acknowledgement.

(e) Connections

Users within a host are identified by a set of internal addresses (or ports). The term 'socket' describes an internal address that is used in conjunction with the network and host addresses appropriate to the lower layers. Each socket therefore provides a unique description of the user access port within a host. When a connection is established it is identified by the sockets in the two communicating TCPs. A local reference number can be established between a TCP and its user to act as a short-cut description of the connection identity.

Connections can be established in either an active or passive manner. In active connection establishment, a user requests the TCP to establish a connection to another user. The connection is identified by two sockets; local and foreign. A local socket is the identity of the connection from the originating TCP perspective. A foreign socket is the identity used at the other end of the connection. Once established, the connection can be used to transfer data between the users.

In passive connection establishment, a user can inform the TCP that it will accept incoming connection requests. The user does this by issuing a Passive Open Request to the TCP. For example, a user that offers a service to other users can issue a Passive Open Request with an unspecified foreign socket, thus permitting TCP to establish a connection from any other user.

A connection record, termed a 'Transmission Control Block' (TCB) is established for a connection. The TCB catalogues a range of variables for a connection, including the socket numbers and pointers to buffers.

18.5.3 Formats

The TCP adds its header to the data transferred from the user, before passing the whole segment to the lower layer (e.g. IP). The format of the TCP segment is illustrated in Fig. 18.7. A brief explanation of the fields in Fig. 18.7 is given in Table 18.4.

The Control Field is sub-divided into six bits:

- an Urgent Pointer Indicator, identifying the need to interpret the Urgent Pointer Field;
- an Acknowledgement Indicator, identifying the need to interpret the Acknowledgement Field;
- a Push Procedure Indicator, indicating the implementation of the push procedure;
- a Reset Bit, indicating a reset function for the connection;
- a Synchronisation Bit, indicating the need to synchronise sequence numbers;
- a Finish Bit, indicating that no more data is to be sent by the user.

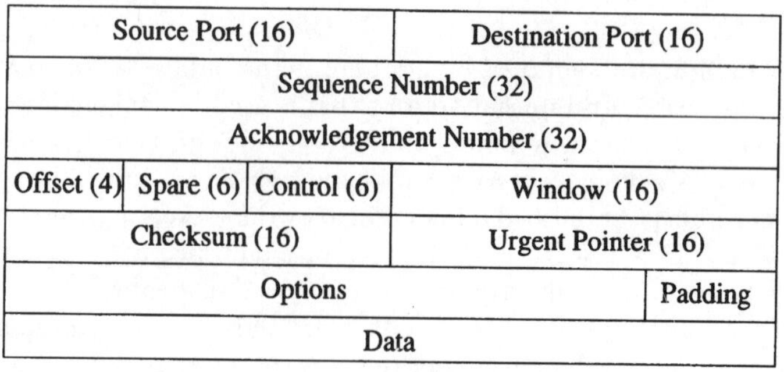

Source Port (16)			Destination Port (16)	
Sequence Number (32)				
Acknowledgement Number (32)				
Offset (4)	Spare (6)	Control (6)	Window (16)	
Checksum (16)			Urgent Pointer (16)	
Options				Padding
Data				

Fig. 18.7   TCP segment format
   ( ) = number of bits

Table 18.4   Explanation of header fields

Field	Description
Source Port Number	address of the source user
Destination Port Number	address of the destination user
Sequence Number	contains the sequence number for error correction. The number generally refers to the first data octet in the segment
Acknowledgement	acknowledges a received segment. It contains the value of the next sequence number that the sender of the segment is expecting to receive
Data Offset	gives the number of 32-bit words in the TCP Header
Control	identifies the presence of a field or gives information on the connection
Window	the number of data octets that the segment sender is willing to accept
Checksum	bits derived by applying the error detection algorithm to the header and data
Urgent Pointer	gives the position of urgent data in the data field
Options	variable length fields
Padding	ensures that the header ends on a 32-bit boundary

Options can have a single-octet format or a multiple-octet format. A single-octet option defines the type of option uniquely. A multiple-octet option has type, length and data fields. The TCP must implement all the options. The options include:

- end of option list, used at the end of each option;
- no-operation, used to align the beginning of an option on a word boundary;
- maximum segment size, indicating the maximum size of a segment that can be received.

18.5.4 Connection establishment

(a) Procedure

Connection establishment is performed using a 'three-way handshake'. The aim of the handshake is to 'synchronise' the sequence numbers between two TCPs, i.e. agree the starting Sequence Numbers for the connection. A basic connection establishment is illustrated in Fig. 18.8. Only some of the segment fields are shown to avoid the Figure becoming too complex.

TCP 1 initiates the connection by sending a Synchronise Segment to TCP 2. The segment contains an Initial Sequence Number (Value 100 in this example), chosen by TCP 1. The Control Field of the segment has its synchronisation bit set to indicate that a synchronisation of sequence numbers is required.

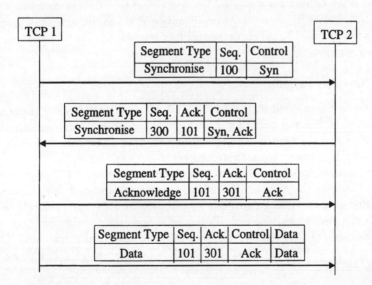

Fig. 18.8 TCP basic connection establishment
  Seq. = Sequence Number, Syn = Synchronise, Ack = Acknowledge

TCP 2 returns a Synchronisation Segment to TCP 1. The segment includes:

- an Initial Receive Sequence Number (Value 300 in this example), chosen by TCP 2;
- an acknowledgement of TCP 1's segment (see below);
- a Control Field showing that the segment includes both Sequence Number Synchronisation and acknowledgement information.

TCP 1 acknowledges the segment from TCP 2. The segment includes:

- a Sequence Number of 101 (i.e. the value of the last segment sent, plus 1);
- an acknowledgement Sequence Number;
- a control bit indicating an acknowledgement.

The connection is now established and the TCPs can exchange data.

(b) Sequence Numbers

The Initial Sequence Number (100 in this case) is chosen by the initiating TCP and is incremented by Value 1 each time TCP 1 sends a segment. The exception to this rule is when a segment is only used for acknowledgement purposes. In this case, the segment sent after the acknowledgement segment has the same Sequence Number Value as the acknowledgement segment (101 in this example).

A segment is acknowledged by including an Acknowledgement Sequence Number in the return segment. The value of the Acknowledgement Sequence Number is set to the next expected Sequence Number. In this example, TCP 2 sends an Acknowledgement Sequence Number of Value 101 to acknowledge the Received Sequence Number of Value 100. Similarly, TCP 1 acknowledges Sequence Number 300 with Value 301.

18.5.5 Connection closure

When a user has finished sending data, it requests the TCP to close the connection. However, the user and TCP at one end of a connection do not know if the user and TCP at the other end of the connection have also finished sending data. Hence, closure is implemented in two stages.

When a user requests a TCP to close a connection the TCP stops sending data and informs the remote TCP of the request. However, if the remote TCP wishes to continue sending data, this is permitted in the closing procedures. An example of a connection closure is illustrated in Fig. 18.9.

In this case, TCP 1 is requested by its user to close the connection. TCP 1 therefore stops sending data and sends a Close Segment to TCP 2. The segment includes:

- the latest Sequence Number appropriate for TCP 1 (410 in this example);
- an acknowledgement of TCP 2's previous segment (Value 619, acknowledged with Value 620);

- a control bit (Finish) that indicates the end of sending data from TCP 1;
- a control bit (Acknowledge) that indicates the presence of an acknowledgement field.

TCP 2 acknowledges receipt of the Close Segment. TCP 2 then determines if it has any more data to send to TCP 1. If TCP 2 had more data to send, then this could be enacted and TCP 1 would still receive the data. However, in this example, TCP 2 decides that there is no more data to send. TCP 2 therefore agrees to close the connection and sends a Close Segment to TCP 1.

Finally, TCP 1 acknowledges receipt of the Close Segment and the connection is regarded as closed at each TCP.

18.5.6 Interface to users

A series of commands and replies are specified for exchanging information between the TCP and its users. The principle is the same as the use of primitives in the OSI Model (explained in Chapter 3). TCP Commands and Replies also include information fields, similar to parameters in the OSI Model. Examples of commands are briefly explained in Table 18.5.

18.6 User Datagram Protocol (UDP)

Like TCP, the User Datagram Protocol (UDP)[5] provides a communications capability between hosts belonging to interconnected networks. However, UDP is designed to minimise signalling overheads by keeping the protocol simple and

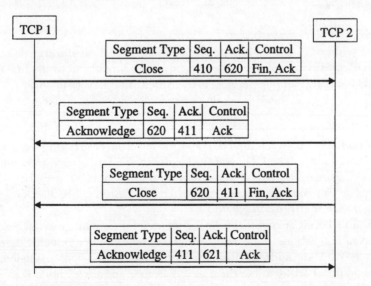

Fig. 18.9 TCP basic connection closure
Seq. = Sequence Number, Ack = Acknowledge, Fin = Finish

Table 18.5   TCP commands

Name	Typical parameters	Description
Active Open	Local port Foreign socket	initiates the TCP procedure to establish and synchronise a connection to the foreign TCP
Passive Open	Local port Foreign socket	requests TCP to listen for incoming connections. The foreign socket can be specific or set to allow any foreign TCP to connect
Send	Local connection name Buffer address Push flag Urgent flag	causes the data in the indicated buffer to be transmitted. The push flag causes immediate transmission. The urgent flag advises the receiving TCP that urgent data is present
Receive	Local connection name Buffer address	allocates the receiving buffer to the connection
Close	Local connection name	initiates the close procedure
Status	Local connection name	seeks the status of the specified connection
Abort	Local connection name	ceases pending send/receive commands and sends a Reset Segment to the foreign TCP

using connectionless procedures. As a result, the delivery of data using UDP is not guaranteed and duplication can occur.

18.6.1 Architecture

UDP fits within the host-to-host layer of the IP Model, as illustrated in Fig. 18.3. UDP makes use of the Internet Layer to provide network functions (typically the IP). It supplies services to the Application Layer and users.

18.6.2 Format

Data is transferred between the UDP and the lower layer (IP in this example) in 'UDP Datagrams'. The UDP adds its header to the data transferred from the user, before passing the whole datagram to the lower layer (e.g. IP). Each datagram header contains address, length and checksum fields. The format of the Datagram is illustrated in Fig. 18.10.

The Source Port is an optional field and identifies the sending user. In connectionless communications, responses are often not required. If a response is given, the Source Port is the address within the host to which the response is delivered. The Destination Port identifies the destination user address within a host. The host itself is identified by the header of the lower layer, e.g. the Destination

Source Port (16)	Destination Port (16)
Length (16)	Checksum (16)
Data	

Fig. 18.10   User datagram format

Address within the IP packet. The Length Field gives the total length of the header and associated data.

The checksum procedure is the same as that used in the TCP. The checksum protects against mis-routed datagrams.

18.6.3 Interfaces

The user passes the data to be transferred to the UDP, together with the source and destination ports and addresses.

18.7 File Transfer Protocol (FTP)

The File Transfer Protocol (FTP)[6] fits within the Application Layer of the IP Model, as shown in Fig. 18.3. The objective of FTP is to transfer data files reliably and efficiently between (a) a user and a server or (b) between servers. The files are stored in the user and server file systems. These file systems vary in nature and FTP is designed to take account of these variations. Thus, the user is relieved of the burden of handling the variations.

FTP makes use of lower layers to transport information, typically TCP in conjunction with IP.

18.7.1 File transfer model

A model for the FTP applied to a user and a server is illustrated in Fig. 18.11. The model has two entities; a user and a server. Communication takes place

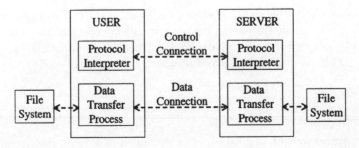

Fig. 18.11   FTP model for user and server

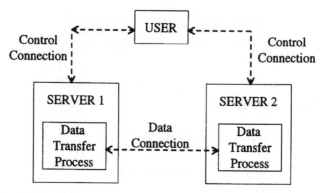

Fig. 18.12 FTP model for two servers

between the entities over two types of connection. The Control Connection covers all aspects of controlling the file transfer, e.g. the user requesting a file to be transferred. The Data Connection provides the path over which the files, etc. are transferred.

For control purposes, each entity has a Protocol Interpreter (PI). The PIs communicate over the Control Connection by means of Commands and Replies. The Commands specify the parameters for the data connection (e.g. the data port to be used and the transfer mode) and the nature of the file system operation (store, retrieve, etc.). Replies are given over the Control Connection in response to the Commands. The user initiates the establishment of the Control Connection.

For data transfer purposes, each entity has a Data Transfer Process (DTP). The DTPs exchange data between the corresponding file systems. In general, the server initiates the establishment of the Data Connection (in response to a user request over the Control Connection).

A similar model applies when a user wishes to transfer data between two servers, as illustrated in Fig. 18.12. Note again the separation of communication over the Control Connection and the Data Connection.

18.7.2 Control Connection

The Control Connection is established by a lower layer (TCP in this case). The User PI is responsible for issuing commands and interpreting the replies. The Server PI interprets the commands, sends replies and directs its Data Transfer Process (DTP) to set up a data connection and transfer data.

(a) Commands

Commands are categorised into:

- Access Control Commands, which determine how a user gains access to the server;

- Transfer Parameter Commands, which relate to the characteristics of the data connection. All data transfer parameters have default values and these commands are only required to change the defaults;
- Service Commands, which define the function requested by the user.

Examples of commands are given in Table 18.6.

(b) Replies

Each command sent over the control connection generates at least one reply. This allows the user to be kept informed of the actions at the server at all times. A reply is coded into a three-digit number and a text statement. The number is primarily intended for interpretation by automatic processes and the text is

Table 18.6 Examples of commands

Command	Description
	Access Control Commands
User Name	identifies the user, e. g. for log-in
Password	password of the user
Change Directory	allows the user to change the directory for file storage or retrieval without re-logging
Reinitialise	resets the data connection, returning parameters to default values. The control connection is left open
Logout	terminates the transaction and closes the control connection
	Transfer Parameter Commands
Transfer Mode	indicates the manner in which data is transmitted
Passive	requests a server to monitor a data port and wait for incoming data, rather than establishing a data connection
File Structure	specifies the structure of the file to be transferred/stored
Representation	specifies the type of coding of data
	Service Commands
Retrieve	causes the server to transfer a copy of a file
Store	requests the server to store data as a file
Delete	deletes a specified file
Status	indicates, for example, the progress in a file transfer

intended to give information to human users. The three digits are presented in order of importance. This allows a user to determine the general response by examining the first digit. If the user requires more information, the second digit can be examined. Finally, if all the information is required, the third digit is examined.

The first digit indicates the overall nature of the response. The following interpretations for digits are specified:

- Digit 1 – Positive Preliminary Reply, indicating that the requested action is being initiated and that the user should expect another reply.
- Digit 2 – Positive Completion Reply, indicating that the requested action has been completed.
- Digit 3 – Positive Intermediate Reply, indicating that the command has been accepted but that more information is needed and the user should send another command.
- Digit 4 – Transient Negative Reply, indicating that the command was not accepted but that the error condition is temporary.
- Digit 5 – Permanent Negative Completion Reply, indicating that the command was not accepted.

The second digit indicates the type of error that has occurred, for example:

- Digit 0 – Syntax error;
- Digit 2 – Control or data connection error;
- Digit 5 – File system error.

The third digit gives greater detail about the error indicated in the second digit. For example, if the second digit is Value 0 (indicating a syntax error), then a third digit of Value 1 indicates that the error was located in the parameter field.

18.7.3 Data connection

(a) Data representation and storage

There are many different ways of coding data, involving combinations of byte-size, alphanumeric compilations, etc. Similarly, data is stored in different file systems in different ways. Thus, when transferring files between systems, it is necessary to take account of these variations. Data representations are specified by the user with a Representation Command.

(b) File structure

In addition to data representation, the FTP allows the user to specify the structure of a file. Three structures are defined:

- File Structure, where the file is considered to be a continuous sequence of bytes without an internal structure;
- Record Structure, where the file consists of sequential records;
- Page Structure, where the file consists of indexed pages.

The Structure Command is used to describe the structure of a file.

(c) Transmission modes

Three transmission modes are defined by the FTP. They are selected by using the Mode Command. The modes are:

- Stream Mode, in which data is transmitted as a stream of bytes without processing;
- Block Mode, in which a file is transmitted in a series of blocks, each having a header giving the number of bytes in the block;
- Compressed Mode, in which replicated data or fillers can be compressed, thus improving transmission efficiency. The original data or filler is sent with an indicator giving the number of replications.

18.8 Simple Network Management Protocol (SNMP)

The Simple Network Management Protocol (SNMP)[7] is designed to provide management capabilities in IP-based networks. SNMP facilitates the inspection and alteration of management information relating to a Network Element by a remote user. As its name implies, one aim of SNMP is to implement a simple approach to network management, avoiding complex procedures. In the future, more management capability will be required. This can be achieved by evolving SNMP. A more likely scenario is the adoption of a more rigorous framework.

18.8.1 Architecture

Network Elements refer to entities like hosts, gateways and servers and the meaning is the same as described in Chapter 10. Management Agents are incorporated within the Network Elements to perform management functions. Network Management Stations execute management applications which monitor and control the Network Elements.

The SNMP is used to transfer management information between a Management Station and a Network Element. A simple architecture illustrating how SNMP fits within the management framework is given in Fig. 18.13.

The Network Management Station and Network Element have Application Entities that drive the management functions and procedures. The Protocol Entities (PEs) transfer the management information necessary to support the Application Entities in their roles. The SNMP refers to the protocol between the

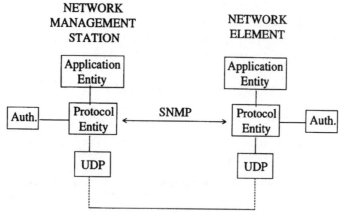

Fig. 18.13   Architecture for management
Auth = Authentication

PEs. Each PE has access to an Authentication Function that validates the identity of a message source.

To keep SNMP simple, it uses the services of the UDP, i.e. simple datagrams are transferred in a connectionless manner.

18.8.2 General message format

Communication between PEs is by the exchange of messages, each of which is included within a UDP datagram. The messages are used to monitor and change management variables. If an Application Entity wishes to determine the value of a variable, the term 'Get' is used to describe this retrieval function. If an Application Entity wishes to change the value of a variable, the term 'Set' is used to describe this alteration function.

The primary procedure used in SNMP is for the Management Station to poll the Network Elements for appropriate information. However, the timing and focus of the polling is influenced by Network Elements providing information in unsolicited messages.

The format of an SNMP message is illustrated in Fig. 18.14. The Version Number gives the version of the SNMP being applied. The Community Name refers to a group of Application Entities.

There are three forms of Protocol Data Unit (PDU). The first form is a polling message, which is sent from the Management Station to the Network Element to seek information or implement a change. In this form, the PDU contains the action that is required. The second form is a Response to the polling

| Version | Community Name | PDU Type + PDU |

Fig. 18.14   SNMP message format

messages. In this form, the PDU contains the response information. The third form refers to the unsolicited messages issued by the Network Element. This form is known a 'Trap' Message. In this form, the PDU contains information supplied by the Network Element.

18.8.3 Polling Protocol Data Units

There are three polling PDUs, namely Get-Request, GetNext-Request and Set-Request. The Get-Request PDU is used by the Management Station PE to request the retrieval of variable names and values for an object. This relates to a monitoring function by the Management Station.

The GetNext-Request PDU is again related to retrieving information, but in this case it relates to variables in a list. The GetNext-Request seeks the name and value of a variable that is one line down from the previous request. This allows a remote user to check each line of a table without knowing the order or contents of the variables within the table.

The Set-Request PDU is used by the Management Station PE to request the setting of variable names and values for an object. This relates to a control function by the Management Station.

The format of a polling PDU is illustrated in Fig. 18.15. The Request ID is a reference number allocated by the Application Entity to facilitate correlation between a response and a reply. The Error Status and Error Index are not relevant to a polling message and are set to zero.

A Variable refers to an instance of a managed object. A Variable Binding gives the name of a variable and its value. A Variable Binding List gives a list of Variable Bindings in a particular PDU.

Request ID	Error Status	Error Index	Variable Bindings	Variable Bindings List

Fig. 18.15 Format of polling PDU

18.8.4 Get-Response PDU

The Get-Response PDU is used by the Network Element to return the result of a request to the Management Station. For example, if a Get-Request Message is received at the Network Element seeking information about an object, the Get-Response Message is used to return the variable and values that apply to that object. If the Get-Response PDU is responding to a Set-Request PDU, the contents of the original PDU are returned as confirmation that the values have been changed to those requested.

The Get-Response PDU has the same format as that illustrated in Fig. 18.15. However, in this case, the Error Status and Error Index Fields have relevance.

The Error Status Field indicates whether or not an error occurred in processing the request. The Error Index gives further information on any errors detected.

18.8.5 Trap PDU

The Trap PDU is used by the Network Element to inform the Management Station of an event. A Trap PDU is sent in an unsolicited manner. The format of the Trap PDU is illustrated in Fig. 18.16.

The Object ID and address indicate the type of object and its location generating the trap. Types of generic trap include:

- ColdStart, signifying that the sending PE is reinitialising itself and changes might result;
- WarmStart, signifying that the sending PE is reinitialising itself without changes;
- LinkDown, signifying that the sending PE recognises the failure in one of the communications links;
- LinkUp, signifying that the sending PE recognises the reinstatement of a communications link;
- Authentication failure, signifying that the sending PE has received a message that is not properly authenticated.

The Time Stamp gives the time elapsed between the last initialisation of the Network Entity and the generation of the trap.

18.8.6 Procedures

High-level procedures for the SNMP describe the construction of a PDU and its analysis at its destination.

(a) Message construction

Upon instruction from the Management Application, the originating PE constructs an appropriate PDU. The PDU is passed to the Authentication Function, together with the community name, its transport address (IP Address plus UDP port) and the destination transport address. The Authentication Function manipulates the data to encode verification information and delivers it back to the PE. The PE formulates a message and passes it to the UDP for transmission to the destination PE.

Object ID	Address	Generic Trap Type	Specific Trap Type	Time Stamp	Variable Bindings	Variable Bindings List

Fig. 18.16 Format of Trap PDU

(b) Message analysis

When the message is received at the destination PE, a simple check is made on the message status. The version number is then verified. If either check fails, the datagram is discarded. The PE now passes the community name, user data and source and destination addresses to the Authentication Function. If the authentication succeeds, the information is passed back to the PE and the PDU is processed accordingly.

18.9 Chapter Summary

18.9.1 Common signalling principles

As the Information Age evolves, the amount of information that is transferred across networks continues to increase substantially and there will be much more focus on data communications. The IP technologies are a key part of this evolution. The structure, organisation and routing approaches used in telecommunications and IP networks are different. However, the signalling systems in the two forms of network have many principles in common. This chapter describes examples of IP signalling at each layer of the IP Model and draws a comparison with the telecommunications field.

18.9.2 IP signalling architecture

IP signalling uses a four-tier architecture consisting of an Application Layer, a Host-to-Host Layer, an Internet Layer and a Link Layer. There is a close relationship with the OSI Model and further work is needed to confirm alignment. The general alignment is:

- the IP Application Layer aligns with the Application, Presentation and Session Layers of the OSI Model;
- the Host-to-Host layer is aligned with the OSI Transport Layer;
- the Internet layer is aligned with the Network Layer;
- the Link Layer is aligned with the OSI Data-Link and Physical Layers.

18.9.3 The Address Resolution Protocol (ARP)

The Address Resolution Protocol (ARP) fits within the Link Layer of the IP Model. The ARP translates an IP address for an outgoing IP packet into the physical address of the destination point of the network. The address correlation is based on a table of translations. The table is populated by a procedure that broadcasts an IP Address and requests the corresponding ARP function to respond with an appropriate physical address.

18.9.4 The Internet Protocol (IP)

The Internet Protocol (IP) fits within the Internet Layer of the IP Model. IP makes use of the Link Layer to provide suitable media, etc. IP offers a routing service to higher layers in a connectionless mode. IP is a simple protocol and it does not guarantee a quality of service to its users.

The format of an IP Packet consists of a Header followed by data. The Header can be followed by Extension Headers that offer additional options, including hop-by-hop analysis, fragmentation, authentication and security.

18.9.5 The Internet Control Message Protocol (ICMP)

The Internet Control Message Protocol (ICMP) is designed to report difficulties in transferring IP Packets. ICMP is an integral part of IP and its implementation is mandatory. ICMP operates between gateway nodes and originating hosts. ICMP is invoked when, for example:

- an IP Packet cannot be delivered;
- an IP Packet is too big for an outgoing link;
- a problem is identified with parameters within the message that prevents processing.

ICMP Messages consist of an IP header with Extension Headers as options. Error Messages include Destination Unreachable, Packet Too Big, Time Exceeded and Parameter Problem.

18.9.6 The Transmission Control Protocol (TCP)

The Transmission Control Protocol (TCP) is a host-to-host protocol that operates in a connection-oriented mode. Connections can be established in either an active or passive manner. In active connection establishment, a user requests the TCP to establish a connection to another user. In passive connection establishment, a user informs the TCP that it will accept incoming connection requests.

Connections are identified by the sockets (combination of addresses) in the two communicating TCPs, with Local Reference Numbers being established as short-cut identities. Connection establishment is performed using a three-way handshake, synchronising the sequence numbers between two TCPs.

A connection is closed in two stages. In Stage 1, a user requests a TCP to close a connection: the TCP stops sending data and informs the remote TCP of the request. However, if the remote TCP wishes to continue sending data, this is permitted in the closing procedures. When the remote TCP has also finished sending data, then Stage 2 of the Closure Procedure is enacted and the connection is released.

TCP transfers information between entities in Segments. The TCP adds its header to the data transferred from the user, before passing the whole segment to the lower layer (e.g. IP). A series of Commands and Replies are specified for exchanging information between the TCP and its users. The principle is the same as the use of Primitives and Parameters in the OSI Model.

In addition to the normal information transfer algorithms, a Push Procedure is defined that initiates immediate transmission of data. Urgent data is indicated by an Urgent Flag. A receiving user has the option of restricting the amount of data transmitted by the sender by defining a window of a particular size.

TCP operates an error detection and correction mechanism. Error detection is based on a checksum field. Error correction is based on acknowledging each octet by the use of sequence numbers.

18.9.7 The User Datagram Protocol (UDP)

The User Datagram Protocol (UDP) is a host-to-host protocol that operates in a connectionless mode. UDP is designed to minimise signalling overheads by keeping the protocol simple. As a result, the delivery of data using UDP is not guaranteed and duplication can occur.

Data is transferred between the UDP and the lower layer (e.g. IP) in UDP Datagrams. The UDP adds its header to the data transferred from the user, before passing the whole datagram to the lower layer. Each datagram header contains address, length and checksum fields.

18.9.8 The File Transfer Protocol (FTP)

The File Transfer Protocol (FTP) fits within the Application Layer of the IP Model. The objective of the FTP is to transfer data files reliably and efficiently between (a) a user and a server or (b) between servers under the control of a user. FTP makes use of lower layers to transport information, typically TCP in conjunction with IP.

Communication takes place between the entities over two types of connection. The Control Connection covers all aspects of controlling the file transfer. The Data Connection provides the path over which the files, etc. are transferred.

The Control Connection is established by a lower layer (e.g. TCP). The User Protocol Interpreter is responsible for issuing commands and interpreting the replies. The Server Protocol Interpreter analyses the commands, sends replies and directs its Data Transfer Process to set up a data connection and transfer data.

Commands are categorised into:

- Access Control Commands, e.g. User Name, Password, Log Out;
- Transfer Parameter Commands, e.g. Transfer Mode, File Structure, Representation;
- Service Commands, e.g. Retrieve, Store, Delete.

Each command sent over the Control Connection generates at least one reply. A Reply is coded into a three-digit number and a text statement. The three digits are presented in order of importance. This allows a user to determine the general response by examining the first digit. The second and third digits are analysed if additional information is required.

The Data Connection is described in terms of Data Representation and Storage, File Structure and Transmission Modes. The Transmission Modes are Stream, Block and Compressed.

18.9.9 The Simple Network Management Protocol (SNMP)

The Simple Network Management Protocol (SNMP) is designed to provide basic management capabilities in IP-based networks. SNMP facilitates the inspection and alteration of management information relating to a Network Element by a remote user.

SNMP is applied between Protocol Elements in a Network Management Station and a Network Element. Each entity has access to an Authentication Function that validates the identity of a message source.

The primary procedure used in SNMP is for the Management Station to poll the Network Elements for appropriate information. However, the timing and focus of the polling is influenced by Network Elements providing information in unsolicited messages called Traps.

An SNMP Message consists of a Version Number, a Community Name and a Protocol Data Unit (PDU). There are three forms of Protocol Data Unit (PDU):

- polling messages, which seek information or implement a change, e.g. Get-Request, Set-Request;
- a Response Message, which provides responses to the polling messages, e.g. Get-Response;
- the unsolicited Trap Messages issued by the Network Element, e.g. ColdStart, LinkDown, Authentication Failure.

Procedures are described to authenticate a message source before PDU processing takes place.

An aim of SNMP is to implement a simple approach to network management, avoiding complex procedures. In the future, more management capability will be required. This can be achieved by evolving SNMP or (more likely) by adopting more rigorous frameworks.

18.10 References

1 Internet Architecture Board, RFC 826: 'Address Resolution Protocol'
2 Internet Architecture Board, RFC 2460: 'Internet Protocol Version 6'
3 Internet Architecture Board, RFC 792: 'Internet Control Message Protocol'
4 Internet Architecture Board, RFC 793: 'Transmission Control Protocol'
5 Internet Architecture Board, RFC 768: 'User Datagram Protocol'
6 Internet Architecture Board, RFC 959: 'File Transfer Protocol'
7 Internet Architecture Board, RFC 1157: 'Simple Network Management Protocol'

Chapter 19
Conclusions

19.1 The Changing Environment

Business practices and social behaviour are changing dramatically, with an increasing reliance being placed upon secure and flexible communications. These demands are being met by the evolution of the Information Age, which is being underpinned by the convergence of the telecommunications, IP, computing and broadcast/media industries. All these industries generate, store, manipulate and transfer information. The amount of data, and the demands for access to it, continue to expand rapidly.

In parallel, customers expect increasing levels of value, choice and control. They require more bandwidth for communication, they are becoming much more global in their outlook and they are demanding greater mobility. There is an increasing trend for companies to adopt more distributed working. Competition amongst operators and suppliers is intense, and many joint ventures are taking place in conjunction with a rise in the number of niche suppliers.

These trends demand ever higher levels of efficiency in the provision of communications.

19.2 Requirements

Networks are designed to provide services to customers with a high level of quality and with maximum value. Networks are built from basic components, e.g. transmission systems and switches. Signalling provides the ability to transfer control information amongst customers and networks, transforming the components into a cohesive and animated force for service provision.

There is a wide variety of signalling systems across a range of industries (telecommunications, IP, etc.). However, all industries require flexibility and evolutionary potential, and modern signalling systems share a range of principles, such as structured architecture. Convergence of the industries will be built upon these foundations, and the principles will apply for the foreseeable future.

The changing environment places demanding requirements upon signalling systems and they need to provide:

- evolutionary potential, with the aim of achieving unimpeded communication between users of the signalling network;
- a high degree of flexibility;
- efficient system and network operation using open interfaces;
- exceptional performance and quality of service, with the ability to support automation and increased management control;
- facilitate efficient interconnection of networks;
- support the global ambitions of operators and customers.

19.3 Architecture

The response to these demanding requirements from a signalling perspective is to build flexibility into the signalling systems. In this context, a structured and flexible architecture for signalling systems is essential to facilitate evolution of the network platforms.

19.3.1 Types of architecture

Ideally, modern signalling systems should have been specified using a common architecture. Unfortunately, various architectures have been adopted as technologies have evolved and as long-term objectives have changed. The main types of architecture are:

- The International Standards Organisation (ISO) uses a 7-Layer Open Systems Interconnection (OSI) Protocol Reference Model. The main focus of the OSI Model is to provide a high degree of openness.
- The original architecture of CCSS7, specified before the need for non-circuit-related (NCR) signalling became widespread, uses a 4-level structure. The main focus of the 4-level structure is to provide services while optimising the use of network resources.
- CCSS7 NCR capabilities use the OSI Model architecture, e.g. the Intelligent Network Application Part (INAP) and the Mobile Application Part (MAP).
- DSS1 uses a 3-layer structure to separate the Physical, Link and Network aspects of access signalling. There are options available for DSS2. If ATM technologies are used, DSS2 uses an ATM Adaptation Layer and an ATM layer. The layers of DSS1 and DSS2 do not align with the OSI Model and there is scope for confusion in the common use of the term 'layer'.
- IP Technologies use a 4-layer structure, having a close relationship with the OSI layers.

Many changes in technology and focus are taking place, including signalling for broadband services, improved processing performance and structured programming for software. With these trends, a change of focus is occurring from network optimisation towards a higher degree of openness. The seven layers in the OSI Model therefore represent the most appropriate architecture for future signalling system evolution.

19.3.2 Common factors

Despite the variance in types, all the architectures adopt a set of principles that provide a high degree of flexibility. These include:

- specifying signalling systems in a series of tiers;
- defining the interfaces between tiers, to allow one tier to evolve without affecting other tiers;
- each tier (plus underlying tiers) offering services to the tiers above in the hierarchy;
- using primitives to communicate between tiers within an entity;
- specifying protocols between corresponding tiers in different entities.

The term 'tier' is used in this book as a common description for the various layers and levels in the differing types of architecture.

19.4 Telecommunications and Internet Signalling

The use of IP technologies is expanding rapidly. Although the routing and addressing techniques used in telecommunications and IP networks are different, there are many common factors from a signalling perspective. Indeed, the biggest differences between the two types of signalling system relate to the quality of service and level of management incorporated.

Telecommunications signalling systems incorporate a very high quality of service and management is regarded as an essential element in its achievement. IP signalling systems offer a lower quality of service, depending more on the user to overcome problems. IP management is similarly low level. These factors make telecommunications more expensive, but extremely reliable. IP signalling systems are cheaper to implement but are susceptible to failures.

The future will see a convergence of these two philosophies. IP will raise its quality of service, especially for business purposes, increasing the cost of implementation. The cost of telecommunications systems will continue to decrease. The result will be the availability of two platforms that can be used to meet customer expectations.

19.5 Standards

The international specification processes conducted by organisations like the ITU-T and groups with a community of interest harness the ideas and views of experts throughout the world, providing standards that are of very high quality. The standards provide the opportunity for great economies of scale in development and deployment costs. However, it is still difficult to produce specifications that are completely unambiguous, and feedback must be used, from trials and implementations, to improve continuously the quality of the specifications.

The production of standards can be a slow business. Specifications are complex, they are at the leading edge of technology and they have to cover a comprehensive range of failure conditions. However, slow standardisation can result in operators developing interim measures in order to meet customer requirements. Thus, the opportunities for economies of scale are missed and non-standard equipment is present as a burden to future evolution.

In many cases a single approach cannot be agreed and options are specified to perform functions in different ways. Options complicate the specification and increase the cost of implementation.

A solution to these issues is to speed up the specification process. Once an idea or new method is considered viable and desirable, a concerted effort should be made to specify it before interim systems are necessary. A more critical examination of options should be performed to reduce the range of variants.

19.6 Vision

Future networks will be based on the provision of transmission capacity (bearers) with unlimited bandwidth and supplied on demand. Transmission capacity will become a commodity item, characterised by low costs, high quality and effective management. The bearers will support a vast range of services, transgressing the current boundaries of telecommunications, computing, IP and broadcast/media and these will provide maximum added value for operators, customers and users. Services will transgress the traditional separation into the fixed and mobile spheres, resulting in a seamless application of services that are independent of the form of access.

The vision for signalling systems is that they will respond to the future needs and they will adapt to the following trends:

- ever increasing customer expectations for quality and value;
- signalling architectures will be based on the OSI Model and its successors;
- technologies will continue to evolve to handle visual communications, distributed intelligence, etc;
- calls will involve numerous signalling connections, both circuit-related and non-circuit-related, and will demand high bandwidth traffic paths;

- non-circuit-related signalling will play a much greater role in communications;
- Signalling will play a greater role in the evolution of management systems.

19.7 Signalling: The Lifeblood

Networks and services are evolving at a rapid pace, with the aim of meeting, and exceeding, the expectations of customers. Signalling, the vitalising influence of networks, is at the heart of this revolution. Without signalling, networks would be inert and passive aggregates of components. Signalling is the bond that provides dynamism and animation, transforming inert components into a living, cohesive and powerful medium. There is a great deal yet to come and signalling will remain at the heart of the evolutionary process.

Signalling: The lifeblood of networks

Appendix 1
Networks

This Appendix gives a brief explanation of networks to assist in the understanding of the role of signalling.

A1.1 Switches

Networks are built in a structured manner, with switches forming a hierarchy. A typical network is shown in Fig. A1.1.

Whereas Fig. A1.1 gives a common structure for networks, the role of each node varies according to the application being considered. Examples are given below for telephony, broadband networks, mobile communications and IP networks.

A1.1.1 Telephony

For traditional telephony, Nodes A, B and C are local exchanges that provide customers with connections to the network. Each local exchange is connected to at least one transit exchange (Nodes D and E) and possibly other local exchanges. Node F is an international gateway and provides access to other countries. Both local and transit exchanges also provide access to other networks. Networks are extremely complex, both in terms of the technology

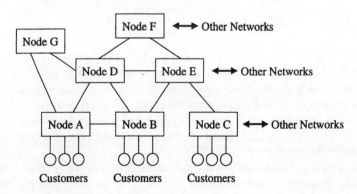

Fig. A1.1   Network structure

needed to provide services and the operational support that is necessary to maintain effective working. Additional network elements are required to support these complexities and Node G is shown as an example of a database storing specialised information.

A1.1.2 Broadband

Future high-speed data services (broadband) will be based on the evolving asynchronous transfer mode (ATM) technologies. In this case, Nodes A, B and C are Edge ATM Nodes that provide access to the network. Nodes D, E and F are Transit ATM Nodes that provide information transfer capability to other Edge Nodes or to other networks, e.g. the Internet.

The term 'broadband' is not strictly defined, but it is generally used to describe communications that use a bandwidth of > 2 Mbit/s. In contrast, communications that use a bandwidth of < 2 Mbit/s are termed 'narrowband'.

A1.1.3 Mobile communications

In cellular mobile networks, Nodes A, B and C are Mobile Switching Centres (MSCs) and they provide access for mobile stations (handsets) via radio station equipment. Node G is a Location Register, which is a database providing information on the location of the mobile stations. Node F is a Gateway MSC, providing interworking with other networks.

A1.1.4 Internet Protocol

In this example, Nodes A, B and C are local exchanges providing customers with access to the Internet. Nodes D and E are Routers, with the role of switching calls. The calls can be routed to Internet Service Providers, supplying services from local switching functions at Node F. Routing can be assisted by a Search Engine at Node G, the role of which is to translate subject areas into network addresses.

A1.2 Transmission Links

The link from a customer to a local exchange is generally carried by a dedicated transmission medium, e.g. a copper cable. For inter-exchange communication, each transmission link provides a number of circuits, as shown in Fig. A1.2. For example, if two optical fibres are provided as transmission media between Exchanges A and B, then each optical fibre provides numerous circuits (typically 64 kbit/s circuits). Each of these circuits can provide a path for traffic (e.g. speech) between the two exchanges.

Transmission media carry information in different ways. For traditional telephony, a circuit is generally dedicated to a call for the whole period of that

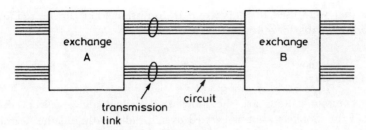

Fig. A1.2   Circuits within transmission links

call. There is a large amount of time during which the circuit is not actively used, e.g. when one customer is listening to another, the return path is quiet. However, the circuit is provided regardless of the real occupancy required.

In ATM-type networks, the physical transmission links are the same as for telephony, but the way in which they are used is different. In this case, the information to be transferred is split into packets called 'cells' and these are transferred over the transmission links. Upon reception at the destination, the cells are re-combined to form the original data stream. In this case, transmission links are shared on a real-time basis. When there are gaps in information flow between customers, the transmission link is used to transmit cells from other customers. For this reason, ATM-type technologies are said to provide a 'virtual circuit'. This means that a circuit is provided, but it is only used by a particular customer when there is information to transfer.

A1.3 Calls

A call refers to the general communication between users and each call can involve a large number of circuits, channels and connections. When a call is being established, it is described as being in the Set-up Phase. Having been established, the call is described as being in the Conversation/Data Phase. During the period in which the call is being released, the call is described as being in the Release or Clear-Down Phase.

Examples of calls are given below to illustrate the principles involved.

(a) Telephony

Consider Fig. A1.1 for telephony. Nodes A and C are local exchanges and Nodes D and E are transit exchanges. Now consider a customer connected to Local Exchange A in Figure A1.1 who wishes to communicate with a customer connected to Local Exchange C. A speech path can be established from Local Exchange A to Local Exchange C, via Transit Exchanges D and E. Hence, a call can be established between the two customers and communication can take place. Upon completion of the communication, the call is released and the circuits that were used in the various transmission links can be relinquished and made available for other customers to use.

(b) Mobile calls

Mobile communications are the same as described above, except that the circuit to the mobile station is carried over a radio path and the terminating node needs to determine the location of the called customer. For an incoming call to a mobile station, Node A uses Node G as a database, termed a Location Register, to identify the location of the called customer. Procedures exist to maintain the record of location of the mobile station and permit roaming.

(c) Asynchronous Transfer Mode (ATM)

The technique for supporting ATM calls is illustrated in Fig. A1.3.

Consider three calls. Call A is taking place between a customer connected to Node 1 and an Internet Service Provider. Call B is taking place between a customer connected to Node 1 and a customer connected to Node 3. Call C is also appropriate to Nodes 1 and 3. All three calls are in the conversation data phase. Fig. A1.3(a) shows the link between nodes 1 and 2 and the following are being transferred between the nodes:

- Cells A2 and A3 for Call A;
- Cell B5 for Call B;
- Cells C3 and C4 for Call C.

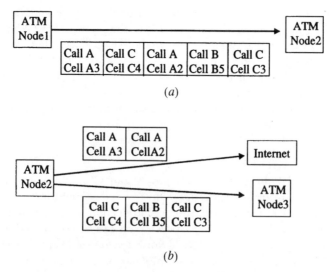

Fig. A1.3   ATM routing

Fig. A.1.3(b) gives a snapshot a short time later, between nodes 2 and 3. Node 2 has received the stream of cells and has routed them to their appropriate destination. Cells A2 and A3 of Call A have been routed to the Internet. Cell B5 for Call B and Cells C3 and C4 for Call C have been routed to Node 3.

It can be seen that the transmission links between Nodes 1 and 2 and between Nodes 2 and 3 are shared by the calls on a real-time basis.

Appendix 2
Channel Associated Signalling

A2.1 Introduction

Channel Associated Signalling (CAS) systems were once the mainstay of telecommunications networks. Over time, they have been replaced by CCS Systems within networks, but some CAS systems are still common for gaining access to networks. Chapter 2 describes the basic tenet of CAS systems, i.e. dedicated signalling capacity is provided for each traffic circuit. This Appendix describes the principles of six categories of CAS system. Examples are given to illustrate the techniques used. Details are well documented elsewhere[1].

One point of terminology needs to be clarified. Signals can be sent in one of two directions. A 'forward signal' is sent in the direction from the calling customer to the called customer. A 'backward signal' is sent in the reverse direction.

A2.2 Loop-Disconnect Signalling

A2.2.1 Loop conditions

In Loop Disconnect (LD) signalling systems, the signalling information pertaining to an analogue traffic circuit is transferred by modifying the electrical conditions applying to the traffic circuit. The status of a direct current (DC) loop on the traffic circuit identifies the information being transferred. Signals from a calling customer are defined by make and break conditions on the loop. Signals from a called customer are transmitted through the network by changing line polarity.

Consider a link between a customer and a local exchange, which is based on a DC loop. When a customer's telephone is in an idle state (i.e. the handset is in the cradle, 'on-hook'), the loop is disconnected (open) and current does not flow. When the calling customer commences a call (by lifting the handset, 'off-hook'), the loop is completed and a direct current flows. The current acts as a forward 'seize' signal at the originating local exchange, indicating that the customer wishes to make a call. The originating local exchange detects the loop from the calling customer and connects appropriate equipment to allow it to receive the dialled digits. As soon as the equipment is allocated by the local exchange, dial tone is returned to the calling customer.

A2.2.2 Dialled digits

The dialled digits are communicated from the calling customer to the originating local exchange as a series of pulses by interrupting (breaking) the loop. Each digit is represented by a corresponding number of pulses. Hence, Digit 1 is represented by one pulse, Digit 2 by two pulses, etc. The normal rate of pulsing is at 10 pulses per second. Thus, to pulse the Digit 0 (10 pulses) takes one second, whereas to pulse Digit 1 takes one tenth of a second. Fig. A2.1 illustrates the principle of transmitting digits.

The format of each pulse varies in national networks, but a typical value is 66.7% break (i.e. open loop) and 33.3% make (i.e. closed loop). In Fig. A2.1, there are three pulses in the first sequence. Thus, the local exchange recognises that Digit 3 has been dialled by the calling customer. There is an 'inter-digit pause' to allow the local exchange to recognise the end of one pulse sequence (i.e. digit) and the start of another. In this sequence, the inter-digit pause is followed by Digit 1. The dialled digits are analysed by the originating local exchange, and the call is routed accordingly to the called customer at the destination local exchange.

A2.2.3 Called customer answer

When the called customer answers by lifting the handset the loop between the destination local exchange and the called customer is closed. The answer state is transmitted through the network to the originating local exchange (i.e. the backward direction) by reversing the polarity of the loop within the network.

A2.2.4 Calling customer cease

When the calling customer ceases the call, by replacing the handset, the on-hook condition is communicated to the originating local exchange by means of a break in the loop. The break to indicate call completion is longer than the break associated with dialled digits. Hence, the originating local exchange can

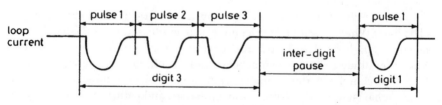

Fig. A2.1   Digit Transmission

determine the difference between dialled digits and a request to clear the call. The request to clear the call is communicated through the network.

A2.2.5 Called customer cease

If the called customer indicates call completion by replacing the handset, the break condition is detected by the destination local exchange and signalled through the network by a reversal of loop polarity (i.e. return to the original loop polarity before answer). However, most telephony-based networks do not commence immediate release of a call in response to a request from the called customer because the calling customer is regarded as controlling the call. Thus, no action is taken by the network until either the calling customer clears or a timer expires.

A2.2.6 Signals

Examples of signals in a typical loop-disconnect signalling system are given in Table A2.1.

A2.2.7 Constraints

The repertoire of signals that can be sent is very limited because of the limited number of conditions that can apply to the line. The application is also limited by the characteristics of the circuit upon which transmission is performed. Line capacitance distorts the shape of the pulse and the distortion increases with the length of the line. Equipment receiving the pulses can only countenance a limited degree of distortion. Thus, the line capacitance restricts the distance over which the signalling system can be used. This is exacerbated by the use of 'single current working', in which the current varies between a positive value and a zero reference. This is illustrated in Fig. A2.2.

Post-dialling delay can be long due to the time to transmit digit trains through the network.

Table A2.1 Typical loop-disconnect signals

Signal	Line condition
Calling customer seizes	Loop to line
Dialled digits	Loop-disconnect pulses
Called customer answers	Reversal of loop polarity within network
Calling customer clears	Disconnection of loop

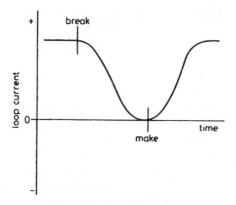

Fig. A2.2   *Single current working*

A2.3 Long-Distance DC Signalling

Long-distance DC (LDDC) signalling systems attempt to overcome the distance limitations of loop-disconnect signalling systems by using 'double current working'. Fig. A2.3 shows that, in double current working, a pulse constitutes a reversal of current, the current varying between a positive and negative value.

This approach of using a symmetrical waveform about a zero reference reduces the impact of pulse distortion due to line capacitance. In addition, double-current working reduces the degree of variance of pulse distortion in conjunction with varying signal level. Thus, the adoption of double current working extends the distance over which the signalling system can be used.

Another advantage of double current working is that more sensitive receive devices can be implemented to detect the current. The implementation of double current working requires the adoption of features additional to those for single current working. Thus, although LDDC signalling systems can be used over greater distances than standard LD systems, LDDC systems are more expensive to implement. Examples of signals used in a typical LDDC signalling system are given in Table A2.2.

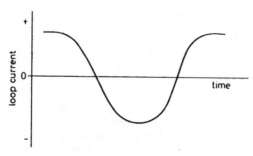

Fig. A2.3   *Double current working*

Table A2.2 Typical LDDC signals

Signal	Line condition
Idle	Negative battery on one leg, earth on the other leg
Calling customer seizes	Loop current reversed
Dialled digits	Loop current reversals at 10 pulses per second
Answer	Earthed loop
Calling customer clears	Reversal of seize condition, i.e. return to idle loop

A2.4 Voice-Frequency Signalling

A2.4.1 Frequencies

Voice-frequency (VF) signalling systems transfer signals by generating one or more tones and transmitting them over the traffic circuit to which the information refers. A tone receiver at the other end of the circuit is used to analyse the information content. The signals share the same 300–3400 Hz frequency band as that used for the traffic circuit. This means that signals cannot be sent during the conversation phase of a call unless expensive filters are provided to prevent the signal being heard by the calling and called customers. During the set-up phase of the call, the traffic circuit is split to avoid the calling customer hearing the signals.

Between exchanges, the signals are treated as normal traffic. Thus, traffic amplifiers are used to maintain signal strength, resulting in a much greater signalling range than is possible in LD and LDDC signalling systems. Tone receivers are permanently allocated to each circuit. Thus, the cost of implementation is relatively high. VF systems can handle both line and selection signalling, but selection signalling is sometimes covered by other means.

A2.4.2 Modes

Several modes of VF signalling exist. They can be categorised into 'continuous compelled', 'continuous non-compelled' and 'pulsed'.

In continuous compelled systems, the information is transferred by sending a continuous frequency (or combination of two frequencies) during the call sequence. The term 'continuous' is used to denote that the frequency is transmitted without interruption, i.e. the frequency is not pulsed. The systems are known as 'compelled' because the application of the frequency is maintained until an acknowledgement is received. The signal meaning is determined by the direction of the signal, the frequency of the signal and the point during the call sequence at which the signal is sent. To illustrate the principles, examples of line signals in the ITU-T Signalling System No. 5[2] are given in Table A2.3.

In continuous non-compelled signalling systems, the information being transferred is denoted in the change of state of the tone. The term 'non-compelled' means that an acknowledgement of a signal is not required before the sending of

356 Telecommunications Signalling

Table A2.3   ITU-T signalling system no. 5 line signals

Signal	Direction	Frequency
Seize	Forward	1
Proceed to send (seize acknowledge)	Backward	2
Answer	Backward	1
Answer acknowledge	Forward	1
Clear forward	Forward	1 + 2 (compound)
Release guard (acknowledgement of clear forward)	Backward	1 + 2 (compound)

Frequency 1 = 2400 Hz, Frequency 2 = 2600 Hz

the signal is stopped. This form of signalling system covers the 2-state on-hook/off-hook conditions by means of tone-on/tone-off. Examples of typical signals using this approach are given in Table A2.4.

In Pulsed VF signalling systems, the information is transferred by timed pulses of tone. Hence, the meaning of the signal is determined by the direction of the signal, the length of the pulse and the point during the call sequence that the signal is sent. The advantages of the pulsed form of VF signalling are that a larger repertoire of signals is available (thus allowing more features), higher signalling levels are possible (due to the non-continuous nature of the signals) and there is less interference with the signals (again due to their non-continuous nature). However, the need to meet the tolerances required for effective signal recognition means that the signalling terminals are relatively complex and hence expensive. Examples of signals in a typical pulsed VF system are given in Table A2.5.

Table A2.4   Typical continuous non-compelled signals

Signal	Forward tone	Backward tone
Idle	on	on
Seizure	off	on
Answer	off	off
Clear forward	on	NA

Table A2.5   Typical pulsed VF signals

Signal	Tone pulse, ms (frequency 2280 Hz)
Seize	70
Digits	60
Answer	250
Clear forward	>700

A2.4.3 Application

VF systems can be used in two modes: 'end-to-end' and 'link-by-link'. In the end-to-end mode, signals of relevance only to originating local exchange and destination local exchange in a connection can be transferred directly over the traffic path, without converting to direct-current conditions at intermediate exchanges. Hence, in end-to-end mode, signals (e.g. the answer signal) can be transferred very quickly.

In the link-by-link mode, each signal is converted to direct-current conditions at each exchange (including intermediate exchanges), thus slowing down the transfer of the signals. Although cumbersome, Link-by-Link signalling is used to facilitate interworking with different signalling systems within the network.

A2.4.4 Signal reception

The use of the same physical path for traffic and signals necessitates the avoidance of imitation of the signals by normal traffic. If specific measures are not taken, the signal receivers could confuse normal traffic with a signal. Several factors are taken into account to avoid imitation:

- The signal frequency is chosen to avoid high energy levels in normal traffic. This factor, in conjunction with other factors, such as filter characteristics, results in frequencies of 2000 to 3000 Hz being suitable.
- A guard circuit is employed such that, if other frequencies are present at the same time as the signal frequency, the guard circuit can over-rule the signal recognition circuit.
- Some signalling systems (especially international systems) use tones of dual frequency rather than single frequency, thus reducing the chances of signal imitation.

A2.5 Outband Signalling

A2.5.1 Frequency

The term 'outband signalling' refers to signalling systems that use frequencies above the normal telephone speech path spectrum. Outband signalling is typically used in frequency division multiplex (FDM) transmission systems. These are systems in which numerous traffic circuits are assembled into blocks for transmission purposes by translating them into different parts of the frequency spectrum. The allocated frequency band per circuit is typically 4 kHz per circuit. The traffic and signalling are transmitted in separate sub-bands, with traffic being carried in the 300-3400 Hz band and signalling in the 3400-4000 Hz band. This allows traffic and signalling to be transferred simultaneously and independently. The traffic and signalling are separated at each exchange by means of appropriate filters.

358   *Telecommunications Signalling*

The frequency recommended by ITU-T[3] for outband signalling is 3825 Hz. Outband signalling is converted to direct-current conditions at each exchange. Hence, it is link-by-link in nature.

A2.5.2 Application

Outband signalling overcomes a number of the inherent disadvantages of VF signalling. Advantages include the ability to signal during traffic and the avoidance of measures to overcome the imitation of signals by normal traffic. A disadvantage of outband signalling is that it can only be applied on transmission systems that permit a wider frequency spectrum than normal non-multiplexed transmission systems. Hence, it is usually limited to appropriate FDM Systems.

A2.5.3 Modes

The nature of outband signalling allows it to be used in numerous modes, including the continuous mode and pulsed mode described for VF signalling. A common implementation is the continuous non-compelled mode, either using tone-on idle or tone-off idle. The influences over whether to use tone-on idle or tone-off idle include transmission system overload considerations and the impact of transmission interruptions. These characteristics vary amongst the numerous types of network, resulting in the use of both forms of signalling.

A2.5.4 Signals

Examples of signals in a typical tone-off idle implementation are given in Table A2.6. It is shown in Table A2.6 that allowance has been made for selection signalling as well as line signalling. Whereas some implementations include selection signalling, other implementations adopt separate selection signalling arrangements.

An example of a tone-on idle system is the line signalling for ITU-T R2 (analogue)[4]. Examples of the signals defined for R2 are given in Table A2.7.

Table A2.6   Typical tone-off idle signals

Signal	Forward tone	Backward tone
Idle	off	off
Seizure	on	off
Digits	off (pulse breaks)	off
Answer	on	on
Clear forward	off	–

Table A2.7 Typical R2 signals

Signal	Forward tone	Backward tone
Idle	on	on
Seizure	off	on
Answer	off	off
Clear forward	on	-

A2.6 Multi-Frequency Inter-register Signalling

A2.6.1 Registers

Registers are used in common-control exchanges to store and analyse routing data. They are provided on a common basis, i.e. a single register provides the routing data for numerous traffic circuits. A register is used by a call only for the time to set up that call. The register is then made available to help set up other calls. The term 'inter-register' refers to the signalling between registers in two different exchanges.

Developments in the common control of exchanges and the need to reduce post dialling delays in networks, led to the implementation of multi-frequency inter-register (MFIR) signalling systems. The term 'multi-frequency (MF)' reflects the use of a variety of frequency tones to denote different signals.

A2.6.2 Selection signalling

MFIR signalling covers selection and network functions. Line functions are covered by an appropriate line signalling system. This allows a simple line signalling arrangement to be maintained, while allowing more complex selection and network functions to be provided. By providing registers on a common basis, the functions can be introduced without unduly increasing the cost per circuit.

A2.6.3 Frequencies

The basic concept in MFIR signalling is that each address digit (or signal) can be coded by means of two frequencies (compounded) taken from a range of N frequencies. Hence, each address digit is represented by a code that is composed of two out of N frequencies. For $N=5$, up to 10 combinations can be coded and this is sufficient to define Digits 0-9. However, $N=6$ gives 15 coding possibilities and thus increases the signal repertoire to a more satisfactory level. Hence, the usual implementation in networks is $N=6$.

A2.6.4 Advantages

The advantages of MFIR signalling include:

- fast transfer of address digits through the network, thus reducing post dialling delay;
- wide repertoire of signals compared with other CAS systems, the repertoire being available for use in the forward and backward directions;
- resilience to distortion, due to signal content being determined by frequency recognition;
- ability to be used in end-to-end mode or link-by-link mode, depending on the particular requirements for a specific network;
- ability to accommodate pulsed, compelled or non-compelled signalling.

A2.6.5 ITU-T signalling system R2

(a) Frequencies

MFIR signalling is very flexible compared with other CAS systems and many variations have been adopted throughout the world, depending on the network characteristics. To illustrate the principles involved, consider the formatting of ITU Signalling System R2[4].

System R2 is a 2-out-of-6 frequencies, compelled, end-to-end signalling system exhibiting a range of forward and backward signals. Each signal is formed by the compound of 2-out-of-6 frequencies, as shown in Table A2.8.

Table A2.8 Signal frequencies in System R2

Signal number	Forward signal frequencies, Hz	Backward signal frequencies, Hz
1	1380 + 1500	1140 + 1020
2	1380 + 1620	1140 + 900
3	1500 + 1620	1020 + 900
4	1380 + 1740	1140 + 780
5	1500 + 1740	1020 + 780
6	1620 + 1740	900 + 780
7	1380 + 1860	1140 + 660
8	1500 + 1860	1020 + 660
9	1620 + 1860	900 + 660
10	1740 + 1860	780 + 660
11	1380 + 1980	1140 + 540
12	1500 + 1980	1020 + 540
13	1620 + 1980	900 + 540
14	1740 + 1980	780 + 540
15	1860 + 1980	660 + 540

(b) Signals

Each frequency combination can have two meanings in both the forward and backward directions. This increases the signal repertoire of the system. In the forward direction, each frequency combination defines a signal in a Group I category and a Group II category. The Group I category is used until a request to change to Group II is received. Typical forward signals are given in Table A2.9.

Backward signals are categorised into Group A and Group B. As for forward signals, each frequency combination can have two meanings: Group A meanings apply until Signal A3 or Signal A5 is returned. Typical backward signals are given in Table A2.10.

(c) Sequence

A typical sequence is for the originating exchange to send the first address digit (using Signals I-1 to I-10) to the destination exchange. This first digit is acknowledged by Signal A-1, which also requests the transfer of the next digit. After the first address digit, the remaining address digits are requested by the destination exchange using Signal A-1. When the last address digit had been received by the destination exchange, it returns a Signal A-3, indicating the need to changeover to Group B and Group II signals. If the called customer is free, a B-6 Signal is returned to the originating exchange.

Table A2.9 Typical forward R2 signals

Signal number	Group I forward signal*	Group II forward signal
1	digit 1	national use
2	digit 2	national use
8	digit 8	data transmission
9	digit 9	customer with priority
15	end of pulsing	national use

Table A2.10 Typical backward R2 signals

Signal	Group A backward signal	Group B backward signal
1	send next digit	national use
3	address complete, changeover to use of Group B and Group II signals	customer busy
4	congestion	congestion
5	send calling customer's category	vacant national number
6	address complete, set up speech conditions	customer line free

A2.7 Signalling in Pulse Code Modulation Systems

A2.7.1 General

Pulse Code Modulation (PCM) is a method of converting information from an analogue form to a digital form for transfer over digital transmission systems. The technique[5] involves sampling the analogue waveform and coding the results in a digital format. Successive sampling allows the analogue waveform to be represented by a series of 8-bit codes. The 8-bit codes from numerous traffic channels are assembled into blocks for transmission by insertion into 'time slots'. The technique is called time division multiplexing (TDM).

The term 'PCM signalling' is in common use, but this is misleading and care must be taken to discriminate between the provision of a transmission medium and provision of corresponding signalling. A PCM system merely provides a digital transmission medium in which signalling capacity is provided. The signalling systems described earlier are carried in the signalling capacity made available by the PCM system.

The bandwidth required to transmit signals is much less than that for traffic, so the signalling for several traffic channels in a PCM system can be handled by a small portion of the bandwidth. The signalling capacity can be used for CAS or CCS. For CAS, the means of identifying to which traffic channel a particular signal refers is to divide the signalling capacity into dedicated bit locations. The means of conveying CCS is to compound the signalling capacity into a signalling channel that is available as and when required.

ITU-T has defined PCM standards for 30-channel and 24-channel systems. The capacity available for signalling in these two standards is different as a result of the differing constraints applied by the PCM standards.

A2.7.2 30-channel PCM systems

In 30-channel PCM systems, the 8-bit codes relating to 30 traffic channels are time division multiplexed into a 'frame'. Each 8-bit code is inserted into a time slot within the frame, as shown in Fig. A2.4. Time Slot 0 is used for alignment purposes and Time Slots 1-15 and 17-31 are used for the encoded traffic relating to the 30 channels. Time Slot 16 is dedicated for the use of signalling. Sixteen frames (Frames 0-15) constitute a 'multi-frame'.

(a) CAS signalling

The tenet of CAS systems is that dedicated signalling capacity is available for each traffic circuit. This is achieved in 30-channel PCM systems by allocating 4 bits in each multi-frame to the signalling for each traffic channel. This is illustrated in Fig. A2.5.

Time Slot 16 in Frame 1 contains the signalling bits for Traffic Channels 1 and 16. Time Slot 16 in Frame 2 contains the signalling bits for Traffic Channels

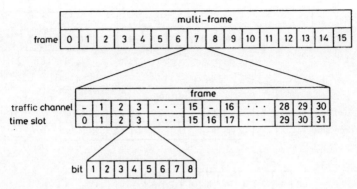

Fig. A2.4   Structure of a 30-channel PCM System

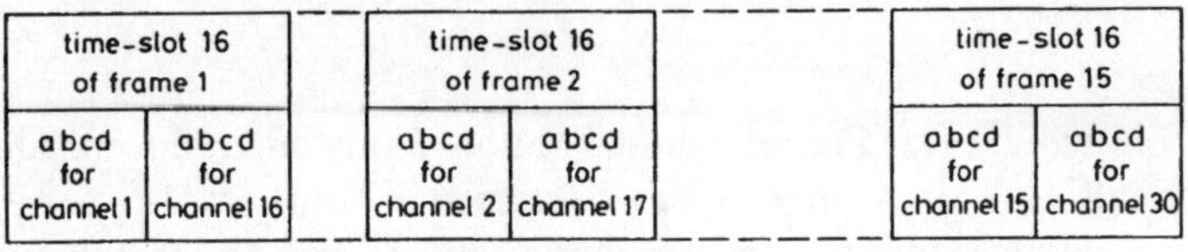

Fig. A2.5   Signalling capacity (30-channel PCM)
abcd = four bits allocated for the signalling for one speech channel

2 and 17. This process is repeated to the end of the multi-frame, in which the Time Slot 16 in Frame 15 contains the signalling bits for Traffic Channels 15 and 30. The next multi-frame repeats this pattern, starting with the signalling bits for Traffic Channels 1 and 16 again.

(b) CCS signalling

For CCS systems, the signalling capacity in Time Slot 16 of successive frames is amalgamated to form a signalling channel. The capacity within the signalling channel is allocated dynamically according to need. For example, if Traffic Channel 1 requires signalling information to be transferred, then the full capacity of the signalling channel is used to transfer the information. When completed, the capacity of the signalling channel is then used for another traffic channel.

A2.7.3   24-channel PCM systems

Fig. A2.6 shows the frame and multi-frame arrangement for the 24-channel PCM standard. In this case, each 8-bit code is again inserted into a time slot, but there are only 24 time slots in a frame. A multi-frame comprises 12 frames. Two blocks of signalling capacity are derived by designating Frames 6 and 12 as signalling frames. Bit 8 of the 8 bits in each time slot in Frames 6 and 12 is 'stolen' for signalling purposes. Hence, Frames 1-5 and 7-11 allow the full 8 bits for

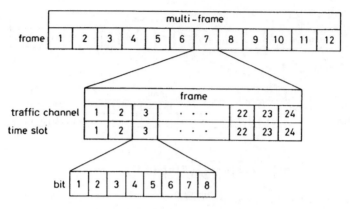

Fig. A2.6 Structure of a 24-channel PCM system

traffic encoding, whereas Frames 6 and 12 allow 7 bits for traffic encoding and 1 bit for signalling. This arrangement is illustrated in Table A2.11.

This bit-stealing technique has a minimal effect upon the traffic quality of the 24 traffic channels, but it is a major constraint on the ability to supply signalling capacity.

For CCS systems, signalling capacity can be provided in two ways. The first realisation is to use the first bit of successive even frames to convey the signalling information. The resultant signalling capacity is amalgamated into a signalling

Table A2.11 Signalling capacity (24-channel PCM)

Frame number	Bit number within each time slot used for:	
	speech	signalling
1	1–8	–
2	1–8	–
3	1–8	–
4	1–8	–
5	1–8	–
6	1–7	8
7	1–8	–
8	1–8	–
9	1–8	–
10	1–8	–
11	1–8	–
12	1–7	8

channel. The second realisation allows a greater signalling capacity to be used by dedicating one of the 24 channels in the PCM system as a signalling channel.

A2.8 References

1 WELCH, S: 'Signalling in telecommunications networks' (Peter Peregrinus, 1981)
2 ITU-T Recommendations Q.140–Q.164: 'Specification of Signalling System No. 5' (ITU, Geneva)
3 ITU-T Recommendation Q.21: 'General Recommendation on Telephone Switching and Signalling Functions' (ITU, Geneva)
4 ITU-T Recommendations Q.350–Q.368: 'Specification of Signalling System R2' (ITU, Geneva)
5 ITU-T Recommendations G.732 and G.733: 'Digital Networks Transmission Systems and Multiplexing Equipment' (ITU, Geneva)

Appendix 3
Evolution of Signalling Systems

A3.1 Introduction

One of the major factors influencing the development of signalling systems in the past was the relationship between signalling and the control functions of nodes. This Appendix traces the evolution of signalling systems in relation to call control technologies.

A3.2 Step-by-step Switches

Early telecommunications networks used analogue step-by-step exchanges. In such systems, the exchange constitutes a large number of discrete switches. The switches have physically moving parts that respond directly to the dialled digits from the calling customer to route the call. The control function, or decision-making, is extremely limited and is located in each discrete switch. In concept, this can be shown as the control and switch function being co-located, as illustrated in Fig. A3.1.

In this type of exchange, when a call is made, the signalling and traffic follow the same path within the node. Step-by-step exchanges are invariably associated with CAS systems. Hence, the signalling and traffic also follow the same path external to the node, i.e. on the transmission link.

A3.3 Separation of Control and Switch Block

The next stage through which exchanges evolved is shown in Fig. A3.2. In this case, the discrete switches of the step-by-step exchange are replaced by an electronic switch block through which calls are routed. The control mechanism in the exchange for setting-up and releasing calls is separated from the switch block. This technique allows much more flexibility in controlling calls and it also reduces costs.

Again, CAS systems are typically associated with this type of exchange. Hence, whereas signalling information is separated from the speech circuit

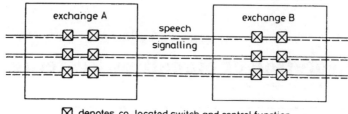

Fig. A3.1 Step-by-step switching

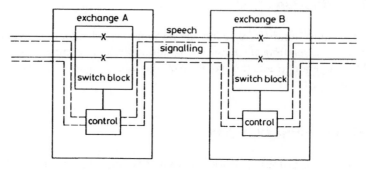

Fig. A3.2 Separate control and switch block

within the exchange, the two are carried on the same path external to the exchange. The speech traffic circuits (denoted by solid lines) are routed by the switch block within the exchange, but the signalling information (denoted by dotted lines) is routed via the control function. Between Exchanges A and B, the signalling and traffic are carried over the same path. This approach was primarily designed to allow optimisation of functions within exchanges, but its effectiveness is constrained by the need to combine the signalling and speech traffic external to the exchange.

A3.4 Common Channel Signalling

With CCS systems, the philosophy is to separate the signalling path from the speech path both within the exchange and external to the exchange, as shown in Fig. A3.3.

Such separation allows optimisation of the control processes, switch block and signalling systems. In a CCS environment, the speech paths are routed by the switch block, as before. However, the signalling (denoted by a dotted line), is routed by a separate path, both internal and external to the exchange. This

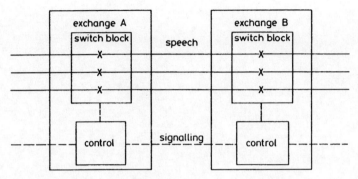

Fig. A3.3 *Common Channel Signalling*

approach allows maximum flexibility in optimising exchange and signalling development.

The approach gains maximum benefit when adopted in parallel with the introduction of digital exchanges and digital transmission systems, CCS systems being particularly efficient in these circumstances. Although CCS systems can be used in an analogue environment, they are a major part in the drive towards all-digital networks. It is in networks offering end-to-end digital communications between customers, e.g. integrated services digital networks (ISDNs), that the full benefits of CCS systems can be unleashed.

A3.5 Non-Circuit-Related Signalling

The key to the next step in the evolutionary path is illustrated by non-circuit-related signalling. In this case, the aim is to separate the signalling and call control to the extent that the signalling system becomes an independent bearer of information that can be used by numerous forms of call control. This situation approaches the aim of unimpeded data transfer between nodes and between customers.

Appendix 4
ITU-T Signalling System No. 6

A4.1 Introduction

ITU-T Signalling System No. 6^1 (CCSS6) was the first CCS system to be implemented internationally. It was originally designed for use in the international network, but some flexibility is included to allow its use in national networks. It is an inter-exchange signalling system that was implemented widely, but it has now been superseded by ITU-T CCSS7.

CCSS6 offers a wide range of features associated with CCS systems, including operation in the quasi-associated mode, error detection and correction mechanisms and re-routing capabilities in fault conditions. However, its main drawback is its limited evolutionary potential, caused by its lack of a tiered architecture. In a dynamic environment, CCSS7 offers far greater flexibility and evolutionary capability, particularly due to its structured architecture.

A4.2 General Description

A4.2.1 Transmission

CCSS6 was originally optimised to be used over analogue transmission paths, but digital paths can also be accommodated. It is operated on a link-by-link basis. Communication between two exchanges is via a datastream over a CCS link.

A4.2.2 Signal units

The information sent over the link is divided into 'Signal Units' of 28 bits in length. All Signal Units are of the same length, the last eight bits being used for error detection. The Signal Units are grouped into 'blocks' of 12, the last Signal Unit in each block being used for acknowledgement purposes. Each Block is allocated a sequence number as part of the error detection and correction mechanism.

Signal Units within a block are either 'Message Signal Units' or 'Synchronisation Signal Units'. Message Signal Units contain call control information relating to the control of telephone calls. Synchronisation Signal Units are sent in the absence of signalling traffic to maintain synchronisation of the link.

A4.2.3 Exchange functions

The functions performed in sending and receiving Signal Units within an exchange are illustrated in Fig. A4.1. The processor is the functional block responsible for call-control within the exchange. The processor therefore recognises the need to establish signalling connections and conduct signalling communication.

The processor formats appropriate messages and passes them to the output buffer. The output buffer stores each message until an appropriate slot is available to transmit the message to the next exchange. When it is the turn for a particular message to be transmitted, it is passed to the coder, which adds eight check bits to the message. The check bits are used by the receiving exchange to detect corruption of the message. In the analogue version of the signalling system, the signal is prepared for analogue transmission by a modulator and delivered to the appropriate outgoing signalling channel. In the digital version, modulation is not necessary.

When receiving Signal Units the decoder accepts the Signal Units from the demodulator (of an analogue transmission system) or from the digital transmission system. The decoder checks each Signal Unit for errors in conjunction with

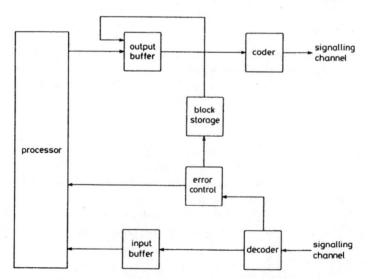

Fig. A4.1   CCSS6 exchange functions (Source: ITU-T Recommendation Q.251)

the eight check bits. If errors are detected, the Signal Unit is discarded. If the Signal Unit is error-free, it is delivered to the input buffer, which stores the Signal Unit until the processor is available to analyse its contents.

Errors detected in Signal Units are overcome by requesting the retransmission of the corrupted Signal Unit. Thus, a copy of each transmitted Signal Unit is stored in 'Block Storage' until receipt of the Signal Unit by the other exchange is acknowledged. The acknowledgements are analysed in 'Error Control'. Successful acknowledgement allows deletion of the Signal Unit stored in Block Storage, whereas unsuccessful acknowledgement invokes the retransmission of the Signal Unit by the outgoing buffer.

A4.3 Formatting Principles

A4.3.1 General

There are two types of Signal Unit. If the signalling information is short enough to fit in a single message, the Signal Unit is termed a 'Lone Signal Unit (LSU)'. If the signalling information is too large to fit within an LSU, then the information is carried in a series of Signal Units, the combination of which is termed a 'Multi-Unit Message' (MUM). A MUM can consist of two to six Signal units. The first Signal Unit in a MUM is termed the 'Initial Signal Unit' and the remaining Signal Units are termed 'Subsequent Signal Units' (SSUs).

A4.3.2 Lone Signal Unit (LSU)

The basic format of the LSU is shown in Fig. A4.2. The Heading Code indicates the general class of signal being transmitted and the Signal Information Field (SIF) distinguishes the particular signal within the class. Hence, the combination of the Heading Code and the SIF defines the specific signal being sent.

The Label defines the speech circuit to which the message refers, reflecting the circuit-related nature of CCSS6. The 11 bits in the label allow up to 2048 speech circuits to be identified. The check bits comprise the error-detection information generated by the coder to detect corruption of messages during transmission from one exchange to another. Example codings for LSUs are shown in Table A4.1.

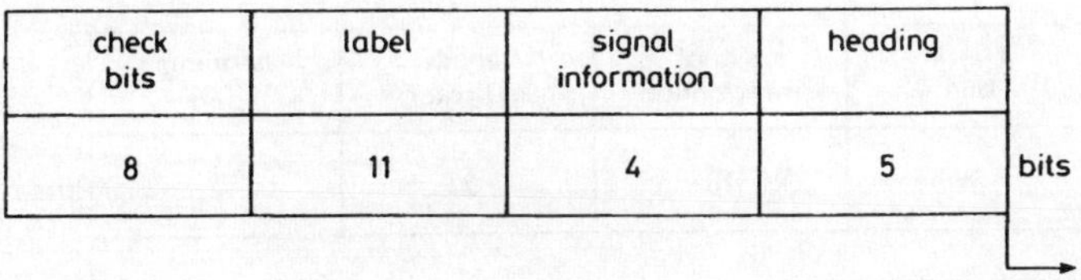

check bits	label	signal information	heading	
8	11	4	5	bits

Fig. A4.2   Format of lone signal unit

Table A4.1   Example lone signal units

Signal	Description	Heading field	Signal-information field
Address Complete	sent by the destination exchange to indicate that enough digits have been received to route a call to the called customer	11011	1010
Answer	sent by the destination exchange to indicate that the called customer has answered the call	11000	0010
Clear Forward	sent by the originating exchange to initiate the release of the call (eg. initiated by the calling customer ceasing the call)	11010	0010

A4.3.3 Multi-Unit Message (MUM)

The basic format of an Initial Signal Unit within a MUM is the same as the LSU. In this case, the SIF identifies the Signal Unit as an Initial Signal Unit and indicates that Subsequent Signal Units will follow. SSUs within a MUM have the general format shown in Fig. A4.3.

The Heading Field is two bits long and is coded 00, thus easing the identification of the Signal Unit as an SSU. The length indicator is two bits long and it indicates the number of SSUs in the MUM, each SSU in a MUM carrying the same length indicator. A Label is not required in an SSU, this information being contained in the LSU.

A4.3.4 Initial Address Message (IAM)

To illustrate the combination of an Initial Signal Unit with SSUs, forming a MUM, Fig. A4.4 shows the general format of an Initial Address Message (IAM). The IAM is the first message to be sent by an exchange to initiate the establishment of a call. The IAM is always a MUM due to the need to include routing information and address digits. The IAM contains between three and six Signal Units.

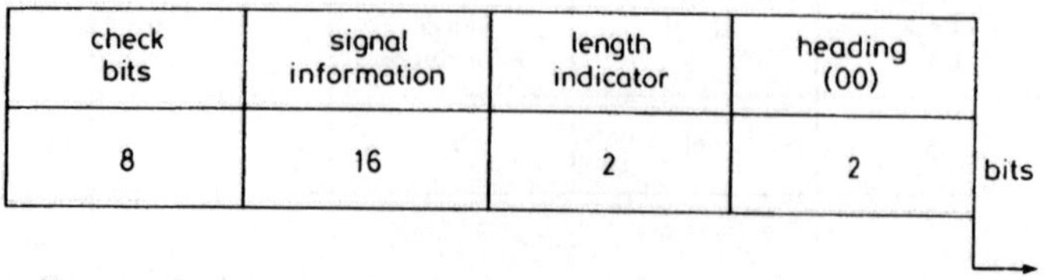

check bits	signal information	length indicator	heading (00)	
8	16	2	2	bits

Fig. A4.3   Format of subsequent signal unit

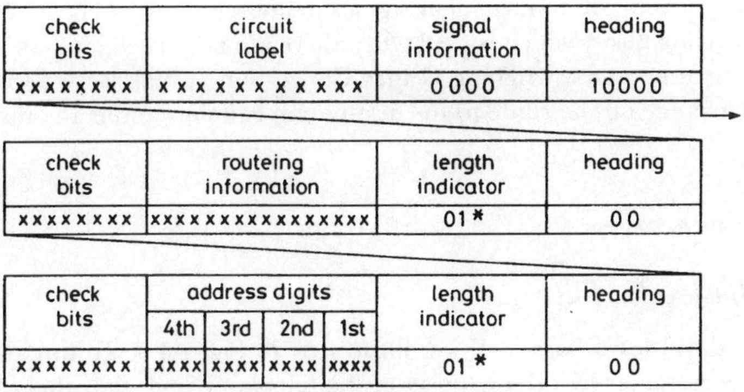

check bits	circuit label	signal information	heading
xxxxxxxx	xxxxxxxxxxx	0000	10000

check bits	routeing information	length indicator	heading
xxxxxxxx	xxxxxxxxxxxxxxx	01*	00

check bits	address digits				length indicator	heading
	4th	3rd	2nd	1st		
xxxxxxxx	xxxx	xxxx	xxxx	xxxx	01*	00

Fig. A4.4   Format of Initial Address Message
x = bit
* code 01 indicates two SSUs in this example

A4.3.5 Signalling system control signals

In addition to signals required for telephony call control, CCSS6 includes messages to ensure that the signalling system itself functions correctly. Such signals are transferred as LSUs and include the 'Acknowledgement Signal Unit' (ACU), the 'Synchronisation Signal Unit' (SYU), the 'System Control Signal Unit' (SCU) and the 'Multi-block Synchronisation Signal Unit' (MBS).

The ACU is used as part of the error-correction mechanism. The ACU indicates to an exchange which has sent a block of Signal Units whether or not the block was corrupted when received by the other exchange. Each block contains 11 Signal Units plus an ACU. Acknowledgement Indicators are used to define the status of the eleven Signal Units in the block that the ACU is acknowledging. A bit value of 0 indicates that the corresponding Signal Unit was received error-free, while a bit value of 1 indicates that an error was detected.

The SYU is used to maintain synchronisation of the signalling channel in the absence of signalling traffic. The SCU is used to carry out control functions on the signalling link. For example, a 'changeover' signal indicates that a particular signalling link has failed and that it is necessary to change to another signalling link to maintain service. The original specification of CCSS6 limited the number of blocks within the error-control loop to eight. This limit was subsequently increased to 256, based on multi-blocks of eight blocks. This gives rise to the need for a multi-block synchronisation (MBS) Signal Unit to enable the monitoring of multi-blocks.

A4.3.6 Management signals

Management signals relate to the management and maintenance of the signalling system and the speech network. Management signals are carried in LSUs or MUMs.

The specification of management signals defines a range of codes. Examples are 'All Circuits Busy', which is used when there are no speech circuits available to the destination and 'Switching Centre Congestion', which is used to indicate that an exchange on the route to the destination cannot handle any more calls.

A4.4 Procedures

A4.4.1 Basic call set-up

The procedures for a basic call are illustrated in Fig. A4.5. To initiate a call, the calling customer lifts the handset of the telephone and dials the address of the called customer. The originating exchange analyses the dialled digits received from the calling customer. When sufficient digits have been received to identify which exchange the call needs to be routed to, the originating exchange formulates an IAM and sends it to the intermediate exchange. The IAM includes the digits dialled so far by the calling customer and also includes information on the type of speech path being provided e.g. whether or not a satellite link is being used as one of the transmission paths.

The IAM is analysed by the intermediate exchange to determine to which exchange the call should be routed. After analysis, an IAM is sent to the destination exchange. Additional address digits received by the originating exchange from the calling customer are passed to the destination exchange in Subsequent

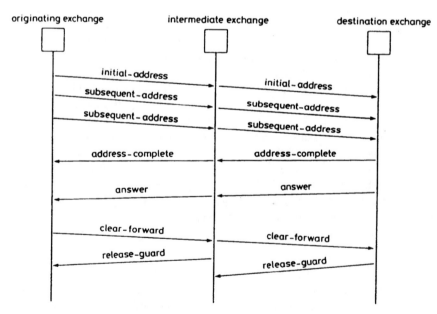

Fig. A4.5   Basic call procedures

Address Messages. When all address digits have been received at the destination exchange, the destination exchange returns an Address Complete Signal and this is passed to the originating exchange.

Ringing current is applied by the destination exchange to the called customer and a ring tone is applied to the speech path by the destination exchange to inform the calling customer of the status of the call.

When the called customer answers, an Answer Signal is sent from the destination exchange to the originating exchange via the intermediate exchange. In most networks, charging commences upon receipt of the Answer Signal at the exchange performing the charging function (typically the originating exchange). Conversation between the calling and called customers can now take place.

A4.4.2 Basic call clearing

When the calling customer clears the call, the originating exchange sends a Clear Forward (CLF) Signal to the intermediate exchange. The intermediate exchange commences clear-down of the speech path and sends a CLF signal to the destination exchange. When clear-down of the speech path is completed at the intermediate exchange, the intermediate exchange returns a Release Guard Signal to the originating exchange. Similarly, the destination exchange commences call clear-down upon receipt of the CLF signal and returns a Release Guard Signal to the intermediate exchange when clear-down is complete.

The originating and intermediate exchanges do not allow the reselection of the relevant speech path for another call until the Release Guard Signal has been received.

A4.4.3 Continuity check

In analogue CAS systems, the signalling path is coincident with the speech path being established for a call. In CCS systems, the physical tie between signalling and speech paths is removed. In early CCS systems, there was concern that the separation of signalling and speech paths could result in error conditions in which signalling messages were not associated with the correct speech path.

In response, CCSS6 specifies the use of a continuity check, during call establishment, to ensure that there is a correct association of signalling and speech paths. This approach is not required in modern CCS systems due to the improved reliability of telecommunications equipment, but it can be provided as an option. The continuity check involves sending a tone down the selected speech path and ensuring that the tone is received correctly. This process confirms that the speech path is coherent and that the correct signalling messages are associated with the speech path. The tone frequency specified by ITU-T is 2000 Hz. If the continuity check for a call on one speech path fails, then an attempt is made automatically to establish the call on another speech path.

A4.4.4 Error control

(a) Error detection

CCSS6 Signal Units contain check bits for error-detection purposes. The bits are generated by applying a polynomial to the remaining 20 bits of the Signal Unit. Upon receipt at the receiving exchange, the check bits are removed and the reverse procedure is applied. If the check is successful, then it is assumed that the message has not been corrupted during transmission and is used accordingly. If the application is unsuccessful, then a transmission error is detected and the error-control function is informed.

The approach adopted for CCSS6 is optimised for detecting short noise bursts in the transmission system. Longer noise bursts, or faults, are detected by circuit failure mechanisms within the transmission system itself.

(b) Error correction

Once an error is detected, it is necessary to correct the corrupted message. This is achieved by re-transmitting the message in question. Consider two exchanges connected by CCSS6, as shown in Fig. A4.6. Assume that Exchange 1 has sent Block A to Exchange 2 and that a copy of Block A is kept in Exchange 1. Block A consists of 11 Call Control Signal Units (Signal Units A.1 - A.11) and an ACU (Signal Unit A.12).

After analysis at Exchange 2, Signal Unit A.5 is recognised as corrupted. In this case, the next available block of Signal Units to be transmitted from Exchange 2 to Exchange 1 is used to inform Exchange 1 of the corruption (Block D in this example). The 12th Signal Unit in Block D is an ACU and the acknowledgement indicators are set to show that Signal Unit A.5 was corrupted. This information is analysed by Exchange 1 causing the retransmission of Signal Unit A.5.

The error-control mechanism used in CCSS6 is limited and can result in an incorrect sequence of messages, duplicated messages and unnecessary retransmissions, particularly during fault conditions. To overcome some of these abnormalities 'reasonableness' checks are incorporated in the signalling system. These require analysis within an exchange to determine whether or not certain message sequences are reasonable. This approach complicates the implementation of the signalling system and uses valuable processing power.

A4.5 Comparison of CCSS6 and CCSS7

CCSS6 is a CCS system exhibiting many attributes that represent a major advance on the capabilities of CAS systems. CCSS6 is a message-based system,

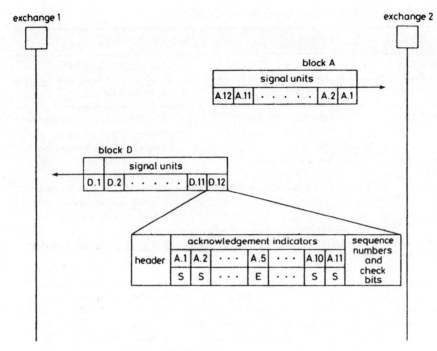

Fig. A4.6   Example of error correction
S = successful transmission
E = error detected

thus allowing a wide repertoire of signals and being compatible with modern technology exchanges. It supports many network characteristics, allowing changeover from faulty signalling links, Signal Transfer Point working and carrying network management and maintenance information. It also provides a basic error-detection and error-correction mechanism to enhance reliability, although the techniques adopted are not comprehensive.

However, CCSS6 has been superseded by CCSS7. This is mainly because of the limited degree of evolutionary potential incorporated within CCSS6. A comparison of the main factors is given in Table A4.2.

A4.6 Summary

CCSS6 was the first CCS system to be implemented internationally. It exhibits many of the advantages of CCS systems, but it has limited evolutionary potential and has been superseded by CCSS7.

Signalling information is included in Signal Units, which are 28 bits in length and are grouped into blocks of twelve. Message Signal Units can either be Lone Signal Units (LSU) or Multi-Unit Messages (MUMs).

380 Telecommunications Signalling

Table A4.2 Comparison of CCSS6 and CCSS7

Feature	CCSS6	CCSS7
designed	1960s	1970s, 1980s and 1990s
environment	optimised for analogue	optimised for digital
architecture	unstructured, limited evolutionary capability	structured to allow evolution: aligns with structure adopted by international organisations
application	optimised for telephony, ie, set-up and release of speech paths	covers telephony, circuit-switched data, ISDN, network features and general data-transfer mechanism
signalling transfer	circuit-related	both circuit-related and non-circuit-related
coding structure	fixed-length signal units	variable-length messages with evolutionary formatting
error control	can lead to unsequenced and duplicated messages: needs reasonableness checks	minimal lost or duplicated messages, even during fault conditions

Lone Signal Units, and the first Signal Unit of a MUM, consist of a Heading Code, Signal Information, a Label and check bits. The combination of the Heading Code and Signal Information defines the meaning of the signal. The Label defines the speech circuit to which the message refers. The check bits are used to check for corruption during transmission of the message.

Subsequent Signal Units (SSUs) contain a Heading Code and check bits. The rest of the SSU format consists of a Length Indicator and a Signalling Information Field (SIF) that is used to transfer, for example, address digits.

In addition to Signal Units for telephony, CCSS6 defines control signals and management signals that are used to operate and maintain the signalling system.

Calls are established by sending an IAM through the network. When sufficient address digits have been received by the destination exchange, an ACM is returned. This is followed by an Answer Message when the called customer answers the call. The call is cleared by using Clear Forward and Release Guard Messages. Procedures are defined to conduct a continuity check of the speech path during the call establishment procedure.

Errors are detected by monitoring the eight check bits in each Signal Unit. Each Signal Unit is acknowledged as having been received successfully or unsuccessfully. If a message is detected as corrupted, it is retransmitted by the sending exchange.

Comparing CCSS6 with CCSS7 highlights the drawback of lack of evolutionary potential. The specification of CCSS6 is not structured in a tiered approach, thus limiting the ability to modify the system. CCSS6 is circuit related only in nature and is optimised for telephony. The coding structure is fixed in length and the error-control mechanism is limited.

A4.7 References

1 ITU-T Recommendations Q.251–Q.300: 'Specifications of Signalling System No. 6' (ITU, Geneva)

Appendix 5
CCSS7 Telephone User Part

A5.1 General

The Telephone User Part (TUP)[1] was defined for use in international and national telephony-based networks (including circuit-switched data). Options are specified to allow for special circumstances in national networks. The TUP is designed to cover all telephone applications, including use over satellites. The TUP is an effective system. However, the fact that it is designed primarily for telephony-based networks constrains its applicability. Working with, for example, non-circuit-related information transfer systems (e.g. Transaction Capabilities) is not as efficient as the ISDN User Part (ISUP) described in Chapter 6. The TUP is still in use, but it is being replaced by the more flexible ISUP.

A5.2 TUP Formats

A5.2.1 Message format

TUP messages are carried by the MTP in Message Signal Units, the general format of which is described in Chapter 4. The format of a TUP Message Signal Unit is shown in Fig. A5.1.

The User Part Information is included in the Signalling Information Field (SIF) of the message. The SIF contains a label, a heading code and one or more information elements (which can be mandatory or optional, according to the type of message). The length of the signal unit can be fixed or variable.

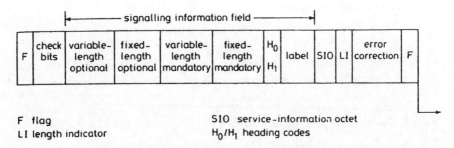

F flag
LI length indicator
SIO service-information octet
H_0/H_1 heading codes

Fig. A5.1   TUP Message Signal Unit

(a) Label

The Label is the information that the TUP uses to identify the traffic circuit to which a message refers. The label consists of the Routing Label and the Circuit Identification Code (CIC). The Routing Label specifies the identity of the originating and destination exchanges. The CIC is a number that identifies the traffic circuit to which the message refers. In the TUP, the CIC includes the 4 bit Signalling Link Selection (SLS) Field. All TUP messages contain a Heading Code, consisting of an H0 field and an H1 field. The H0 field identifies to which general category (message group) a message belongs. For example, a message containing address information (e.g. digits dialled by the calling customer) belongs to the category 'Forward Address'.

The meaning of the H1 field depends on the complexity of the message. For a simple message, the H1 field contains enough information to define fully the meaning of the message. For example, if a telephone conversation has been completed and the calling customer clears (i.e. releases) the call, the H1 field consists of information instructing exchanges to clear the traffic path that is being used.

For a complex message, the H1 field contains information that specifies the format of the remainder of the message. For example, in a message setting-up a call, the H1 field indicates the type of information that is included in the rest of the SIF.

(b) Sub-fields

After the Label and Heading Codes, the SIF generated by the TUP is divided into a number of sub-fields. Mandatory sub-fields are compulsory for a given message type and they appear in all messages of that type. Optional sub-fields can be present, but are not compulsory, for a given message type.

Mandatory and optional sub-fields can be of fixed length or variable length. Fixed length sub-fields have the same number of bits in each message of a given type. Variable length sub-fields can vary in size for a message of a given type. The length of a variable length sub-field is specified by a 'Length Indicator'. The order of sub-field transmission in the SIF is:

(i)   fixed length mandatory;
(ii) variable length mandatory;
(iii) fixed length optional;
(iv) variable length optional.

A wide range of message formats is defined by ITU-T. To illustrate the principles involved, examples of formats of commonly used messages are described in the following sections

A5.2.2 Initial Address Message (IAM)

The Initial Address Message (IAM) is the first message to be sent during the set-up of a call. It contains the required address (e.g. digits dialled by the calling customer) and other information needed for routing purposes. The format of the IAM is shown in Fig. A5.2.

The IAM belongs to a message group termed 'Forward Address' and the H0 field is coded accordingly (0001). The IAM is a complex message: the H1 code (0001) identifies the message as an IAM and determines the format of the rest of the SIF. The Calling Party Category indicates the type of customer initiating the call, e.g. ordinary customer or operator. The message indicators give a variety of information pertinent to the call, e.g. whether or not a satellite is being used to provide one of the transmission links. The address digits dialled by the calling customer are included in the IAM after the message indicators. A sub-field precedes the digits to indicate the number of address digits included in the message.

A5.2.3 Address Complete Message (ACM)

The Address Complete Message (ACM) is sent by the destination exchange to the originating exchange to indicate that all the address digits required to identify the called customer have been received. The format is shown in Fig. A5.3.

The ACM belongs to the group of messages termed 'Successful Backward Set-up Information'. The H0 is coded 0100 in accordance with the message group. The H1 code of 0001, in conjunction with the H0 code, identifies the message as an ACM. The message indicators give general information about the characteristics of the established call, e.g. whether or not echo suppressors are involved with the traffic path.

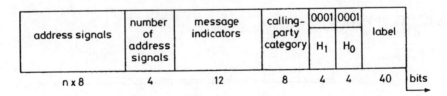

address signals	number of address signals	message indicators	calling-party category	0001 H_1	0001 H_0	label
n x 8	4	12	8	4	4	40 bits

Fig. A5.2 Format of Initial Address Message
Source: (ITU-T Recommendation Q.723)

message indicators	0001 H_1	0100 H_0	label
8	4	4	40 bits

Fig. A5.3 Format of Address Complete Message
Source: (ITU-T Recommendation Q.723)

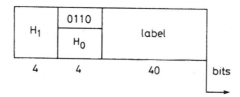

Fig. A5.4   Format of Call Supervision Message

A5.2.4 Answer Signal (ANS)

The Answer (ANS) Signal is sent by the destination exchange to indicate that the call has been answered by the called customer. The ANS Signal belongs to the group of messages termed 'Call Supervision', with the general format shown in Fig. A5.4. The ANS is a simple message: the message consists only of the H0/H1 combination, without further indicators. The H0 code for the call supervision message is 0110 and the H1 code for a normal answer message is 0000.

A5.2.5 Clear Forward (CLF) Signal

The Clear Forward (CLF) Signal is sent by the originating exchange when the calling customer indicates that the call has been completed. The CLF is a Call Supervision signal with the same general format as the Answer Signal. Again, the CLF is a simple message and the HO/H1 combination is sufficient to define the message. For CLF, the H1 code is 0100.

A5.3 TUP Procedures

A5.3.1 Basic call set-up

A basic call set-up sequence under normal operating conditions is shown in Fig. A5.5. The originating exchange initiates the call by sending an IAM to the intermediate exchange. The direction in which the IAM is transmitted determines the forward direction of the call, the opposite direction being the backward direction.

The IAM contains all the information relating to the characteristics of the required traffic path. It may also contain all the address digits required to identify the called customer (en bloc operation). Alternatively, the IAM may contain only sufficient address digits to route the call to the intermediate exchange (overlap operation). In both cases, sending an IAM results in the seizure of an appropriate traffic path between the originating and intermediate exchanges. Seizure indicates that the traffic path is reserved.

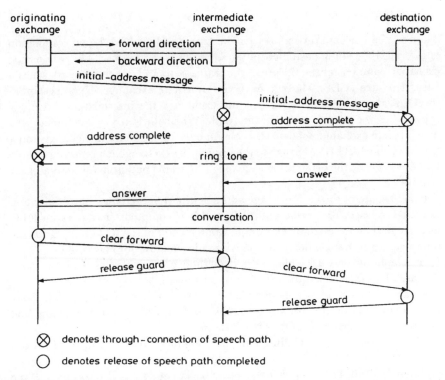

Fig. A5.5 *Basic call set-up and release*

The intermediate exchange analyses the received IAM and selects an appropriate traffic path to the destination exchange. The IAM is forwarded to the destination exchange over an appropriate signalling link.

In en bloc operation, the destination exchange returns an Address Complete Message (ACM) to confirm receipt of the called customer's number. In overlap operation, only some of the address digits are included in the IAM. In this case, the originating exchange supplies the extra digits in one or more Subsequent Address Messages (not shown in Fig. A5.5). When all address digits are received at the destination exchange, the ACM is returned to the originating exchange. In both the en bloc and overlap methods of operation, the receipt of the ACM at the originating exchange causes it to connect through the traffic path to the calling customer.

While returning the ACM to the originating exchange, the destination exchange starts ringing the called customer. The destination exchange also returns ring tone over the allocated traffic path to the calling customer. When the called customer answers the telephone, ringing and ring tone are stopped and an Answer Signal is returned to the originating exchange. It is usual for the originating exchange to commence charging upon receipt of the Answer Signal. At this stage, conversation may take place.

A5.3.2 Call-release

The basic call-release procedures are illustrated in Fig. A5.5. Upon completion of the call, the calling customer initiates the release procedure. The originating exchange commences the release of the traffic path and, when released, sends a Clear Forward (CLF) Message to the intermediate exchange. Upon receipt of the CLF Message from the originating exchange, the intermediate exchange commences the release of the traffic path. When the release is complete, the intermediate exchange sends a CLF message to the destination exchange and a Release Guard (RLG) Message to the originating exchange. Upon receipt of the RLG message at the originating exchange, the traffic path is made available for new traffic.

Upon receipt of the CLF message at the destination exchange, the destination exchange releases the traffic path. When this is complete, it returns an RLG message to the intermediate exchange. Upon receipt of the RLG message at the intermediate exchange, the traffic path is made available for new traffic.

In telephony networks, the normal method of operation is 'Calling Party Release'. This means that the calling customer is deemed to control the call. This approach differs from circuit-switched data networks, in which a 'First Party Release' method of operation is generally used, in which both the calling and called customers have the authority to release a call.

The TUP operates the Calling Party Release method of operation. However, there is a need to safeguard the calling customer from unintentionally prolonging a call. For this purpose, a clear-back sequence is also specified. If the called customer ends the conversation, the destination exchange generates a Clear Back (CLB) Message and sends it to the intermediate exchange. The intermediate exchange passes the message back to the originating exchange without analysis. After receipt of the CLB Message, the originating exchange commences a timing period of 1-2 minutes. If the calling customer clears down during this period, then the clear forward sequence is enacted immediately. If no indication is received from the calling customer, the originating exchange waits until expiration of the timing period before commencing the clear forward sequence.

A5.3.3 Abnormal conditions

The TUP covers many conditions to ensure that appropriate action is taken by the network to maintain service in abnormal circumstances. Examples are 'Reset', 'Dual Seizure' and 'Unexpected Messages'.

(a) Reset

The Reset Procedure is used when the status of a traffic circuit, or a number of circuits, becomes unclear to an exchange (e.g. due to memory mutilation within the exchange). In this case, procedures are defined for the exchanges at each end of a circuit to reset the circuit to the idle condition. If a small number of circuits is affected, Reset Messages are used for each circuit. If a large number of circuits is affected, a Group Reset Message may be applied to reset all the circuits with one message.

(b) Dual seizure

Dual seizure is when two interconnected exchanges choose the same traffic circuit for two different calls at approximately the same time. The risk of dual seizure is minimised by applying different algorithms for choosing traffic paths at each end of the transmission link. To resolve the contention when a dual seizure does occur, each circuit is nominally controlled by one exchange. The IAM sent by the controlling exchange for a particular circuit is processed normally by the non-controlling exchange. The IAM sent by the non-controlling exchange for the circuit is disregarded by the controlling exchange. Upon detection of dual seizure at the non-controlling exchange, an automatic repeat attempt is initiated for another traffic path.

(c) Unexpected messages

Rules are defined to cover the actions to be taken at each exchange upon receipt of an unexpected message. For example, if an exchange receives a Clear Forward Message relating to an idle circuit, the exchange will acknowledge the signal with a Release Guard Message. Receipt of many other unexpected messages causes the return of a Reset Message.

A5.4 TUP Supplementary Services

A5.4.1 General

The TUP defines procedures and formats for supplementary services to provide features additional to basic call set-up and release. Specifying the procedures for supplementary services is very complex due to:

- the need to specify the procedures in sufficient detail without imposing undue constraints on the network operator;
- the complex nature of some supplementary services;
- the need to cover abnormal conditions and failures at each stage of each supplementary service;
- the changing nature of many supplementary services;
- the impact of combining supplementary services.

Detailed procedures are given in the specification[2] and are not discussed here. The implementation of some supplementary services is similar to the ISUP, explained in Chapter 6. Below is a brief description of some of the more common supplementary services specified for the TUP.

A5.4.2 Closed User Group (CUG)

The Closed User Group (CUG) Service allows customers to form groups or clubs, providing facilities and placing restrictions upon members.

The CUG facility is implemented using an interlock code. During call set-up, a validation check is performed to ensure that the calling and called customers belong to the same CUG, as verified by the interlock code. The data to verify interlock codes can be held at local exchanges (decentralised version of CUG) or at dedicated points in the network (centralised version of CUG). Only the decentralised version of CUG is specified for the TUP.

A5.4.3 Calling line identification

This service allows the presentation of the calling customer's telephone number to the called customer and is very similar to that for the ISUP.

A5.4.4 Called line identification

This service is used by the calling customer to confirm to which telephone number a call is connected and is very similar to that for the ISUP.

A5.4.5 Redirection of calls

The Redirection Facility allows calls to a particular telephone number to be redirected to another predetermined number. Procedures are included to allow a called customer to reject a redirected call. To avoid numerous redirections, resulting in a continuous loop, only one redirection is allowed per call. The procedures are similar to those for the ISUP.

A5.4.6 Digital connectivity

A calling customer can request that the traffic path to be established through the network should be digital. An indicator is included in the IAM to allow analysis at each exchange. If it is not possible to provide digital equipment for each connection, the call is rejected and the originating exchange is informed that the call cannot be completed.

A5.5 Summary

The TUP is an effective system, but the fact that it is designed primarily for telephony-based networks constrains its applicability and it is being replaced by the more flexible ISUP. The TUP Signalling Information Field consists of a

Label, a Heading Code and sub-fields. The Label consists of the Routing Label and a Circuit Identification Code (CIC). The Heading Code defines the class of message and, for simple messages, defines the message type. For complex messages, the Heading Code defines the format of the rest of the message. The sub-fields can be mandatory or optional. In both cases, sub-fields can be of fixed or variable length.

Traffic circuits are established by sending an IAM which includes the address of the called customer. An Address Complete Message indicates that sufficient address information has been supplied and an Answer Message indicates that the called customer has answered the call. The call can be cleared by the calling customer initiating a clear forward sequence.

Supplementary services are defined which are applicable to telephony-based networks (including circuit-switched data).

A5.6 References

1 ITU-T Recommendations Q.721–Q.725: 'Telephone User Part' (ITU, Geneva)
2 ITU-T Recommendation Q.724: 'Signalling Procedures' (ITU, Geneva)

Appendix 6
Inter-PABX Signalling

A6.1 Introduction

As mentioned in Chapter 13, private network signalling systems evolved to take advantage of the feature-rich nature of PABXs before the advent of virtual private networks (VPNs). A successful system was the Digital Private Network Signalling System (DPNSS). This Appendix gives a brief outline of such a private network signalling system. The system is based on three layers, similar to those of DSS1. This Appendix focuses on the Layer 3 aspects.

A6.2 Private Network

In private networks, the switching functions are performed by PABXs, as illustrated in Fig. A6.1. In this case, users are connected to PABXs 1 and 2. PABXs 3, 4 and 5 form a transit network.

A6.3 Format

A typical message format contains a Message Type Field and one or more Information Element Fields, as illustrated in Fig. A6.2.
 The Message Type indicates the category of the message, e.g.:

- Service Request;
- Incoming Call Indication;
- Call Accepted;
- Release Request.

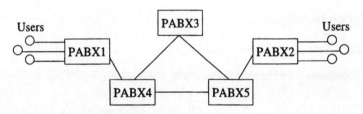

Fig. A6.1 Private network switching

Information Element	Information Element	Information Element	Message Type

Fig. A6.2 *Typical message format*

Table A6.1 *Typical Information Elements*

Information Element	Description
Service Indicator	type of service, e.g. telephony
Calling Line Category	type of line, e.g. ordinary extension
Originating Line Identity	number of calling user
Destination Line Identity	number of called user

In some cases, the Message Type is sufficient to define the action that is required. In other cases, Information Elements are provided to give additional information. Typical Information Elements are given in Table A6.1.

Private network signalling systems, in common with their public counterparts, were designed to be as flexible as possible and to cater for evolution. Thus, if new features are required, it is undesirable to have to update all the PABXs. The impact of this evolution is limited by including an indicator in the Information Element to inform the receiving PABX how to deal with the call if the Information Element is not understood. Three cases are defined:

- Mandatory – if this is not recognised the receiving PABX should abort the call and return the information to the calling PABX.
- Optional – if this is not recognised, the call may still proceed, but the calling PABX is informed that some information has been ignored.
- Informative – if this is not recognised, the call may still proceed and the calling PABX is not informed.

The cases can vary according to the type of PABX used. For example, an Information Element that is mandatory for a terminating PABX can be made Informative for a transit PABX. Thus, the transit PABX can continue to operate despite new features being added to a terminating PABX.

A6.4 Procedures

A6.4.1 Basic call

Typical procedures for a basic call set-up are illustrated in Fig. A6.3.

User A indicates that a call is required by dialling appropriate digits. PABX1 selects a circuit to be used and sends a Service Request Message to

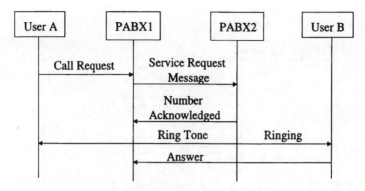

Fig. A6.3 *Procedures for basic call*

PABX2. If User B is free, PABX2 returns a Number Acknowledged Message to indicate that the call is accepted and applies ring tone towards User A and ringing current to User B. When User B answers, an Answer Indication is returned to PABX1 and conversation can take place. If User B is busy, the call is cleared by PABX2 issuing a Clear Request Message and PABX1 confirming clearance with a Clear Indication Message.

A6.4.2 Supplementary services

As an example, consider the procedure for the completion of a call to a busy subscriber (CCBS), also called Call Back (Fig. A6.4). Upon detection of a busy tone, User A can initiate the CCBS service by sending a Call Back Request to PABX1. PABX1 sets up a virtual connection (i.e. a signalling connection without a physical circuit for speech) with Service Request 1. This includes an Information Element that indicates that PABX2 is requested to call back when User B is free. PABX2 accepts or rejects the request and includes the response in a Clear Request Message, which is acknowledged by a Clear Indication. At this point, the virtual connection is released. If PABX2 accepts the request, User B is monitored until the line becomes free.

When User B becomes free, PABX2 initiates Service Request 2 to inform PABX1 of the status. This is again a virtual connection. The clearing sequence confirms that User A is free and initiates the blocking of User B from other calls.

PABX1 then initiates Service Request 3, but this time the request is for a physical speech circuit to be established to PABX2. The Service Request Message includes information that the call is a Call Back. Upon receipt of the Number Acknowledged Message, ringing is applied to User A's line. Upon answer, PABX2 is requested to ring User B and the call continues as for a basic call.

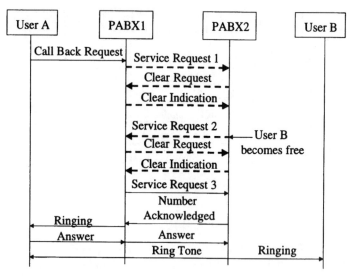

Fig. A6.4 Procedures for CCBS

A6.5 Conclusion

The principles outlined in this Appendix have been used to provide private network signalling for some years. This eventually spurred the specification of PSS1 (QSIG), explained in Chapter 13. The role for such systems will continue until public networks can provide virtual private networks with greater functionality and at a lower cost.

Glossary

Term	Abb	Explanation	Context
Access	–	The link between a customer and a network node	general
Access Significance	–	Relevance of a message to the originating and terminating accesses	DSS
Access Unit	AU	The point at which access is gained to a packet-switched public data network	DSS
Acknowledged Operation	–	A Layer 2 form of working in which each frame is numbered, thus permitting error control	DSS
Active Connection	–	A user requests the establishment of a TCP connection	IP (TCP)
Address Resolution Protocol	ARP	An IP Link Layer Protocol that translates IP addresses into physical addresses	IP
Application	–	An activity or process that a user wishes to perform	general
Application Context	AC	A combination of Application Service Elements and their relationship	IN
Application Entity	AE	The functions that Applications use to communicate	OSI
Application Management	–	The exchange of management information between signalling points	CCSS7 (OMAP)
Application Part	AP	A set of OSI Layer 7 functions	OSI
Application Service Element	ASE	A function, or a group of functions, within an Application Entity	OSI
Architecture	–	The structured approach to the specification of signalling systems	general

Term	Abb	Explanation	Context
Associated Mode of Operation		The transfer of messages between directly connected network nodes	CCS
Association Control Service Element	ACSE	An Application Service Element used to establish an association between two Application Layers	OSI
Asynchronous Transfer Mode	ATM	A form of packet (or cell) switching	general
B Channel	–	An access traffic channel	DSS
Backward Message	–	A message sent towards the calling customer	general
Base Station System	BSS	Provides the radio functions in a Public Land Mobile Network	CCSS7 (MAP)
Basic Access	BA	A means of connecting a customer to a local node in an ISDN using two traffic channels and a signalling channel	DSS
Basic Error Correction		A form of correcting errors in messages	CCSS7 (MTP)
Bearer	–	The type of medium used to transfer information	general
Block	–	Aggregate a small amount of information into a single package	IP (FTP)
Block Mode	–	The transmission of data in a series of blocks with header	IP (FTP)
Blocking	–	Removing a circuit from service	CCSS7
Broadband	B	High bandwidth, typically greater than 2 Mbit/s	general
Broadcast Working	–	The sending of signalling information to a range of customer terminals	DSS
Buffer	–	A unit that stores messages in preparation for transmission	general
Call Deflection	CD	The diversion of a call after alerting the called user	DSS
Call Forwarding		A supplementary service in which a call is diverted from one called customer to another	CCSS7 (ISUP)

Call Gapping	–	A method of reducing the number of calls to a destination	general
Call Rearrangement Procedure	–	A procedure that allows a customer to suspend a call, make changes to the terminal being used and then resume the call	DSS
Call Reference	CR	A number that is used to identify a particular call	general
Called Line Identification	–	The identity of the called customer	general
Calling Line Identification	CLI	The identity of the calling customer	CCSS7
Calling Party Release	–	A method of operation in which the calling customer controls, and usually releases, a call	general
Capability Set	CS	Attributes defined in terms of interfaces and services	IN
Card Acceptor	–	A network or service operator that gives service to a charge card holder	CCSS7
Card Issuer	–	A network or service operator that issues a charge card	CCSS7
Cell Loss Priority	–	Discriminates between high and low priority cells	IP
Centralised Intelligence	–	Call control vested in one node	IN
Changeback	–	A procedure for reversing changeover	CCSS7 (MTP)
Changeover	–	A procedure for transferring signalling from one link to another	CCSS7 (MTP)
Channel Associated Signalling	CAS	A type of signalling in which signalling capacity is dedicated for use by a traffic circuit	general
Check Bits	–	A field within a signal unit used for error control purposes	CCS
Circuit	–	A physical traffic path	general
Circuit-Related Signalling	CR	Signalling in which messages are identified by the number of the traffic circuit to which the messages refer	general

(continued)

Term	Abb	Explanation	Context
Circuit Validation Test	CVT	A procedure that is used to check data consistency	CCSS7 (OMAP)
Circuit-Switched Data	–	The use of traffic circuits that are set up and dedicated to a data call	general
Client	–	A remote user making use of the services of a server	general
Closed User Group	CUG	A supplementary service in which customers form clubs	general
Codeset	–	A group of information elements within which the codings are uniquely defined	DSS
Common Management Information Protocol	CMIP	Used to exchange information between Application Layer Entities	CCSS7 (OMAP)
Common Services	–	Available to all users	CCSS7 (MAP)
Common-Channel Signalling	CCS	A type of signalling in which signalling capacity is allocated from a common pool when required by a particular call	general
Community	–	A group of Application Entities	IP (SNMP)
Component	Comp	An element of information used to request a remote node to perform an action or to return the result of an action	CCSS7 (TC)
Component-Sub Layer	–	A sub-layer responsible for requesting an action to be performed and returning the results of actions	CCSS7 (TC)
Compressed Mode	–	The transmission of files in which replicated data is compressed	IP (FTP)
Configuration Management	CM	Controlling resources within a network	general
Connection-Oriented	–	A type of information transfer that is accomplished by establishing a formal relationship between entities	general
Connection Section	–	A part of a connection between two nodes	SCCP

Connectionless	–	A type of information transfer that is accomplished without establishing a formal relationship between entities	general
Constructor	–	A complex type of information element that is structured in a recursive manner	CCSS7 (TC)
Control Connection	–	A connection for transferring control messages	IP (FTP)
D Channel	–	An access signalling channel	DSS
Data Connection	–	A connection for transferring data	IP (FTP)
Data Form Message	–	A message used to carry information in connection-oriented signalling	CCSS7 (SCCP)
Data Transfer Process	DTP	The signalling functions for a data connection	IP (FTP)
Data User Part	DUP	A user part defining procedures and formats for circuit-switched data	CCSS7
Database	–	A node storing information	general
Data-Link Connection	–	A connection at Layer 2 between a local node and a customer terminal	DSS
Deblock	–	Split a package of information into its original format	IP (FTP)
Destination Point Code	DPC	The point code of the node to which a message is destined	CCSS7 (MTP)
Detection Point	DP	The juncture in call processing at which an IN call is triggered	IN
Dialogue	–	A form of connection	general
Digital Connectivity	–	A calling customer request to establish a call using digital nodes	CCSS7 (TUP)
Digital Private Network Signalling System	DPNSS	A signalling system defined for inter-PABX use	general
Digital Subscriber Signalling System No. 1	DSS1	A common channel signalling system used between customers and ISDN nodes	general
Direct-Dialling In	DDI	A feature allowing a call to be made to a specific extension on a PABX	general

(*continued*)

Term	Abb	Explanation	Context
Disconnect Frame	–	A frame used to clear a connection between a customer and a local node	DSS
Disconnected Call	–	A call in which a traffic circuit is cleared but not yet ready to be used for another call	general
Distributed Intelligence	–	Call control spread over several nodes	general
Dual Seizure	–	An attempt by two entities to select the same speech circuit	general
Dynamic Modelling	–	Modelling of the set-up and clear-down of connections	general
en bloc	–	Sending address information in one batch	general
Entity	E	A node, user, program, etc.	OSI
European Association for Standardising Information & Communications Systems	ECMA	A European standards organisation	general
European Telecommunications Standards Institute	ETSI	A European standards organisation	general
Extension Header	–	Provides optional information	IP (IP)
Extensive Primitive	–	Used in two-way information flow	OSI
Fallback	–	The provision of a second choice connection type	general
Fault Management	FM	Detecting and correcting abnormal network operations	general
File Structure	–	Ways in which files are structured	IP (FTP)
File Transfer Protocol	FTP	An IP Application Layer Protocol for transferring files between a user and a server or between servers	IP (FTP)
Fixed Communications	–	Communications when the users are of fixed location	general
Flag	–	A defined end point or indicator	general
Flow	–	A sequence of packets that is treated in a special manner	IP (IP)
Flow Control	–	A mechanism to restrict the amount of data transferred	general

Format	–	The coding structure applicable to messages	CCS
Forward Message	–	A message sent towards the called customer	general
Frame	–	A unit of signalling information	general
Frame in a PCM System	–	A collection of 8-bit codes relating to 24 or 30 traffic channels	general
Frame-Check Sequence	–	The check bits used by DSS	DSS
Frequency Division Multiplex	FDM	A means of assembling traffic circuits, using the frequency spectrum, for transmission purposes	general
Functional Protocol	–	A means of providing supplementary services by relying on customer terminal participation	DSS
Gateway Mobile Switching Centre	GMSC	A Mobile Switching Centre interworking with other networks	MAP
Generic Functional Transport	GFT	A means of transferring information transparently between PABXs	PN
Get	–	A message type used for retrieving information	IP (SNMP)
Global Significance	–	Relevance of a message to all CCS portions of a call	DSS
Global Title	GT	A form of address that needs to be translated to facilitate routing	general
Global Virtual Network Service	GVNS	A means of implementing a Virtual Private Network internationally	general
Handover	–	A procedure to facilitate a customer moving from one area to another	MAP
Header	–	A field within a message defining the class of message	CCS
High-Level Data-Link Control	HDLC	A protocol defined by the International Standards Organisation	general
Home Location Register	HLR	A database to which a mobile station is assigned upon subscription	MAP

(*continued*)

404 *Telecommunications Signalling*

Term	Abb	Explanation	Context
Hop	–	The link between two IP nodes	IP
Host	–	A node or computer that originates or terminates IP traffic	IP
I-Format Frame	–	A type of frame used to transfer information	DSS
In Call Modification	–	A procedure allowing the modification of the characteristics of a speech circuit during a call	CCSS7 (ISUP)
Incoming Signalling System	I/C	A function, used in interworking of CCS systems, receiving a set-up message	CCS
Information Element	IE	A field(s) within a message	general
Integrated Services Digital Network	(ISDN)	A network providing digital connections between customers	general
Intelligent Network	IN	A network using centralised intelligence	general
Intelligent Network Application Part	INAP	A user of Transaction Capabilities for communicating within an Intelligent Network	IN
Inter-Exchange Signalling	–	Signalling between exchanges within networks	general
Intermediate Service Part	ISP	Layers 4 to 6 of Transaction Capabilities	CCSS7 (TC)
Internal Management	–	Management functions specified as part of the signalling system	OMAP
International Mobile Equipment Identity	IMEI	Equipment number of a mobile terminal	CCSS7 (MAP)
International Mobile Subscriber Identity	IMSI	Number of a mobile terminal used within the network	CCSS7 (MAP)
International Standards Organisation	ISO	An international standards organisation responsible for defining communications environments	general
International Telecommunications Charge Card	ITCC	A supplementary service in which calls are billed to a user's account	general

International Telecommunications Union-Telecommunications Standardisation	ITU-T	A subsidiary of the United Nations focusing on telecommunications standards	general
International Telegraph & Telephone Consultative Committee	CCITT	An old term for the International Telecommunications Union	general
Internet Architecture Board	IAB	A standards organisation for IP Technologies	IP
Internet Control Messaging Protocol	ICMP	An IP Internet Layer Protocol that reports difficulties in delivering packets	IP
Internet Protocol	IP	(a) a signalling type used in the Internet (b) a general term to describe Internet technologies	IP
Inter-Nodal Signalling	–	Signalling between nodes within networks	general
Invoke	–	A component used to request an action to be performed	CCSS7 (TC)
IP Model	–	An architecture model for IP protocols	IP
ISDN Signalling Control Part	–	A form of ISUP in which call control functions are separated from traffic circuit control functions	CCSS7
ISDN User Part	ISUP	A user part defining procedures and formats for use in ISDNs	CCSS7
ITU-T Signalling System No. 7	CCSS7	A common channel signalling system used between network nodes	general
Keypad Protocol	–	A means of providing supplementary services by generating alphanumeric codes	DSS
Label	–	A field within a signal unit defining a speech circuit	CCSS7
Layer	–	A tier in the architecture of signalling systems	general
Level	–	A tier defined for use by CCSS7 circuit-related applications	CCSS7

(continued)

Term	Abb	Explanation	Context
Line Signalling	–	Signalling dealing with setting-up and clearing-down traffic circuits	general
Local Reference	LR	A number allocated by a node to identify a connection	general
Local Significance	–	Relevance of a message to either the originating or the terminating access	DSS
Location Area	–	An area within which a mobile station can roam without updating a location register	CCSS7 (MAP)
Loop Disconnect	LD	A signalling system in which signals are transferred by modifying the status of a direct current loop	general
Management Information Service	MIS	The overall management function within a network	general
Management Inhibit	–	A procedure facilitating maintenance or testing	CCSS7 (MTP)
Message	–	A unit of information in common channel signalling systems	CCS
Message Signal Unit	MSU	A signal unit used to carry user-part information	CCSS7
Message Transfer Part	MTP	Levels 1 to 3, responsible for successfully transferring messages from one node to another, even under failure conditions	CCSS7 (MTP)
Mobile Communications	–	Communication when users can change locations during a call	general
Mobile Station	MS	A mobile user terminal	CCSS7 (MAP)
Mobile Station ISDN Number	MSISDN	Address of a mobile terminal dialled by a calling user	CCSS7 (MAP)
Mobile Switching Centre	MSC	A switch with functions particular to a mobile network	CCSS7 (MAP)
Mode of Operation	–	The means of routing messages through a signalling network	CCS
MTP Routing-Verification Test	MRVT	A procedure that is used to verify MTP routing data	CCSS7 (OMAP)
Multi-Frame	–	A collection of 12 or 16 frames in PCM systems	general

Multi-Frequency Inter-Register Signalling System	MFIR	A signalling system using compounded frequencies (2 out N frequencies) to transfer signals	general
Multimedia		Using several types of connection, call or service	general
Multiple Association Control Function	MACF	A set of rules for the use of functions in multiple communications	OSI
Narrowband	N	Low bandwidth, typically 2 Mbit/s or less	general
National Number Group (Code)	NNG	The number that is normally associated with a geographic area when dialling a trunk call	general
Network Element	NE	An element of a network, e.g. a switch	general
Network Node Interface	NNI	An interface between nodes	general
Network Service Data Unit	NSDU	Blocks of data in non-circuit-related applications	CCSS7 (SCCP)
Network Service Part	NSP	The combination of the SCCP and MTP	CCSS7
Node	–	A basic element of a network, e.g. an exchange	general
Non-Associated Mode of Operation		A means of transferring messages between network nodes that are not directly connected	CCS
Non-Circuit-Related Signalling	NCR	Signalling in which messages are identified by a reference number, thus allowing messages to be transferred in the absence of traffic circuits	general
Object	–	An element of a network, e.g. a signalling termination	general
Off-Air Call Set Up	–	The allocation of a radio channel after the called customer answers	CCSS7 (MAP)
Off-Line Applications		Actions that do not need to be performed in a short timescale	general
Open Systems Interconnection	OSI	Connecting entities, networks, etc. using standard interfaces	OSI
Openness	–	Providing access to a wide range of applications and making common interfaces	general

(continued)

Term	Abb	Explanation	Context
Operation	–	A message that causes an action to take place at a remote location	general
Operations and Maintenance ASE	OMASE	An ASE designed for managing networks	CCSS (OMAP)
Operations System	–	A support system, e.g. a network management system	general
Operations, Maintenance and Administration Part	OMAP	A user of TC designed to manage a signalling network	CCSS7 (OMAP)
Originating Point Code	OPC	The point code of the node sending a message	CCSS7 (MTP)
Outband Signalling System	–	Signalling in which signals are transferred in frequency spectra additional to those employed by traffic circuits	general
Outgoing Signalling System	O/G	A function, used in interworking of CCS systems, sending a set-up message	CCS
Overlap	–	The sending of address information in more than one batch	general
Packet Data	PD	Data divided into cells, each cell being routed independently through the network	general
Packet Handler	PH	Equipment within nodes capable of handling packet data	general
Parameter	–	Information contained in message fields or primitives	general
Pass Along	–	A form of end-to-end signalling	CCSS7 (ISUP)
Passive Connection	–	A user indicates availability to receive a TCP connection	IP (TCP)
Payload	–	Data to be transferred within a message	general
Performance Management	PM	Gathering statistics on network behaviour	general
Personal Identification Number	PIN	A code identifying an individual	general
Point Code	PC	The unique identity of a signalling point	CCSS7 (MTP)

Glossary 409

Term	Abbr.	Definition	Category
Pointer	–	A field used to indicate the location of another field within a message	CCS
Point-to-Point Working	–	Exchange of signalling between a local node and a specific customer terminal	DSS
Portable	–	Can be used in several scenarios	general
Post Dialling Delay	PDD	The time between a calling customer completing dialling and receiving a tone or announcement	general
Pre-Arranged End	–	A procedure in which a dialogue is deemed to have ended unless indicated otherwise	CCSS7 (TC)
Preventive-Cyclic Retransmission Error Correction	–	A form of correcting errors in messages	CCSS7 (MTP)
Primary Rate Access	PRA	A means of connecting a customer to a local node in an ISDN using 30 traffic channels and a signalling channel	general
Primitive	–	A simple type of information element	CCSS7 (TC)
Primitive	–	A unit of information passed between adjacent tiers within an entity	OSI
Primitive Constituent	–	Part of the means of defining the interworking of CCS systems using OSI primitives	CCS
Private Automatic Branch Exchange	PABX	A switch owned or leased by a customer	general
Private Integrated Services Network Exchange	PINX	A PABX designed for an ISDN environment	PN
Private Network	PN	Network elements owned or leased by a customer	PN
Private Signalling System No. 1	PSS1	A signalling system for use in private networks defined by ETSI (also known as QSIG)	PN
Private Telecommunications Network Exchange	PTNX	A PABX designed for a digital environment	PN

(continued)

Term	Abb	Explanation	Context
Procedure Constituent	–	Part of the means of defining the interworking of CCS systems using procedures	CCS
Procedures	–	The logical sequence of events that can occur during calls	general
Process	–	A series of inter-related events	general
Protocol	–	The primitives, procedures and formats for peer-to-peer communication	general
Protocol	–	The procedures, formats and functions for communication	IP
Protocol Class	–	Types of service offered by the SCCP	CCSS7 (SCCP)
Protocol Interpreter	PI	The signalling functions for a control connection	IP (FTP)
Public Land Mobile Network	PLMN	A network providing mobile communications	general
Pulse Code Modulation	PCM	A method of converting information to a digital form for transmission over digital transmission systems	general
Push Procedure	–	A procedure whereby a user requests an immediate transfer of data	IP (TCP)
Q.3	–	A management interface defined by ITU-T	CCSS7 (OMAP)
QSIG	–	A signalling system for use in private networks defined by ETSI (also known as Private Signalling System No. 1)	PN
Quasi-Associated Mode of Operation	–	A form of the non-associated mode of operation in which the network pre-determines the routing of messages	CCS
Real-time Applications	–	Actions that need to be performed in a short timescale, e.g. during call establishment	general
Reassembly	–	The recombination of small blocks of data	general
Recommendation	–	A specification produced by the ITU	general

Redirection	–	A supplementary service in which a call is diverted from one called customer to another	CCSS7 (TUP)
Reference Number	RN	A number used to identify a particular transaction	general
Relay Point	SPR	A signalling point relaying SCCP messages between other nodes	CCSS7 (SCCP)
Released Call	–	A call in which a traffic circuit is cleared-down and ready for use on another call	CCS
Remote Operations Application Service Element	ROSE	An Application Service Element designed to facilitate the implementation of actions remotely	OSI
Representation	–	Ways of coding data	IP
Reset	–	A procedure to overcome an uncertain status of a speech circuit	CCS
Restart Procedure	–	A procedure in which traffic circuits are returned to an idle condition	CCS
Resume	–	A feature allowing cessation of the suspend feature	CCSS7 (ISUP)
Roam	–	The ability of a mobile user to move from one location to another	CCSS7 (MAP)
Router	–	A node that forwards IP traffic between hosts	IP
Routing Label	–	A field identifying the originating and destination nodes of a message	CCSS7
Routing Number	–	An ISDN number used for routing purposes	PN
SCCP Method	–	A form of end-to-end signalling	CCSS7 (ISUP)
Search Engine	–	A node that translates a subject request into an IP address	IP
Segment	–	The unit of data transfer in TCP	IP
Segmentation	–	The splitting of data into small blocks	general
Seize	–	The act of reserving a resource, e.g. reserving a speech circuit	CCS

(*continued*)

Term	Abb	Explanation	Context
Selection Signalling	–	Signalling to transfer address digits	general
Sequence Numbers	SN	Numbers used in error-control mechanisms to detect loss of messages	CCS
Server	–	A computer or node providing services to a remote user	general
Service	–	The functions performed by a number of layers for use by higher layers	OSI
Service Access Point	SAP	The point at which services are offered to higher layers	OSI
Service Control Function	SCF	The logic and processing for services	IN
Service Creation Environment	SCE	Facilitates flexible and speedy derivation of services	IN
Service Feature	–	Elements of a service	IN
Service Key	–	Describes a form of service	IN
Service Logic	–	A set of routines and rules to implement services	IN
Service Management	SM	Provides management capabilities for services	general
Service Switching Function	SSF	The ability to trigger a request for an IN call	IN
Set	–	A message used to alter variables	IP (SNMP)
S-Format Frame	–	A type of frame used in the operation of an access signalling channel	DSS
Shift	–	A move from using one set of procedures &/or formats to another	general
Short Message Service	SMS	A service in which a mobile station sends/receives alphanumeric messages	CCSS7 (MAP)
Signal Transfer Point	STP	A node that routes messages from one point to another using Levels 1 to 3	CCSS7
Signal Unit	SU	A block of signalling information	CCSS7
Signalling	–	The transfer of control information between customers and networks, within networks and between customers	general

Signalling Activity	–	A measure of the amount of signalling traffic	general
Signalling Connection	–	A signalling relationship established between entities	general
Signalling Connection Control Part	SCCP	Used, in conjunction with the MTP, to provide a network service (i.e. Layers 1 to 3)	CCSS7 (SCCP)
Signalling Identifier	SID	Identifier for a signalling instance	general
Signalling Link	–	The direct interconnection of two signalling points	CCS
Signalling Link Activation	–	A procedure preparing a signalling link for service	CCSS7 (MTP)
Signalling Link Management	–	A form of signalling network management for signalling links	CCSS7 (MTP)
Signalling Link Restoration	–	A procedure restoring a signalling link into service	CCSS7 (MTP)
Signalling Message Handling	–	Level 3 functions covering the routing of messages through a network	CCSS7 (MTP)
Signalling Point	SP	A network node capable of operating CCSS7	CCSS7
Signalling Relation	–	An ability for two network nodes to exchange signalling information	general
Signalling Route	–	A collection of signalling links	CCSS7 (MTP)
Signalling Route Management	–	A form of signalling network management distributing status information	CCSS7 (MTP)
Signalling Traffic Management	–	A form of signalling network management reconfiguring signalling traffic	CCSS7 (MTP)
Signalling-Network Functions	–	The Level 3 functions of the MTP	CCSS7 (MTP)
Signalling-Network Management	–	Level 3 functions controlling the signalling network	CCSS7 (MTP)
Simple Network Management Protocol	SNMP	An IP Application Layer Protocol that provides basic management capabilities	IP (SNMP)
Single Association Control Function	SACF	A set of rules for the use of functions in a communication	OSI
Single Association Object	SAO	A collection of functions and rules used by an Application to communicate with a peer over a single association	OSI

(continued)

Term	Abb	Explanation	Context
Socket	–	A set of addresses that identify a TCP connection	IP (TCP)
Specialised Resource Function	SRF	A range of resources for use in an IN call, e.g. collect digits	IN
Specific Services		Available only to some users	CCSS7 (MAP)
Specification and Description Language	SDL	A method of defining the procedures used in systems	general
Static Modelling	–	The modelling of an existing signalling association	general
Stream Mode	–	The transmission of files in a series of bytes	IP (FTP)
Structured Dialogue	–	A dialogue in which an explicit association between two nodes is established	CCSS7 (TC)
Subscriber Identity Module	SIM	A removable card that stores user information in a mobile station	CCSS7 (MAP)
Supplementary Services	SS	Features and facilities additional to a basic service	general
Suspend	–	A feature allowing temporary suspension of a connection	CCSS7 (ISUP)
Synchronise	–	A procedure to establish the sequence numbers for a TCP connection	IP (TCP)
Telecommunications Management Network	TMN	An ITU-T framework for the management of networks and services	general
Telephone User Part	TUP	A user part defining procedures and formats for telephony	CCSS7 (TUP)
Terminal End Point Identifier	TEI	A number identifying a specific terminal	DSS
Tier	–	A self-contained group of functions, forming part of the architecture of a signalling system	CCS
Time-Division Multiplexing	TDM	A means of assembling numerous traffic circuits, using time-slots, for transmission purposes	general
Transaction Capabilities	TC	A generic protocol used to transfer non-circuit-related information between network nodes	CCSS7 (TC)

Glossary 415

Term	Abbr.	Definition	Category
Transaction Capabilities Application Part	TCAP	Part of TC within Layer 7 of the OSI model	CCSS7 (TC)
Transaction Identity	TID	A number used to identify to which dialogue a message pertains	CCSS7 (TC)
Transaction Sub-Layer	–	Part of Transaction Capabilities responsible for establishing and maintaining a connection	CCSS7 (TC)
Transfer Allowed	–	A procedure reversing Transfer Prohibited	CCSS7 (MTP)
Transfer Prohibited	–	A procedure preventing access to a Signal Transfer Point	CCSS7 (MTP)
Transit Control	–	Functions in a transit node	general
Transmission Control Protocol	TCP	A host-to-host connection-oriented protocol	IP (TCP)
Trap	–	An unsolicited message from a server	IP (SNMP)
U-Format Frame	–	A type of frame used to transfer information in unacknowledged operation	DSS
Unacknowledged Operation	–	A Layer 2 form of working in which frames are not numbered, obviating the provision of error control	DSS
Unconfirmed Service	–	Delivery of information not validated	general
Unidirectional Message	–	A message used in a connectionless or unstructured dialogue	CCS
Unitdata	–	A message or primitive used in one-way information flow	general
Unreasonable Information	–	Unexpected or unrecognised message or information	general
Unstructured Dialogue	–	A dialogue in which an explicit association between two nodes is not established	CCSS7 (TC)
User	–	A person, computer or program making use of services	general
User Datagram Protocol	UDP	A host-to-host connectionless protocol	IP (UDP)

(continued)

Term	Abb	Explanation	Context
User Part	UP	Level 4, defining the meaning of messages and the logical sequence of events that occur during calls	CCSS7
User-Network Interface	UNI	An interface between terminal equipment and a network node	general
User-To-User Signalling	UU	Signalling between customers that is not analysed by network nodes	CCS
Variable Mandatory Fields	–	Fields within messages that are compulsory and of variable length	CCS
Version	–	The identity of a variant, e.g. in a signalling system's evolution	general
Virtual Circuit	–	Traffic path in a packet network	general
Virtual Private Network	VPN	The emulation of a Private Network in a Public Network	general
Visitor Location Register	VLR	A database with which a mobile station associates to facilitate roaming	CCSS7 (MAP)
Voice Frequency Signalling System	VF	A signalling system in which signals are transferred as tones within traffic circuits	general
Wide Area Network	WAN	A data network for use over a wide geographical area	general

Abbreviations

AAL	ATM Adaptation Layer
AC	Application Context
ACK	Acknowledgement
ACM	Address Complete Message
ACSE	Association Control Service Element
ACU	Acknowledgement Signal Unit
AE	Application Entity
AL	Alignment
ANS	Answer
AP	Application Part
APDU	Application Protocol Data Unit
ARP	Address Resolution Protocol
ASE	Application Service Element
ATM	Asynchronous Transfer Mode
AU	Access Unit
B	Broadband
BA	Buffer Allocation
BA	Basic Access
BC	Bearer Control
BIB	Backward Indicator Bit
B-ISDN	Broadband ISDN
B-ISUP	Broadband ISDN User Part
BSN	Backward Sequence Number
BSS	Base Station System
B-STP	Broadband Signal Transfer Point
BT	Beginning Tag
C	Confirmed
C/R	Command/Response
CAS	Channel Associated Signalling
CC	Call Control
CC	Connection Confirm
CC	Credit Card
CCBS	Completion of Call to a Busy Subscriber
CCF	Call Control Function
CCITT	International Telegraph & Telephone Consultative Committee
CCS	Common Channel Signalling

CCSS6	Common Channel Signalling System No. 6
CCSS7	Common Channel Signalling System No. 7
CD	Call Deflection
CDMA	Code Division Multiple Access
CFB	Call Forwarding on Busy
CFNR	Call Forwarding on No Reply
CFU	Call Forwarding Unconditional
CIC	Circuit-Identity Code
CLB	Clear Back
CLF	Clear Forward
CLI	Calling Line Identity
CLIP	Calling Line Identification Presentation
CLIR	Calling Line Identification Restriction
CLP	Cell Loss Priority
CM	Configuration Management
CMIP	Common Management Information Protocol
Comp	Component
Con	Confirmation
CPCS	Common Part Convergence Sub-Layer
CPI	Common Part Identifier
CR	Circuit Related
CR	Connection Request
CR	Call Reference
CRC	Cyclic Redundancy Check
CS	Capability Set
CS	Convergence Sub-Layer
CUG	Closed User Group
CVI	Circuit Validation Test Indication
CVT	Circuit Validation Test
DC	Direct Current
DCS	Digital Communications System
DDI	Direct Dialing In
DP	Detection Point
DPC	Destination Point Code
DPNSS	Digital Private Network Signalling System
DSE	Dialogue Service Element
DSS1	Digital Subscriber Signalling System No. 1
DSS2	Digital Subscriber Signalling System No. 2
DTID	Destination Transaction Identity
DTP	Data Transfer Process
DUP	Data User Part
E	Entity
ECMA	European Association for Standardising Information and Communications Systems

ET	End Tag
ETSI	European Telecommunications Standards Institute
FDM	Frequency Division Multiplex
FE	Functional Entity
FIB	Forward Indicator Bit
FISU	Fill-in Signal Unit
FM	Fault Management
FSN	Forward Sequence Number
FTP	File Transfer Protocol
GF	Generic Function
GFC	Generic Flow Control
GFT	Generic Functional Transport
GMSC	Gateway Mobile Switching Centre
GSM	Global System for Mobile Communications
GT	Global Title
GVNS	Global Virtual Network Service
HDLC	High-Level Data-Link Control
HEC	Header Error Control
HLR	Home Location Register
I/C	Incoming
IAA	Initial Address Message Acknowledgement
IAB	Internet Architecture Board
IAM	Initial Address Message
ICMP	Internet Control Message Protocol
ID	Identity
IDU	Interface Data Unit
IE	Information Element
IMEI	International Mobile Equipment Identity
IMSI	International Mobile Subscriber Identity
IN	Intelligent Network
INAP	Intelligent Network Application Part
Ind	Indication
IP	Internet Protocol
ISCP	ISDN Signalling Control Part
ISDN	Integrated Services Digital Network
ISO	International Standards Organisation
ISP	Intermediate Service Part
ISUP	ISDN User Part
ITCC	International Telecommunications Charge Card
ITU	International Telecommunications Union
ITU-T	ITU - Telecommunications Standardisation
IWF	Interworking Function
L	Length
LA	Local Area

LD	Loop Disconnect
LDDC	Long Distance Direct Current
LI	Length Indicator
LR	Local Reference
LSSU	Link Status Signal Unit
LSU	Lone Signal Unit
M	Mandatory
M/M	Multi-Media
MACF	Multiple Association Control Function
MAP	Mobile Application Part
MBS	Multi-Block Synchronisation Signal Unit
MFIR	Multi-Frequency Inter-Register
MID	Multiplexing Identification
MIS	Management Information Service
MRVA	MTP Routing Verification Acknowledgement
MRVR	MTP Routing Verification Result
MRVT	MTP Routing Verification Test
MS	Mobile Station
MSC	Mobile Switching Centre
MSISDN	Mobile Station ISDN Number
MSU	Message Signal Unit
MTP	Message Transfer Part
MUM	Multi-Unit Message
N	Narrowband
NCR	Non-Circuit Related
NE	Network Element
NEM	Network Element Management
N-ISDN	Narrowband ISDN
N-ISUP	Narrowband ISDN User Part
NM	Network Management
NNG	National Number Group
NNI	Network Node Interface
NPI	Numbering Plan Indication
NSDU	Network Service Data Unit
NSP	Network Service Part
O	Optional
O/E	Odd/Even
O/G	Outgoing
OM	Operations and Maintenance
OMAP	Operations, Mantenance and Administration Application Part
OMASE	Operations and Maintenance Application Service Element
OPC	Originating Point Code
OSI	Open Systems Interconnection
OTID	Originating Transaction Identity

PABX	Private Automatic Branch Exchange
PAD	Padding
PC	Point Code
PCM	Pulse Code Modulation
PCS	Personal Communications Service
PD	Packet Data
PDD	Post Dialling Delay
PDU	Protocol Data Unit
PE	Physical Entity
PE	Protocol Entity
PH	Packet Handler
PI	Protocol Interpreter
PIN	Personal Identification Number
PINX	Private Integrated Services Network Exchange
PLMN	Public Land Mobile Network
PM	Performance Management
PN	Private Network
PRA	Primary Rate Access
PSPDN	Packet Switched Public Data Network
PSS1	Private Signalling System No. 1
PSTN	Public Switched Telephony Network
PT	Payload Type
PTNX	Private Telecommunication Network Exchange
QSIG	Signalling System for the Q Reference Point
R2	ITU-T Signalling System R2
REL	Release
Req	Request
Resp	Response
RLC	Release Complete
RLG	Release Guard
RLSD	Released
RN	Reference Number
ROSE	Remote Operations Service Element
RSN	Receive Sequence Number
SABME	Set Asynchronous Balanced Mode Extended
SACF	Single Association Control Function
SAM	Subsequent Address Message
SAO	Single Association Object
SAP	Services Access Point
SAPI	Service Access Point Identifier
SAR	Segmentation and Reassembly
SCCP	Signalling Connection Control Part
SCE	Service Creation Environment
SCF	Service Control Function

SCP	Service Control Point
SCU	System Control Signal Unit
SDF	Service Data Function
SDL	Specification and Description Language
SDU	Service Data Unit
Seq	Sequence
SIB	Service Independent Building Block
SID	Signalling Identifier
SIF	Signalling Information Field
SIM	Subscriber Identity Module
SIO	Service Information Octet
SLDC	Service Logic and Data Control
SLS	Signalling Link Selection
SM	Short Message
SM	Service Management
SMF	Service Management Function
SMS	Short Message Service
SN	Sequence Number
SNMP	Simple Network Management Protocol
SNP	Sequence Number Protection
SP	Signalling Point
SPR	Signalling Point with SCCP Relay Functionality
SRF	Specialised Resource Function
SRTS	Synchronous Residual Time Stamp
SRVT	SCCP Routing Verification Test
SS	Supplementary Service
SSCS	Service Specific Part Convergence Sub-Layer
SSF	Service Switching Function
SSN	Send Sequence Number
SSP	Service Switching Point
SSRC	Service Switching and Resource Control
SSU	Subsequent Signal Unit
ST	Segment Type
STP	Signal Transfer Point
SU	Signal Unit
Syn	Synchronise
SYU	Synchronisation-Signal Unit
TC	Transit Control
TC	Transaction Capabilities
TCAP	Transaction Capabilities Application Part
TCB	Transmission Control Block
TCP	Transmission Control Protocol
TDM	Time Division Multiplexing
TDMA	Time Division Multiple Access

TE	Terminal Equipment
TEI	Terminal End-Point Identifier
TID	Transaction Identity
TMN	Telecommunications Management Network
TMSI	Temporary Mobile Station Identity
TPIE	Transaction Portion Information Element
TRA	Traffic Restart Allowed
TUP	Telephone User Part
U	Unconfirmed
UDP	User Datagram Protocol
UI	Un-numbered Information
UMTS	Universal Mobile Telecommunications System
UNI	User-Network Interface
UP	User Part
UU	User-to-User
UUS	User-to-User Signalling
VC	Virtual Channel
VCI	Virtual Channel Identifier
VF	Voice Frequency
VLR	Visitor Location Register
VP	Virtual Path
VPC	Virtual Path Connection
VPCI	Virtual Path Connection Identifier
VPI	Virtual Path Identifier
VPN	Virtual Private Network
WAN	Wide Area Network

Index

A interface 248
AAL parameters 268, 284
AAL primitives 255
AAL services 255
abort command 326
abort message 123
abort service 137
access 8, 22, 279
access control commands 328
access significance 209
access unit 219
acknowledged operation 51, 194, 199
acknowledgement field 322
acknowledgement signal unit 375
action indicator field 282
action service 185
activate service 142, 151
active call supplementary services 234
active connection establishment 321
active open command 326
adaptive clock method 256
Address Complete Message (ACM) 92, 269, 298, 374, 385
address field 193
address information 9
Address Resolution Protocol (ARP) 315
alerting message 206, 233, 283, 298
algorithm SIB 162
allocate handover number service 139
Answer (ANS) Message 9, 92, 269, 298, 374, 395
answer signal 352, 353, 355, 358, 359, 386
Application 41
Application Context 137, 163
Application Entity (AE) 43, 182, 263, 331
Application Layer 43, 312
Application Part 47
Application Process 41, 182
Application Protocol Data Unit (APDU) 45
Application Service Element (ASE) 43, 163, 182, 263

application type of management 176, 181
architecture 29, 177, 312
ARP request/response packet 315
assigning node 271
associated mode 24, 248
associated signalling 287
Association Control Service Element (ACSE) 43
Assured Transfer 257
Asynchronous Transfer Mode (ATM) 4, 247, 346, 348
ATM Adaptation Layer (AAL) 253
ATM cell rate 268
ATM layer 251
ATM primitives 253
ATM traffic descriptor field 284
authentication failure trap 334
authentication function 332
authentication header 317
authorisation code 237
automatic answering terminal 214
automatic TEI assignment 194

B channel 191
backward call indicator 94
backward direction 65
backward GVNS parameter 241
Backward Sequence Number/ Indicator Bit 65
backward signal/direction 351, 386
base station system/controller 134
base transceiver station 134
basic access 191
basic call processing ASE 165
basic end procedure 128
basic error correction 65
bearer capability 210, 284, 302
bearer connection control part 39
bearer control ASE 263
Begin Message 122, 306
B-ISUP message/primitive mapping 274
bit-stealing technique 364
block 371
block mode 331

block storage 373
blocking/deblocking 101, 273, 254
broadband access signalling 279
broadband bearer capability 268, 284
broadband high layer information 285
Broadband ISDN User Part (B-ISUP) 47, 263
broadband network 247, 346
broadband signalling relations 247
broadband STP 248
broadcast industry 4
broadcast operation 50, 192, 212, 286
BSS APDU parameter 138
buffer 14, 372

cable television 4
call 7, 347
call accepted message category 393
call back 395
call clearing message 205, 352
call completion 106
call control 16, 158
call control ASE 263
call control part 39
call deflection 105
call establishment messages 205
call forwarding 104
call gap 166, 170
call handling service 141
call history parameter 270
call ID data parameter 165
call independent supplementary services 235
call information messages 205
call offering 104
call proceeding message 206, 233, 283, 298
call progress message 100, 298
call reference 99, 204, 235, 282
call related supplementary services 234
call waiting 106
called party address 84
called party number 94, 210, 268
calling line category field 394
Calling Line Identification Presentation (CLIP) 103
Calling Line Identification Restriction (CLIR) 103
calling party address 85
calling party number 94, 210
calling party release 388
calling/called line identification 390
call-rearrangement procedure 218
cancel location service 138

Capability Set (CS) 160
card acceptor 109
card issuer 109
cause field 212
cause indicator 269, 284
cause parameter 102
CCSS6/7 comparison 378
cell 249, 251
cell loss priority field 253
centralised intelligence 157
change directory command 329
changeback 70
changeover 70
changeover signal 375
channel associated signalling 9, 351
    disadvantages 10
    principle 9
    types 10, 351
channel identification 210, 284
charge SIB 162
charging ASE 165
charging service 109
check bits 14, 196, 318, 327
Circuit Identification Code (CIC) 91, 180, 384
Circuit Related (CR) 11, 13
circuit switching 3, 346
Circuit Validation Test (CVT) 184, 188
class of service 79, 86, 116
Clear Back signal 9, 388
Clear Forward signal 9, 356, 358, 359, 374, 386
Clear Request Message 395
clear signal 353, 355
client request 163
client/server response pair 163
close command 326
close service 137
Closed User Group (CUG) 108, 390
Code Division Multiple Access (CDMA) 132
code field 318
coder 372
codeset 207, 235, 286
ColdStart trap 334
command 325
command frame 193
common channel signalling
    advantages 16
    modes 24
    principle 11, 368
    requirements 19
    types 21

Common Management Information
    Protocol (CMIP) 185
common MAP service 137
common part of AAL 250
communications capabilities 41
community name 332
community of interest 108
compare SIB 162
compatibility 31, 213, 267, 279, 288
compatibility, DSS2 types 288
compelled mode 355
complete call service 140
Completion of Call to a Busy Subscriber
    (CCBS) 395
component 116
component portion of TC 119
component sub-layer of TC 116
compressed mode 331
computing industry 4
conference calling 107
configuration management 175, 238
confirmation primitive 34
confirmed MAP service 136
confusion message 102
congestion control 222
congestion level parameter 270
connect ASE 165
connect message 98, 206, 234, 283, 298
connect operation 166
connection 321
Connection Confirm (CC) message 82,
    100
connection element identifier parameter
    268
connection establishment ASE 165
connection establishment phase 79
connection identifier field 284
connection oriented 42, 77
connection reference 52
connection release phase 79
Connection Request Message 82, 100
connection section 86
connectionless 42, 77
constructor information element 119
contents field 119
contiguous multi-rate 222
Continue Message 122
continuity check 20, 377
continuous mode 355
control connection 328
control field 194, 322
control information 7
controlled rerouting 71

convergence 4
convergence sub-layer of AAL 250
co-ordination function 232
coupling function 87
credit card calling 169
customer care system 177
cyclic retransmission 66

D channel 191
data connection 330
Data Form 1 and 2 82
Data link connection 51
Data Link Layer 42
data offset 322
data parameter 255
data primitive 281
data representation, storage, and
    transfer 330
data transfer phase 79
database 3, 12, 48, 304
D-channel 49
de-activation service 151
decoder 372
delete command 329
destination address 317
destination line identity 394
destination local reference 86
destination options header 317
Destination Point Code (DPC) 60, 68
destination port 322, 326
destination signalling identifier 269
Destination Transaction Identity
    (DTID) 122
Destination Unreachable message 319
Detection Point (DP) 165
dialled digits 352, 353, 355, 358
dialogue control 125
dialogue for TC 117
dialogue portion for TC 119
Dialogue Service Element (DSE) 232
digital communications system 132
digital connectivity 390
Digital Private Network Signalling
    System (DPNSS) 393
Digital Subscriber Signalling System
    No.1 (DSS1) 21, 49
Digital Subscriber Signalling System
    No.2 (DSS2) 21, 279
digital transmission link 10, 346
direct access to VPN 237
Disconnect command 195
Disconnect message 206, 234, 283,
    298

428 *Index*

disconnected 216
display 210, 284
distributed functional plane 162
distributed intelligence 157
double current working 354
DSS1 data link layer 192
DSS1 network layer 203, 294
DSS1 physical layer 191
DSS1/DSS2 comparison 283
DSS2 interworking with narrowband 289
DSS2 primitives 281
DSS2 support of narrowband 288
dual seizure 102, 271, 389
dual significance 209
dynamic modelling 264

echo control 101, 318
en bloc operation 96, 212, 286, 297, 386
encapsulating security payload header 317
End message 122, 306
end of option list 323
end point 99
end primitive 118
end-to-end 18, 99, 357
end-to-end significance 281
end-to-end transit delay 285
entity 53, 293
erase service 142, 151
error index field 333
error recovery 16, 64, 280, 320, 378
error status field 333
establish primitive 281
establish temporary connection operation 166
ethernet 315
event-report service 185
evolution 18, 367
executive intrusion 237
extension field 193, 282, 284
extension header 317
extensive primitive 34

facility indication 107
facility information element 225
facility message 235
fallback 100
fault management 175
feature activation information element 224
feature key management protocol 223
field 13

file structure 331
file structure command 329
File Transfer Protocol (FTP) 327
fill in signal unit 62
finish bit 321
first party release 98, 388
fixed length sub-fields 384
fixed mandatory parameter 83
flag 13, 63, 193, 282, 320
flow and flow label 316
flow control 51, 66, 320
forced rerouting 70
forced retransmission 66
foreign socket 321
format 16
format constituent 296
forward access signalling service 139
forward address group 385
forward call indicator 94
forward GVNS parameter 240
Forward Sequence Number/Indicator Bit 65
forward short message service 140
forward signal/direction 351, 386
fragment header 317
frame 51, 193
frame check sequence 196
frame for PCM 362
frame relay 219
freephone service 167, 305
Frequency Division Multiplex (FDM) 357
functional entity 158, 238
functional information element 212
functional protocol 224

gateway MSC 134
gateway node 318
generic address 166
generic flow control field 252
generic functional procedures 231
generic name 34
generic notification 106
get function 332
get PDUs 333
global functional plane 161
global significance 209, 281
Global System for Mobile Communications (GSM) 132
global title 88, 124
Global Virtual Network Service (GVNS) 230
Group A and B signals 361

Index   429

group blocking/unblocking 101
Group I and II signals 361
GVNS access procedure 241
GVNS functional model 239
GVNS transfer/routing procedure 241

handover number 138
handover service 138
handset 131
header 251, 316
header error control field 253
heading code 373, 383
high layer compatibility 210, 284
High Level Data-Link Control (HDLC) Protocol 192
hold function 225
Home Location Register (HLR) 134
hop limit 319
hop-by-hop header 317
host 312
host address 321
host-to-host layer 312

ID function 184
identifier 13
I-format 194
IN conceptual model 161
Incoming Call Indication message category 393
incoming signalling system 299
indication primitive 34
indirect access to VPN 237
information 4, 14
information age 1, 311
information command 195
Information Element 119, 393
information flows for IN 163
Information Message 223
Informational Message 318
informative indicator 394
Initial Address Message (IAM) 92, 268, 298, 301, 374, 385
initial detection point 166
initial signal unit 373
insert subscriber data service 143
instruction field 284
Integrated Services Digital Network (ISDN) 3, 90, 191, 219, 294, 369
Intelligent Network (IN) 157
Intelligent Network Application Part (INAP) 113, 161
inter-digit pause 352

inter-exchange 8
Interface Data Unit (IDU) 257
interlock code 108
Intermediate Service Part (ISP) 46, 115
internal management 176
international charge card 109
International Mobile Equipment Identity (IMEI) 133
International Mobile Subscriber Identity (IMSI) 133
international standards 21
International Standards Organisation (ISO) 36
International Telecommunications Union (ITU) 5, 21
International Telegraph and Telephone Consultative Committee (CCITT) 21
Internet 3, 311
Internet Control Message Protocol (ICMP) 318
Internet layer 313
Internet Protocol (IP) 3, 5, 315
Internet Protocol network 346
Internet Protocols 311
inter-nodal 8, 22
inter-operability 311
inter-PABX signalling 230, 393
inter-register 359
interrogation service 142, 152
interworking 31, 280, 293
invoke component 124, 306
invoke identity 124
invoke primitive 118, 258
invoke service 142
IP address 315
IP architecture 312
IP format 316
IP layers 312
IP packet/service 315
IP terminology 311
IP Version 6 316
IP/OSI model relationship 313
ISDN customer number 88
ISDN Signalling Control Part (ISCP) 39
ISDN User Part (ISUP) 39, 90, 294

keypad facility information element 223
keypad protocol 223

label 373, 384
layer 36, 40, 312

430 *Index*

legacy 31
length field for TC 119
Length Indicator (LI) 62
level 36, 37
line signalling 9
link efficiency 23
link layer 313
link set activation 72
Link Status Signal Unit 62
link-by-link 357
LinkDown trap 334
linked identity 124
LinkUp trap 334
local reference number 78, 321
local significance 209
local socket 321
location area 135
location register 113
location service 137
location updating 144
locking shift procedure 208
logical object 163
logout command 329
lone signal unit 373
long distance DC signalling 354
loop disconnect signalling 10, 351
loop routing 317

maintenance control 273
maintenance control ASE 263
managed object 179, 333
managed object model 180
management 4, 23
management agent 331
Management Information Service (MIS) 179
management inhibiting 71
management of CCSS7 175
management signals 375
mandatory fixed part 91
mandatory indicator 394
mandatory sub-fields 384
mandatory variable part 91
MAP dialogue 135
MAP service model 135
maximum segment 323
maximum transmission unit 319
measurement 181
media 4
message 11,13
message body 318
message discrimination and distribution 69

message group 384
message handling 38, 68
message mode 257
message routing 68
message signal unit 62, 372, 383
Message Transfer Part (MTP) 38, 59
message type 205, 210, 235, 281, 393
message type code 81, 91, 267
message type tag 120
meta signalling 250
miscellaneous messages 205
Mobile Application Part (MAP) 113, 133
mobile call 347
mobile communications 131
mobile country code 133
mobile network 346
mobile network code 133
mobile station 131
mobile station identification number 133
Mobile Station ISDN Number (MSISDN) 133
mobile supplementary services 141
Mobile Switching Centre (MSC) 133
mobile technologies 132
mobile terminal 134
mode 24
MTP restart 71
MTP Routing Verification Test (MRVT) 184, 187
multi-block synchronisation signal unit 375
multi-frame for PCM 362
multi-frequency inter-register signalling 359
multimedia 3
multi-mode handset 132
multi-party 107
Multiple Association Control Function (MACF) 44, 163
multiple invocation procedure 126
multiple network 311
multi-rate procedures 222
multi-unit message 374
multi-vendor 159

name 83
Narrowband ISDN User Part (N-ISUP) 39, 90
National Number Group (NNG) code 305
nature of connection 94
network 1, 7, 345

Index 431

network address 321
network element 177, 331
Network Element Management (NEM) 177
network interface function 263
network layer 42
network layer connection 232
network management 2, 4, 175
network management station 331
Network Node Interface (NNI) 250
network reconfiguration 16
network service 46, 77, 88, 114
Network Service Data Unit (NSDU) 77
Network Service Part (NSP) 77
network signalling 10
next header field 316
nodal functions 263
non-associated mode 24
non-associated signalling 287
non-assured transfer 257
non-automatic answering terminal 214, 297
non-automatic TEI assignment 194
Non-Circuit-Related (NCR) 12, 14, 369
non-compelled 65, 320, 355
non-contiguous multi-rate 222
non-locking procedure 209
non-P format 256
no-operation 323
notice service 137
notification indicator field 286
notify message 206, 283
number acknowledged message 395
number identification 103

object attribute, notification and information 179
object ID 184, 334
off-air call set up 147
off-line usage 114
off-net 237
OMAP model 181
OMASE user 183
on-net 237
open interface 311
open service 137
open systems interconnection 36
openness 30
operation 45, 116, 163
operation code information element 125
Operations and Maintenance ASE (OMASE) 182
operations system 177

Operations, Maintenance and Administration Application Part (OMAP) 114, 177
option 54
optional indicator 394
optional parameter 83, 92
optional part 91
optional sub-fields 384
options field 322
originating host 318
originating line identity field 394
Originating Point Code (OPC) 60, 68
originating service provider 240
Originating Transaction Identity (OTID) 122
origination signalling identifier 269
OSI model 36, 40
OSI/IP model relationship 313
outband signalling 10, 357
outgoing signalling system 299
overlap 96, 215, 286, 386

PABX signalling 230, 393
packet data 3, 50, 219, 311
packet handler 219
packet size 210
Packet Switched Public Data Network 219
packet too big message 319
packet via B channel 220
packet via circuit switched access 219
packet via D channel 220
padding 322
page service 143
page structure 331
parameter 33
parameter problem message 319
pass-along 99
passive command 329
passive connection establishment 321
passive open command 326
passive open request 321
password command 329
payload 251
payload length 316
payload type field 252
PCM signalling 362
peer-to-peer 35, 248, 312
performance management 175
permanent negative completion reply 330
Personal Communications Services (PCS) 132

432 *Index*

personal computer 4
Personal Identification Number (PIN) 109
P-format 256
physical address 315
physical entity 159, 163
physical layer 42
physical plane 162
platform 1, 5, 19
play announcement operation 166
Point Code 68
pointer 83, 91, 256, 320
point-to-multi point connection 258, 279
point-to-point 50, 192, 212, 258, 279
polling message 332
polling protocol data unit 333
portable 47, 114
positive acknowledgement 66, 320
positive reply 330
positive/negative acknowledgement 65
post-dialling delay 10, 17, 353
pre-arranged end 137
pre-arranged end procedure 128
prepare handover service 139
presentation layer 43, 313
preventive cyclic retransmission error correction 66
primary account number 109
primary rate access 192
primitive 33
primitive constituent 294
primitive information element 119
principles of signalling 7
Private Automatic Branch Exchange (PABX) 22, 229, 393
Private Integrated Services Network Exchange (PINX) 231
private network 229
Private Signalling System No.1 (PSS1) 230
Private Telecommunication Network Exchange (PTNX) 230
problem diagnostic parameter 137
procedure 16
procedure constituent 295
proceed to send signal 356
process access request service 143
process access signalling service 139
process co-ordinator 151
processing model 162
progress message 105, 206, 273, 283
prompt and collect operation 166
propagation delay 101, 273

protocol 35, 312
protocol class 79, 84
protocol control function 231
Protocol Data Unit (PDU) 251, 332
protocol discriminator 204, 235, 282
Protocol Entity (PE) 331
Protocol Interpreter (PI) 328
protocol reference model 36
provider abort service 137
PSS1 protocol model 231
Public Land Mobile Network (PLMN) 131
Public Switched Telephony Network (PSTN) 133
pulse 352
Pulse Code Modulation (PCM) 362
pulsed mode 355
push procedure 320

Q Interface 248
Q Reference Point 231
Q.3 Interface 178
QSIG 230
quasi-associated mode 25, 248
queue SIB 162

rapid service creation 159
real time usage 114
reason parameters 137
reasonableness check 378
reassemble 78
receive command 326
receive not ready frame 194
receive ready command 195
receive ready/not ready instruction 222
receive sequence number 194, 200
received loss priority 253
receiver 320
recommendation 21
record structure 331
redirection 105, 390
reference model 36
reference number 14
register 359
register service 142, 152
reinitialise command 329
reject command 195
reject component 124
Relay Point (SPR) 86
Release (REL) Message 95, 206, 234, 269, 286, 298
Release Complete Message 82, 206, 234, 283, 298

release guard signal 356, 377, 388
release method parameter 137
release primitive 281
release request message category 393
released 216
released message 82
reliability 18, 19, 30, 59
Remote Operations Service Element (ROSE) 44, 116
reply 325
representation command 329
request ID 333
request primitive 34
reset 102, 273, 321, 388
response frame 193
response message 332
response primitive 34
restart procedure 218
result last/not-last primitive 118
resume message 100, 206, 273, 283
retransmission 65
retrieve command 329
retrieve function 225
return error component 124
return result component 125, 306
roam 134, 144
route trace 186
router 3, 312
routing control function 78
routing field 252
routing header 317
routing label 64, 68, 267, 384
routing number 242

satellite 4, 131
SCCP end-to-end 99
SCCP Routing Verification Test (SRVT) 184, 188
SCF Activation ASE 165
search engine 3
search for mobile service 143
security 19
segment 78, 319
segment type field 260
segmentation 96, 118, 127, 270
segmentation and reassembly 254, 250, 318
seize signal (examples) 9, 351, 353, 356, 359
selection signalling 9
send command 326
send end signal service 139

send handover report service 139
send ID service 138
send information on SMS service 140
send information service 140
send routing information service 140
send sequence number 194, 200
sender 320
separation of control and switch 367
sequence count procedure 256
sequence number 63, 320, 324
sequence number field 256, 260
server 3, 327
service 33
service access code 237
service access point 51
Service Access Point Identifier (SAPI) 51, 194
service class 79, 86
service commands 329
service control function/point 159
Service Creation Environment Function (SCEF) 159
Service Data Function/Point 109, 159
Service Data Unit (SDU) 251
service feature 160
Service Independent Building Block (SIB) 160, 162
service indicator field 394
service information octet 63, 69
service key 167
service logic 162
service logic and data control 239
service management 2, 159, 177
Service Management ASE 165
Service Management Function/Point 159
service plane 161
service profile 223
service provider field 34
service request message category 393
service specific part 250
service switching and resource control 239
Service Switching Function/Point 159
session layer 43, 313
set function 332
Set-Up message 206, 233, 285, 298, 301
set-asynchronous balanced-mode extended (SABME) command 195
set-request PDU 333
S-format 194
shift information element 208
short message service 141, 150

signal imitation 357
signal information element 211, 284
signal information field 373
signal primitive 258
Signal Transfer Point 25, 40
signal unit 62, 371
signal unit error rate monitor 67
signalling 5, 7
signalling activity 15
signalling channel 11
signalling connection 86
Signalling Connection Control Part (SCCP) 46, 77
Signalling Identifier (SID) 264
signalling in PCM systems 362
Signalling Information Field (SIF) 64, 81, 91, 383
signalling link 37, 62
signalling link activation/ deactivation 72
signalling link management 71, 176
signalling link restoration 72
signalling link selection 69, 79, 384
signalling network functions 38, 67
signalling network management 38
signalling path 12
signalling point 24, 37
signalling relation 24
signalling route management 72, 176
signalling route set test 72
Signalling System No.5 355
Signalling System No.6 21, 371
Signalling System No.7 21
Signalling System R2 358, 360
signalling termination 17
signalling traffic flow control 71
signalling traffic management 67
Simple Network Management Protocol (SNMP) 331
Single Association Control Function (SACF) 44, 163, 263
Single Association Object (SAO) 44, 163
single current working 353
single ended 160
single invocation procedure 126
single octet information element 205
single point of control 160
socket 321
source address 316
source clock frequency recovery 256
source local reference 83
source port 322, 326
specialised resource control ASE 165

Specialised Resource Function/Point 159
specialised resource report operation 166
specific MAP service 136
specific name 34
specification 16
Specification and Description Language (SDL) 295
speech clipping 18
static modelling 265
status command 326
status parameter 256
step-by-step switch 367
stimulus information element 212
stimulus procedures 223
store command 329
STP screening table 181
stream mode 331
streaming mode 257
structure 18, 29, 32, 177, 312
structure parameter 256
structured dialogue 117, 128
submitted loss priority 253
Subscriber Identity Module (SIM) 134
Subsequent Address Message (SAM) 96, 387
subsequent signal unit 373
sub-system number 87
successful backward group 385
supplementary services control function 232
suspend message 100, 206, 273, 283
switch 7, 345
synchronisation 14, 321
synchronisation signal unit 372
synchronise segment 323
synchronous residual time stamp procedure 256
system control signal unit 375

tag field 119
target location area ID 138
TCP/IP 312
Telecommunications Management Network (TMN) 5, 177
Telephone User Part (TUP) 383
telephony call/network 345
telephony over IP 5
television 4
temporary mobile station number 138
terminal 1, 24

Index   435

Terminal End-Point Identifier (TEI) 51, 194, 197
terminal equipment 247
terminating access indicator 241
terminating network routing number 241
three-way handshake 323
tier 32
Time Division Multiple Access (TDMA) 132
Time Division Multiplex (TDM) 362
time exceeded message 319
time sequence diagram 295
time stamp 334
toll free service 305
traffic class 316
traffic management ASE 165
trailer 251
Transaction Capabilities (TC) 46, 113, 306
Transaction Capabilities Application Part (TCAP) 46, 116
Transaction Capabilities User 113
transaction portion 119
transaction sub-layer 117
transfer allowed 72
transfer command 329
transfer prohibited 72
transfer request primitive 274
transient negative reply 330
transit control 240
translate SIB 162
translation procedure for IN 166
transmission control block 321
Transmission Control Protocol (TCP) 319
transmission link 7, 346
transmission medium requirements 94, 302
transmission mode 331
transport layer 42
transport message 319
transport service 279
trap message 333
trap PDU 334
TUP supplementary services 389
Type A/B Service 160
type field 318
Types 1 and 2 information element 205
Types A and B node 102
Types A, B and C call 240

U interface 248

UDP datagram 326
U-format 195
unacknowledged operation 51, 194, 196
unblocking 101
unconfirmed MAP service 136
unexpected message 102, 389
unidirectional message 121
unitdata 34, 82, 118, 281
Universal Mobile Telecommunications System (UMTS) 132
unreasonable information 102
unrecognised information ASE 263
unstructured dialogue 117, 128
update location service 138
urgent pointer indicator 321
user 8, 40, 203, 327
user abort service 137
User Datagram Protocol (UDP) 325
user group field 240
user information message 206, 283
user name command 329
user notification 218
User Part (UP) 39, 90
user signalling bearer service 221
user-abort/error primitive 118
User-Network Interface (UNI) 249
user-reject primitive 118
user-to-user 18, 110, 221

variable binding and list 333
variable length information element 205
variable length sub-fields 384
variable mandatory parameter 83
verify SIB 162
Version 6 316
version compatibility 149
version number 316, 332
virtual channel 53, 229, 279, 287
Virtual Channel Identifier (VCI) 252, 287
virtual circuit 251
virtual network 236
Virtual Path Connection Identifier (VPCI) 271, 287
virtual path consistency check 273
Virtual Path Identifier (VPI) 252, 287
virtual private network 229
Visitor Location Register (VLR) 134
voice frequency signalling 10, 355

WarmStart trap 334
Wide Area Network (WAN) 311
window 322